Theoretical Ecology

Theoretical Ecology
Principles and Applications

EDITED BY

Robert M. May,
Department of Zoology, University of Oxford,
Oxford, UK

AND

Angela R. McLean,
Department of Zoology, University of Oxford,
Oxford, UK

OXFORD
UNIVERSITY PRESS

OXFORD
UNIVERSITY PRESS

Great Clarendon Street, Oxford OX2 6DP

Oxford University Press is a department of the University of Oxford.
It furthers the University's objective of excellence in research, scholarship,
and education by publishing worldwide in

Oxford New York

Auckland Cape Town Dare es Salaam Hong Kong Karachi
Kuala Lumpur Madrid Melbourne Mexico City Nairobi
New Delhi Shanghai Taipei Toronto

With offices in

Argentina Austria Brazil Chile Czech Republic France Greece
Guatemala Hungary Italy Japan Poland Portugal Singapore
South Korea Switzerland Thailand Turkey Ukraine Vietnam

Oxford is a registered trade mark of Oxford University Press
in the UK and in certain other countries

Published in the United States
by Oxford University Press Inc., New York

British Library Cataloguing in Publication Data

Data available

Library of Congress Cataloging in Publication Data

Data available

Typeset by Newgen Imaging Systems (P) Ltd., Chennai, India
Printed in Great Britain
on acid-free paper by
Antony Rowe Ltd., Chippenham, Wiltshire

ISBN 978–0–19–920998–9 (Hbk)   978–0–19–920999–6 (Pbk)

10 9 8 7 6 5 4 3 2 1

Contents

Acknowledgements

The aims and scope of this book are set out in the beginning of the first chapter (unimaginatively labelled Introduction). So these prefacing comments are confined to acknowledging some of the help we and the other authors have received in putting this book together.

The two of us are deeply indebted to the other 21 authors who have contributed to the book, both for the work they did and for their exemplary adherence to a rather fast production schedule. Individual authors have wished to thank both funding agencies and helpful colleagues who gave assistance of various kinds. This would have been an impressively long list, but we unkindly decided against including it.

We must, however, recognize the generosity of Merton College and the Zoology Department at the University of Oxford, and particularly their respective heads, Dame Jessica Rawson and Professor Paul Harvey. They made it possible to bring the authors and others together for a 2-day conference, in which the sweep of material in this book was exposed to discussion and constructive criticism. It helped shape the book.

Our thanks are also owed to enthusiastic and helpful people at Oxford University Press, particularly the commissioning editor, Ian Sherman, the production editor, Christine Rode, and the copy-editor, Nik Prowse. R.M.M.'s assistant, Chris Bond, was her usual invaluable self, helping in every facet of the enterprise with unflappable competence.

Sadly, one of the authors—Geoff Kirkwood—unexpectedly died the week after the gathering in Oxford. Everyone remembers him with affection, and his shadow lies on the book. He will be missed.

A.R. McLean and R.M. May
22 September 2006

Contributors

John R. Beddington, Division of Biology, Faculty of Natural Sciences, RSM Building, Imperial College London, SW7 2BP, UK. E-mail: j.beddington@imperial.ac.uk

Michael B. Bonsall, Department of Zoology, Tinbergen Building, University of Oxford, Oxford OX1 3PS, UK. E-mail: michael.bonsall@zoo.ox.ac.uk

Gordon Conway, Centre for Environmental Policy, 4th Floor, RSM Building, Imperial College, South Kensington, London SW7 2AZ, UK. E-mail: g.conway@imperial.ac.uk

Tim Coulson, NERC Centre for Population Biology and Division of Biology, Imperial College London, Silwood Park Campus, Ascot, Berkshire SL5 7PY, UK. E-mail: t.coulson@imperial.ac.uk

Michael J. Crawley, Department of Biological Sciences, Imperial College London, Silwood Park, Ascot, Berkshire SL5 7PY, UK. E-mail: m.crawley@imperial.ac.uk

Andy Dobson, Ecology and Evolutionary Biology, Princeton University, Princeton, NJ 08544, USA. E-mail: dobber@princeton.edu

H. Charles J. Godfray, Department of Zoology, Tinbergen Building, University of Oxford, Oxford OX1 3PS, UK. E-mail: charles.godfray@zoo.ox.ac.uk

Bryan Grenfell, Biology Department, 208 Mueller Laboratory, Pennsylvania State University, University Park, PA 16802, USA. E-mail: grenfell@psu.edu

Michael P. Hassell, Department of Biological Sciences, Imperial College London, Silwood Park, Ascot, Berkshire SL5 7PY, UK. E-mail: m.hassell@ic.ac.uk

Anthony R. Ives, Department of Zoology, University of Wisconsin, Madison, WI 53706, USA. E-mail: arives@wisc.edu

Matthew Keeling, Department of Biological Sciences and Mathematics Institute, University of Warwick, Gibbet Hill Road, Coventry CV4 7AL, UK. E-mail: m.j.keeling@warwick.ac.uk

Jeremy T. Kerr, Canadian Facility for Ecoinformatics Research (CFER), Department of Biology, University of Ottawa, Box 450, Station A, Ottawa, ON, K1N 6N5, Canada. E-mail: jkerr@uottawa.ca

Heather Kharouba, Canadian Facility for Ecoinformatics Research (CFER), Department of Biology, University of Ottawa, Box 450, Station A, Ottawa, ON, K1N 6N5, Canada. E-mail: hkar075@uottawa.ca

Geoffrey P. Kirkwood, Division of Biology, Faculty of Natural Sciences, RSM Building, Imperial College London, SW7 2BP, UK

Robert M. May, Department of Zoology, Tinbergen Building, University of Oxford, Oxford OX1 3PS, UK. E-mail: robert.may@zoo.ox.ac.uk

Angela R. McLean, Department of Zoology, Tinbergen Building, University of Oxford, Oxford OX1 3PS, UK. E-mail: angela.mclean@zoo.ox.ac.uk

Sean Nee, Institute of Evolutionary Biology, School of Biological Sciences, University of Edinburgh, West Mains Road, Edinburgh EH9 3JT, UK. E-mail: sean.nee@ed.ac.uk

Martin A. Nowak, The Program for Evolutionary Dynamics, Faculty of Arts and Science, One Brattle Square, Harvard University, Cambridge, MA 02138, USA. E-mail: nowak@fas.harvard.edu

Karl Sigmund, Faculty for Mathematics, University of Vienna, Nordbergstrasse 15, A-1090 Vienna, Austria. E-mail: karl.sigmund@univie.ac.at

George Sugihara, Scripps Institution of Oceanography, University of California, San Diego, 9500 Gilman Drive, La Jolla, CA 92093 0202, USA. E-mail: gsugihara@ucsd.edu

David Tilman, Department of Ecology, Evolution and Behavior, University of Minnesota, St. Paul, MN 55108, USA. E-mail: tilman@umn.edu

Will R. Turner, Center for Applied Biodiversity Science, Conservation International, 1919 M St. NW Suite 600, Washington, DC 20036, USA. E-mail: w.turner@conservation.org

David S. Wilcove, Ecology and Evolutionary Biology and Princeton Environmental Institute and Woodrow Wilson School of Public and International Affairs, Princeton University, Princeton, NJ 08544, USA. E-mail: dwilcove@princeton.edu

CHAPTER 1

Introduction

Angela R. McLean and Robert M. May

In this introductory chapter, we indicate the aims and structure of this book. We also indicate some of the ways in which the book is not synoptic in its coverage, but rather offers an interlinked account of some major developments in our understanding of the dynamics of ecological systems, from populations to communities, along with practical applications to important problems.

Ecology is a young science. The word ecology itself was coined not much more than 100 years ago, and the oldest professional society, the British Ecological Society, is less than a century old. Arguably the first published work on ecology was Gilbert White's *The Natural History of Selborne*. This book, published in 1789, was ahead of its time in seeing plants and animals not as individual objects of wonder—things to be assembled in a cabinet of curiosities—but as parts of a community of living organisms, interacting with the environment, other organisms, and humans. The book has not merely remained in print, but has run steadily through well over 200 editions and translations, to attain the status of the fourth most published book (in the sense of separate editions) in the English language. The following excerpt captures White's blend of detailed observation and concern for basic questions.

Among the many singularities attending those amusing birds, the swifts, I am now confirmed in the opinion that we have every year the same number of pairs invariably; at least, the result of my inquiry has been exactly the same for a long time past. The swallows and martins are so numerous, and so widely distributed over the village, that it is hardly possible to recount them; while the swifts, though they do not all build in the church, yet so frequently haunt it, and play and rendezvous round it, that they are easily enumerated. The number that I constantly find are eight pairs, about half of which reside in the church, and the rest in some of the lowest and meanest thatched cottages. Now, as these eight pairs—allowance being made for accidents—breed yearly eight pairs more, what becomes annually of this increase? and what determines every spring, which pairs shall visit us, and re-occupy their ancient haunts?

This passage is unusual in giving quantitative information about the population of swifts in Selborne two centuries ago, a small exception to the almost universal absence of population records going back more than a few decades. It is even more remarkable for its clear articulation of the central question of population biology: what regulates populations? Interestingly, the swift population of Selborne these days is steadily around 12 pairs, which in ecological terms is not much different from eight, even though much of their environment has changed—entries to the church tower all wired-off to keep out squirrels, and the gentrified cottages no longer low and mean with their thatch, when it remains, neatly wired down (Lawton and May, 1983). Interpreted generously, these population data on Selborne's swifts could be seen as one of ecology's longest time series, so it is sobering to realize there is still no agreed explanation of what actually regulates the swifts' numbers.

Moving on from Gilbert White, the first half of the twentieth century saw some more explicitly mathematical models aimed at understanding the dynamical behaviour of populations. Notable examples include Ross' work on malaria, with its first introduction of the basic reproductive number, R_0, discussed in later chapters of this book, and Lotka and Volterra's indication of the inherently oscillatory properties of prey–predator systems. Despite this, ecology seems to us to have

1

remained a largely observational and descriptive subject up to the decade of the 1960s. Witness two of the most influential texts of that time: Andrewartha and Birch (1954), an excellent book but explicitly antithetic to theory in the form of anything resembling a mathematical model; Odum (1953), arguably foreshadowing aspects of 'systems ecology' with its insightful focus on patterns of energy flow in ecosystems, but with the emphasis descriptive rather than conceptual.

For evolutionary studies as well as for ecological ones, we think the 1960s saw a change in the zeitgeist. For evolution, much of the stimulus derived from Bill Hamilton's conceptual advances. For ecology, it was the reframing by Evelyn Hutchinson (1965) and his student Robert McArthur (1972; see also MacArthur and Wilson, 1967) of old questions in more explicitly analytic ways; one could perhaps say, rephrasing them in the idiom of theoretical physics. How similar can species be, yet persist together? What tends to govern the number of species we see on an island, and how does this number depend on the size and isolation of the island? Gilbert White's question of population abundance was revisited—and expanded beyond the sterile controversies of the 1950s about whether populations typically are governed by tight density dependence or fluctuate greatly under the influence of environmental factors—to ask the more precise dynamical question of why do some populations remain relatively steady, others show regular cycles, and yet others fluctuate wildly? Given the observed patterns of relative abundance of the different species in particular communities, what are the underlying causes? What is the relation between the complexity of a food web (variously defined) and its ability to withstand disturbance, natural or human created?

These more deliberately conceptual or theoretical approaches differed from early work, in our view, in that they went beyond the codification of descriptive material, and the search for patterns within such codification, to ask questions about underlying mechanisms. To ask questions about why, rather than what. Mathematics enters into such studies, essentially as a tool for thinking clearly. In pursuing a 'why' or 'what if' question about a complicated situation, it can be helpful to ask whether particular factors may be more important than others, and to see if such insight or guesswork does indeed provide testable explanations. Mathematical models can be precise tools for doing this, helping us to make our assumptions explicit and unambiguous, and to explore 'imaginary worlds' as metaphors for such hypothetical simplicity underlying apparent complexity. The 1970s saw much activity of this kind in ecological research, helped in part by basic advances in our understanding of nonlinear dynamical systems and by the advent of increasingly powerful and user-friendly computers.

In particular, the phenomenon of deterministic chaos received wide recognition in the 1970s. The finding that very simple and purely deterministic laws or equations can give rise to dynamical behaviour that not merely looks like random noise, but is so sensitive to initial conditions that long-term prediction is effectively impossible, has huge implications. It ends the Newtonian dream that if the system is simple (very few variables) and orderly (the rules and parameters exactly known), then the future is predictable. The 'law' can be as trivial as $x(t+1) = \lambda x(t) \exp[-x(t)]$, with λ a known and unvarying constant, but if λ is big enough then an error of one part in one million in the initial estimate of $x(0)$ will end up producing a completely wrong prediction within a dozen or so time steps. Interestingly, it is often thought that chaotic phenomena found applications in ecology after others had developed the subject. In fact, one of the two streams which brought chaos centre stage in the 1970s derived directly from ecological research on models for a single population with discrete, non-overlapping generations. These models were first-order difference equations; the other strand was Lorenz's metaphor for convective phenomena in meteorology, involving more complex—although still relatively simple—three-dimensional differential equations.

Advances in computing have also been of great help in all areas of ecology: statistical design of experiments; collecting and processing data; and, coming to the present book, developing and exploring mathematical models for both simple and complicated ecological systems. There are, however, some associated dangers, which deserve passing mention. The understanding derived from computer studies of complicated models can sometimes be substantially less complete than that gained from the analytic methods of classical applied mathematics and theoretical physics. The early days of computers—mechanical calculators—saw them used by theoretical physicists in conjunction with analytic approximations, to explore previously intractable problems. The result, however, was that at every step there was preserved an intuitive understanding of the relation between the underlying assumptions and the results. In contrast, many scientists who today use computers to explore increasingly complex mathematical models have little formal background in mathematics, or have forgotten what they were once taught. Most of this work is interesting and excellent. But, absent any degree of intuitive understanding of how the input assumptions about the system's biology relate to the consequent output, we need to be wary (May, 2004). Too often, an 'emergent phenomenon' means little more than 'I've no clue what is going on, but it looks kinda interesting'. Happily, there are very few examples of this in ecology. More particularly, throughout the present book we aim, wherever possible, to provide intuitive understanding of the lessons learned from mathematical models.

Be all this as it may, there has been a marked rise in theoretical ecology as a distinct sub-discipline over the past three decades or so. Many of the practitioners are not to be found in the field or laboratory; a greater number, however, find their experimental contributions in field and/or laboratory to be inextricably interwoven with their theoretical and mathematical contributions. Ecology has come a long way from the 1970s, when a few empirical ecologists resented outsiders, who had not paid their dues of years of toil in the field, presuming to mathematize their problems (often sweeping aside arguably irrelevant, but certainly beloved, details in the process). Others perhaps welcomed the intrusion too uncritically.

The end result, however, is seen clearly by comparing today's leading ecology texts with those of the 1950s and 1960s. In the latter, you will find very few equations. Today, in contrast, you will find a balanced blend of observation, field and laboratory experiments, and theory expressed in mathematical terms. The comparison, for example, between the first edition of Begon, Townsend and Harper (1986) and the earlier Andrewartha and Birch (1954) or Odum (1953) is pronounced. We think this marks a maturation of the subject, although there undeniably remain large and important areas where there are still more questions than answers.

1.1 This book and its predecessors

This book (TEIII) is essentially a greatly transmogrified version of one first published in 1976 (TEI), and followed with substantial changes in 1981 (TEII; this was not a perfunctory update, but had three chapters completely re-written by different authors, two new chapters added, and all others revised; TEI's 14 chapters involved 11 authors, TEII's 16 chapters had 13 authors, of whom nine were from TEI). This new version, 25 years on, has 15 chapters by 23 authors, only three of whom are veterans of TEII.

Like the previous two, this book is not a basic undergraduate ecology text, but equally it is not a technical tome for the front-line specialist in one or other aspect of theoretical ecology. Rather, the book is aimed at upper-level undergraduate, postgraduate, and postdoctoral students, and ecological researchers interested in broadening aspects of the courses they teach, or indeed of their own work. As such, we think it fair to claim that TEI and TEII in their own time played a part in the above-mentioned transition in the general subject of ecology, where earlier texts, in which mathematical content was essentially absent, contrast markedly with today's, where theoretical approaches—sometimes explicitly mathematical and sometimes not—play an important part, although no more than a part, of the presentation of the subject. Some of our acquaintances, indeed, still use the earlier volumes as supplements to their undergraduate courses. TEII, although out of print, still trades actively on the online bookseller Amazon.

This book, on the other hand, differs from the previous two by virtue of these changes in how the subject of ecology is defined and taught. Much of the material in TEI and TEII would now, 25 years and more on, be seen as a routine part of any basic ecology text. Other bits, of course, are just out of date, overtaken by later advances.

One essential similarity with its predecessors is that the present book does not aim at synoptic coverage. Instead, it attempts first (in Chapters 2–9) to give an account of some of the basic principles that govern the structure, function, and temporal and spatial dynamics of populations and communities. These chapters are not tidily kept to uniform length; we think the dynamics of plant populations have probably received less attention than those of animal populations, and so have encouraged the authors in this area to go into somewhat greater detail. Conversely, we recognise that there are important and interesting areas of theoretical ecology—aspects of macroecology, or energy flows in ecosystems, for example—which are not covered here. By the same token, the 'applied' chapters are a selection from the larger universe of interesting and illuminating possibilities. In short, advances over the past quarter century have seen significant growth in field and laboratory studies, along with major theoretical advances and practical applications. Any book on 'theoretical ecology' simply has much more ground to cover—many more subdisciplines and specialized areas—than was the case for TEII. The result is inevitably that the present book has more gaps and omissions than its predecessors; inclusions and exclusions are bound to be more quirky. A charitable interpretation would be that, just as the gates to Japanese temples, *tori*, have deliberate imperfections to avoid angering the gods, so too we have avoided the sublime. The real reason is a mixture of our own interests, and a feeling that enough is enough.

1.2 What is in the book

Previous editions of this text began with a chapter on the evolutionary forces which shape the behaviour of individuals on a stage set by specific environmental and ecological factors, and then show how such individual behaviour ultimately determines the demographic parameters—density-dependent birth and death rates, movement patterns, and so on—governing the population's behaviour in space and over time.

The past three decades have seen extraordinary advances in our understanding of the behavioural ecology and life-history strategies of individuals (e.g. Krebs and Davies, 1993). On the one hand, this is a formidable field to cover concisely, but on the other hand, only in a relatively few corners does this work deal directly with deducing the overall dynamics of a population from the behavioural ecology of its constituent individuals. There are some interesting examples of phenomena whose understanding unavoidably requires bringing the two together—for instance, odd aspects of brood parasitism where you cannot understand the population dynamics without understanding the evolution of individual's behaviour, and conversely (Nee and May, 1993)—but they are few, and seem to have evoked little interest so far. A good review of some other open questions at the interface between natural selection and population dynamics is by Saccheri and Hanski (2006). Resource managers get by, and seem to be content, with treating the parameters in population models as phenomenological constants, fitted to data.

One really big problem, however, which is in many ways as puzzling today as it was to Darwin, is how large aggregations of cooperating individuals (where group benefits are attained for a relatively small cost to participating individuals, but where the whole thing is vulnerable to cheats who take the benefits without paying the cost) can evolve and maintain themselves. Relatively early work by Hamilton and Trivers pointed the way to a solution of this problem for small groups of closely related individuals. But much of this work is so restricted as to defy application to large aggregations of human or other animals. The past few years have, however, seen a diverse array of significant advances in this area, and we thought it would be better to begin with a definitive review of this underpinning topic, which is still wide open to further advances. Hence Chapter 2, *How populations cohere: five rules for cooperation*.

1.2.1 Basic ecological principles

The next two chapters deal with single populations. In Chapter 3 Coulson and Godfray distil the essence of several recent monographic treatments of one or other aspect, to discuss how density-dependent or nonlinear effects, interacting to various degrees with demographic and environmental stochasticity, can result in relatively steady, or cyclic, or erratically fluctuating population dynamics. They also sketch progress that has been made in looking at the 'flipside of chaos', namely the question of whether, when we see apparently noisy time series, we are looking at 'environmental and other noise' or a deterministic but chaotic signal. This survey is woven together with illustrative accounts of field studies and laboratory experiments. In Chapter 4 Nee widens the discussion of population dynamics to look at some of the complications which arise when a single population is spatially distributed over many patches. Foreshadowing later chapters on conservation biology and on infectious diseases, he emphasizes that you do not have to destroy all of a population's habitat to extinguish it. Widening the survey to include two populations interacting as competitors, predator–prey or mutualists, Nee further indicates other aspects of the dynamics of such so-called metapopulations which may seem counter-intuitive.

The next three chapters expand on interacting populations. Bonsall and Hassell first survey the dynamical behaviour of prey–predator interactions. This chapter takes for granted some of the by-now familiar material presented in TEII, giving more attention to the way spatial complexities contribute to the persistence of such associations (and also noting that such spatial heterogeneity can even be generated by the nonlinear nature of the interactions themselves, even in an homogeneous substrate). Crawley gives an overview of the dynamics of plant populations, interpreting 'plants' broadly to emphasize the range of different considerations which arise as we move from diatoms to trees. This chapter also discusses plant–herbivore interactions as an important special case of predators and prey. Competitive interactions are discussed by Tilman in Chapter 7, drawing together theoretical advances with long-term and other field studies.

Chapters 8 and 9 deal with the theoretical ecology of communities. Ives' chapter might have been called *Complexity and stability* in the 1970s (not *Diversity and stability*; diversity was not a much-used term then—the word *diversity* does not appear in the index to May's *Stability and Complexity in Model Ecosystems* (1973a), and although it does appear in the indexes for TEI and TEII, it clearly means simply numbers of species). Ives carefully enumerates the varied interpretations which have been placed on the terms complexity/diversity and stability. He goes on to give a thumbnail sketch of the way ideas have evolved in this area, guided by empirical and theoretical advances, and concludes by presenting models which illustrate how the answers to questions about community dynamics can depend on precisely how the questions are framed. In Chapter 9 May, Crawley, and Sugihara survey a range of recent work on 'community patterns': the relative abundance of species; species–area relations; the network structure of food webs; and other things. This survey, which in places is a bit telegraphic, seeks to outline both the underlying observations and the suggested theoretical explanations, including null models (old and new) and scaling laws.

1.2.2 Applications to practical problems

The next five chapters turn to particular applications of these theoretical advances. Grenfell and Keeling (Chapter 10) deal with the dynamics and control of infectious diseases of both humans and other animals. They begin by explaining how basic aspects of predator–prey theory apply here, with particular emphasis on the infection's basic reproductive number, R_0. Recent applications to the outbreak of foot-and-mouth disease among livestock in the UK are discussed in some detail, although other examples could equally well have been chosen (HIV/AIDS, SARS, H5N1 avian flu). Grenfell and Keeling emphasize the essential interplay between massively detailed computations (the foot-and-mouth disease outbreak was modelled at the level of every farm in Britain, an extreme example of an individual-level approach to a population-level phenomenon) and basic dynamical understanding of what is going on, based on simple models.

In Chapter 11 Beddington and Kirkwood give an account of the ecology of fisheries and their practical management. This chapter explains how the dynamics of fish populations—as single species or in multispecies communities—interacts with practical policy options (quotas, tariffs, licenses, etc.), in ways which can be complicated and sometimes counter-intuitive. This is an area in which science-based advice can be in conflict with political considerations, sometimes in ways which have interesting resonance with the problems discussed in Nowak and Sigmund's opening chapter on the evolution of cooperation. In passing, we observe that a vast amount of interesting ecological data, and also of excellent theoretical work, is to be found in the grey literature associated with the work of bodies like the International Council for the Exploration of the Seas (ICES) or the Scientific Committee of the International Whaling Commission (IWC); it is unfortunate that too little of this makes its way into mainstream ecological meetings and scientific journals. We think Chapter 11 is particularly interesting for the way it reaches into this grey literature.

The term Doubly Green Revolution was coined by Gordon Conway, one of the three continuing authors from TEII (along with Hassell and May). Here, in Chapter 12, he surveys the triumphs and problems of the earlier Green Revolution, which has doubled global food production on only 10% additional land area over the past 30 years or so. Looking to the future, he suggests how new technologies offer the potential to feed tomorrow's population, and to do so in a way where crops are adapted to their environment (as distinct from past practice, where too often the environment was wrenched to serve the crops by fossil-fuel energy subsidies). Conway stresses that engagement and empowerment of local people is essential if this Doubly Green Revolution is to be realized, which again harks back to Nowak and Sigmund.

Chapter 13 by Dobson, Turner, and Wilcove deals directly with conservation biology, surveying some of the factors which threaten species with extinction, indicating possible remedial actions, but also noting some of the economic and political realities that can impede effective action. Chapter 14, on *Climate Change and Conservation Biology*, by Kerr and Kharouba, amplifies one particularly important threat to the survival of species, namely the effects that climate change are likely to have on species' habitats and ranges.

The concluding Chapter 15 offers a selective and opinionated review of some of the major environmental threats that loom for us and other species over the coming few centuries. The emphasis is on issues where ecological knowledge can provide a guide to appropriate action, or to areas where current lack of ecological understanding is a handicap. One thing is sure: the future for other living things on planet Earth, not just humans, depends on our understanding and managing ecosystems better than we have been doing recently.

CHAPTER 2

How populations cohere: five rules for cooperation

Martin A. Nowak and Karl Sigmund

Subsequent chapters in this volume deal with populations as dynamic entities in time and space. Populations are, of course, made up of individuals, and the parameters which characterize aggregate behavior—population growth rate and so on—ultimately derive from the behavioral ecology and life-history strategies of these constituent individuals. In evolutionary terms, the properties of populations can only be understood in terms of individuals, which comes down to studying how life-history choices (and consequent gene-frequency distributions) are shaped by environmental forces.

Many important aspects of group behavior—from alarm calls of birds and mammals to the complex institutions that have enabled human societies to flourish—pose problems of how cooperative behavior can evolve and be maintained. The puzzle was emphasized by Darwin, and remains the subject of active research today.

In this book, we leave the large subject of individual organisms' behavioral ecology and life-history choices to texts in that field (e.g. Krebs and Davies, 1997). Instead, we lead with a survey of work, much of it very recent, on five different kinds of mechanism whereby cooperative behavior may be maintained in a population, despite the inherent difficulty that cheats may prosper by enjoying the benefits of cooperation without paying the associated costs.

Cooperation means that a donor pays a cost, c, for a recipient to get a benefit, b. In evolutionary biology, cost and benefit are measured in terms of fitness. While mutation and selection represent the main forces of evolutionary dynamics, cooperation is a fundamental principle that is required for every level of biological organization. Individual cells rely on cooperation among their components. Multicellular organisms exist because of cooperation among their cells. Social insects are masters of cooperation. Most aspects of human society are based on mechanisms that promote cooperation. Whenever evolution constructs something entirely new (such as multicellularity or human language), cooperation is needed. Evolutionary construction is based on cooperation.

The five rules for cooperation which we examine in this chapter are: kin selection, direct reciprocity, indirect reciprocity, graph selection, and group selection. Each of these can promote cooperation if specific conditions are fulfilled.

2.1 Kin selection

The heated conversation took place in an unheated British pub over some pints of warm bitter. Suddenly J.B.S. Haldane remarked, 'I will jump into the river to save two brothers or eight cousins.' The founding father of population genetics and dedicated communist in his spare time never bothered to develop this insight any further. The witness of the revelation was Haldane's eager pupil, the young John Maynard Smith. But given John's high regard for entertaining stories and good beer, can we trust his memory?

The insight that Haldane might have had in the pub was precisely formulated by William Hamilton. He wrote a PhD thesis on this topic, submitted a long paper to the *Journal of Theoretical Biology*, and spent much of the next decade in the Brazilian jungle. This was one of the most important papers in evolutionary biology in the second half of the twentieth century (Hamilton, 1964a, 1964b). The theory was termed kin selection by Maynard Smith (1964). The crucial equation is the following. Cooperation among relatives can be favored by natural selection if the coefficient of genetic relatedness, r, between the donor and the recipient exceeds the cost/benefit ratio of the altruistic act:

$$r > c/b \qquad (2.1)$$

Kin-selection theory has been tested in numerous experimental studies. Indeed, many cooperative acts among animals occur between close kin (Frank, 1998; Hamilton, 1998). The exact relationship between kin selection and other mechanisms such as group selection and spatial reciprocity, however, remains unclear. A recent study even suggests that much of cooperation in social insects is due to group selection rather than kin selection (Wilson and Hölldobler, 2005). Note that kin selection is more likely to work in quite small groups; in large groups, unless highly inbred, the average value of r will be tiny.

2.2 Direct reciprocity

In 1971, Robert Trivers published a landmark paper entitled 'The evolution of reciprocal altruism' (Trivers, 1971). Trivers analyzed the question how natural selection could lead to cooperation between unrelated individuals. He discusses three biological examples: cleaning symbiosis in fish, warning calls in birds, and human interactions. Trivers cites Luce and Raiffa (1957) and Rapoport and Chammah (1965) for the Prisoner's Dilemma, which is a game where two players have the option to cooperate or to defect. If both cooperate they receive the reward, R. If both defect they receive the punishment, P. If one cooperates and the other defects, then the cooperator receives the sucker's payoff, S, while the defector receives the temptation, T. The Prisoner's Dilemma is defined by the ranking $T > R > P > S$.

Would you cooperate or defect? Assuming the other person will cooperate it is better to defect, because $T > R$. Assuming the other person will

defect it is also better to defect, because $P > S$. Hence, no matter what the other person will do it is best to defect. If both players analyze the game in this rational way then they will end up defecting. The dilemma is that they both could have received a higher payoff if they had chosen to cooperate. But cooperation is irrational.

We can also imagine a population of cooperators and defectors and assume that the payoff for each player is determined by many random interactions with others. Let x denote the frequency of cooperators and $1 - x$ the frequency of defectors. The expected payoff for a cooperator is $f_C = Rx + S(1 - x)$. The expected payoff for a defector is $f_D = Tx + P(1 - x)$. Therefore, for any x, defectors have a higher payoff than cooperators. In evolutionary game theory, payoff is interpreted as fitness. Successful strategies reproduce faster and outcompete less successful ones. Reproduction can be cultural or genetic. In the non-repeated Prisoner's Dilemma, in a well-mixed population, defectors outcompete cooperators. Natural selection favors defectors.

Cooperation becomes an option if the game is repeated. Suppose there are m rounds. Let us compare two strategies, always defect (ALLD), and GRIM, which cooperates on the first move, then cooperates as long as the opponent cooperates, but permanently switches to defection if the opponent defects once. The expected payoff for GRIM versus GRIM is nR. The expected payoff for ALLD versus GRIM is $T + (m - 1)P$. If $nR > T + (m - 1)P$ then ALLD cannot spread in a GRIM population when rare. This is an argument of evolutionary stability. Interestingly, Trivers (1971) quotes 'Hamilton (pers. commun.)' for this idea.

A small problem with the above analysis is that given a known number of rounds it is best to defect in the last round and by backwards induction it is also best to defect in the penultimate round and so on. Therefore, it is more natural to consider a repeated game with a probability w of having another round. In this case, the expected number of rounds is $1/(1 - w)$, and GRIM is stable against invasion by ALLD provided $w > (T - R)/(T - P)$.

We can also formulate the Prisoner's Dilemma as follows. The cooperator helps at a cost, c, and the other individual receives a benefit, b. Defectors do not help. Therefore we have $T = b$, $R = b - c$, $P = 0$, and $S = -c$. The family of games that is described by the parameters b and c is a subset of all possible Prisoner's Dilemma games as long as $b > c$. For the repeated Prisoner's Dilemma, we find that ALLD cannot invade GRIM if

$$w > c/b \qquad (2.2)$$

The probability of having another round must exceed the cost/benefit ratio of the altruistic act (Axelrod and Hamilton, 1981; Axelrod, 1984). Notice, however, the implicit assumption here that the payoff for future rounds is not discounted (i.e. distant benefits count as much as present ones). In evolutionary reality, this is unlikely. We can address this by incorporating an appropriate discount factor in w (May, 1987), but note, from eqn 2, that this makes cooperation less likely.

Thus, the repeated Prisoner's Dilemma allows cooperation, but the question arises: what is a good strategy for playing this game? This question was posed by the political scientist, Robert Axelrod. In 1979, he decided to conduct a tournament of computer programs playing the repeated Prisoner's Dilemma. He received 14 entries, of which the surprise winner was tit-for-tat (TFT), the simplest of all strategies that were submitted. TFT cooperates in the first move, and then does whatever the opponent did in the previous round. TFT cooperates if you cooperate, TFT defects if you defect. It was submitted by the game theorist Anatol Rapoport (who is also the co-author of the book *Prisoner's Dilemma*; Rapoport and Chammah, 1965). Axelrod analyzed the events of the tournament, published a detailed account and invited people to submit strategies for a second championship. This time he received 63 entries. John Maynard Smith submitted tit-for-two-tats, a variant of TFT which defects only after the opponent has defected twice in a row. Only one person, Rapoport, submitted TFT, and it won again. At this time, TFT was considered to be the undisputed champion in the heroic world of the repeated Prisoner's Dilemma.

But one weakness became apparent very soon (Molander, 1985). TFT cannot correct mistakes.

The tournaments were conducted without strategic noise. In a real world, trembling hands and fuzzy minds cause erroneous moves. If two TFT players interact with each other, a single mistake leads to a long sequence of alternating defection and cooperation. In the long run two TFT players get the same low payoff as two players who flip coins for every move in order to decide whether to cooperate or to defect. Errors destroy TFT.

Our own investigations in this area began after reading a News and Views article in *Nature* where the author made three important points: first, he often leaves university meetings with a renewed appreciation for the problem of how natural selection can favor cooperative acts given that selfish individuals gain from cheating; second, strategies in the repeated Prisoner's Dilemma should not be error-free but subjected to noise; third, evolutionary stability should be tested not against single invaders but against heterogeneous ensembles of invaders (May, 1987). This was the motivation for the following work.

In 1989, we conducted evolutionary tournaments. Instead of inviting experts to submit programs, we asked mutation and selection to explore (some portion of) the strategy space of the repeated Prisoner's Dilemma in the presence of noise. The initial random ensemble of strategies was quickly dominated by ALLD. If the opposition is random, it is best to defect. A large portion of the population began to adopt the ALLD strategy and everything seemed lost. But after some time, a small cluster of players adopted a strategy very close to TFT. If this cluster is sufficiently large, then it can increase in abundance, and the entire population swings from ALLD to TFT. Reciprocity (and therefore cooperation) has emerged. We can show that TFT is the best catalyst for the emergence of cooperation. But TFT's moment of glory was brief and fleeting. In all cases, TFT was rapidly replaced by another strategy. On close inspection, this strategy turned out to be generous tit-for-tat (GTFT), which always cooperates if the opponent has cooperated on the previous move, but sometimes (probabilistically) even cooperates when the opponent has defected. Natural selection had discovered forgiveness (Nowak and Sigmund, 1992).

After many generations, however, GTFT is undermined by unconditional cooperators, ALLC. In a society where everybody is nice (using GTFT), there is almost no need to remember how to retaliate against a defection. A biological trait that is not used is likely to be lost by random drift. Birds that escape to islands without predators lose the ability to fly. Similarly, a GTFT population is softened and turns into an ALLC population.

Once most people play ALLC, there is an open invitation for ALLD to seize power. This is precisely what happens. The evolutionary dynamics run in cycles: from ALLD to TFT to GTFT to ALLC and back to ALLD. These oscillations of cooperative and defective societies are a fundamental part of all our observations regarding the evolution of cooperation. Most models of cooperation show such oscillations. Cooperation is never a final state of evolutionary dynamics. Instead it is always lost to defection after some time and has to be re-established. These oscillations are also reminiscent of alternating episodes of war and peace in human history (Figure 2.1).

A subsequent set of simulations, exploring a larger strategy space, led to a surprise (Nowak and Sigmund, 1993). The fundamental oscillations were interrupted by another strategy which seems to be able to hold its ground for a very long period of time. Most surprisingly, this strategy is based on the extremely simple principle of win-stay, lose-shift (WSLS). If my payoff is R or T then I will continue with the same move next round. If I have cooperated then I will cooperate again, if I have defected then I will defect again. If my payoff is only S or P then I will switch to the other move next round. If I have cooperated then I will defect, if I have defected then I will cooperate (Figure 2.2).

If two WSLS strategists play each other, they cooperate most of the time. If a defection occurs accidentally, then in the next move both will defect. Hereafter both will cooperate again. WSLS is a simple deterministic machine to correct stochastic noise. While TFT cannot correct mistakes, both GTFT and WSLS can. But WSLS has an additional ace in its hand. When WSLS plays ALLC it will discover after some time that ALLC does not retaliate. After an accidental defection, WSLS will switch to permanent defection. Therefore, a population of WSLS players does not drift to ALLC. Cooperation based on WSLS is more stable than cooperation based on TFT-like strategies.

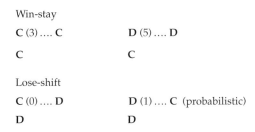

Win-stay

C (3) C   D (5) .... D

C   C

Lose-shift

C (0) .... D   D (1) .... C (probabilistic)

D   D

Figure 2.2 Win-stay, lose-shift (WSLS) embodies a very simple principle. If you do well then continue with what you are doing. If you are not doing well, then try something else. Here we consider the Prisoner's Dilemma payoff values $R = 3$, $T = 5$, $P = 1$, and $S = 0$. If both players cooperate, you receive three points, and you continue to cooperate. If you defect against a cooperator, you receive five points, and you continue to defect. But if you cooperate with a defector, you receive no points, and therefore you will switch from cooperation to defection. If, on the other hand, you defect against a defector, you receive one point, and you will switch to cooperation. Your aspiration level is three points. If you get at least three points then you consider it a win and you will stay with your current choice. If you get less than three points, you consider it a loss and you will shift to another move. If $R > (T + P)/2$ (or $b/c > 2$) then WSLS is stable against invasion by ALLD. If this inequality does not hold, then our evolutionary simulations lead to a stochastic variant of WSLS, which cooperates after a DD move only with a certain probability. This stochastic variant of WSLS is then stable against invasion by ALLD.

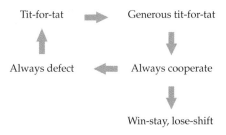

Figure 2.1 Evolutionary cycles of cooperation and defection. A small cluster of tit-for-tat (TFT) players or even a lineage starting from a single TFT player in a finite population can invade an always defect (ALLD) population. In fact, TFT is the most efficient catalyst for the first emergence of cooperation in an ALLD population. But in a world of fuzzy minds and trembling hands, TFT is soon replaced by generous tit-for-tat (GTFT), which can re-establish cooperation after occasional mistakes. If everybody uses GTFT, then always cooperate (ALLC) is a neutral variant. Random drift leads to ALLC. An ALLC population invites invasion by ALLD. But ALLC is also dominated by win-stay, lose-shift (WSLS), which leads to more stable cooperation than TFT-like strategies.

The repeated Prisoner's Dilemma is mostly known as a story of TFT, but WSLS is a superior strategy in an evolutionary scenario with errors, mutation, and many generations (Fudenberg and Maskin, 1990; Nowak and Sigmund, 1993).

In the infinitely repeated game, WSLS is stable against invasion by ALLD if $b/c > 2$. If instead $1 < b/c < 2$ then a stochastic variant of WSLS dominates the scene; this strategy cooperates after a mutual defection only with a certain probability. Of course, all strategies of direct reciprocity, such as TFT, GTFT, or WSLS can only lead to the evolution of cooperation if the fundamental inequality (eqn 2.2) is fulfilled.

2.3 Indirect reciprocity

Whereas direct reciprocity embodies the idea of you scratch my back and I scratch yours, indirect reciprocity suggests that you scratch my back and I scratch someone else's. Why should this work? Presumably I will not get scratched if it becomes known that I scratch nobody. Indirect reciprocity, in this view, is based on reputation (Nowak and Sigmund, 1998a, 1998b, 2005). But why should you care about what I do to a third person?

The main reason why economists and social scientists are interested in indirect reciprocity is because one-shot interactions between anonymous partners in a global market become increasingly frequent and tend to replace the traditional long-lasting associations and long-term interactions between relatives, neighbors, or members of the same village. Again, as for kin selection, it is a question of the size of the group. A substantial part of our life is spent in the company of strangers, and many transactions are no longer face to face. The growth of online auctions and other forms of e-commerce is based, to a considerable degree, on reputation and trust. The possibility to exploit such trust raises what economists call moral hazards. How effective is reputation, especially if information is only partial?

Evolutionary biologists, on the other hand, are interested in the emergence of human societies, which constitutes the last (up to now) of the major transitions in evolution. In contrast to other eusocial species, such as bees, ants, or termites, humans display a large amount of cooperation between non-relatives (Fehr and Fischbacher, 2003). A considerable part of human cooperation is based on moralistic emotions, such as anger directed towards cheaters or the warm inner glow felt after performing an altruistic action. Intriguingly, humans not only feel strongly about interactions that involve them directly, they also judge actions between third parties as evidenced by the contents of gossip. There are numerous experimental studies of indirect reciprocity based on reputation (Wedekind and Milinski, 2000; Milinski et al., 2002; Wedekind and Braithwaite, 2002; Seinen and Schram, 2006).

A simple model of indirect reciprocity (Nowak and Sigmund, 1998a, 1998b) assumes that within a well-mixed population, individuals meet randomly, one in the role of the potential donor, the other as potential recipient. Each individual experiences several rounds of this interaction in both roles, but never with the same partner twice. A player can follow either an unconditional strategy, such as always cooperate or always defect, or a conditional strategy, which discriminates among the potential recipients according to their past interactions. In a simple example, a discriminating donor helps a recipient if her score exceeds a certain threshold. A player's score is 0 at birth, increases whenever that player helps and decreases whenever the player withholds help. Individual-based simulations and direct calculations show that cooperation based on indirect reciprocity can evolve provided the probability, q, of knowing the social score of another person exceeds the cost/benefit ratio of the altruistic act:

$$q > c/b \qquad (2.3)$$

The role of genetic relatedness that is crucial for kin selection is replaced by social acquaintanceship. In a fluid population, where most interactions are anonymous and people have no possibility of monitoring the social score of others, indirect reciprocity has no chance. But in a socially viscous population, where people know each other's reputation, cooperation by indirect reciprocity can thrive (Nowak and Sigmund, 1998a).

In a world of binary moral judgements (Nowak and Sigmund, 1998b; Leimar and Hammerstein,

2001; Fishman, 2003; Panchanathan and Boyd, 2003; Brandt and Sigmund, 2004, 2005), there are four ways of assessing donors in terms of first-order assessment: always consider them as good, always consider them as bad, consider them as good if they refuse to give, or consider them as good if they give. Only this last option makes sense. Second-order assessment also depends on the score of the receiver; for example, it can be deemed good to refuse help to a bad person. There are 16 second-order rules. Third-order assessment also depends on the score of the donor; for example, a good person refusing to help a bad person may remain good, but a bad person refusing to help a bad person remains bad. There are 256 third-order assessment rules. We display four of them in Figure 2.3.

With the scoring assessment rule, cooperation, C, always leads to a good reputation, G, whereas

Reputation of donor and recipient

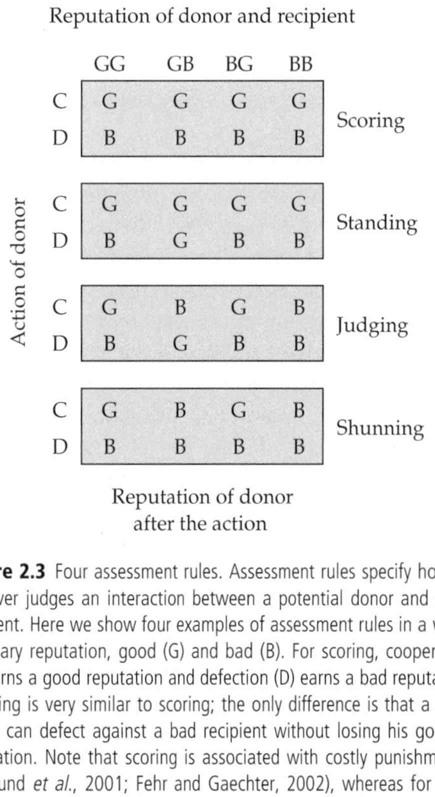

Figure 2.3 Four assessment rules. Assessment rules specify how an observer judges an interaction between a potential donor and a recipient. Here we show four examples of assessment rules in a world of binary reputation, good (G) and bad (B). For scoring, cooperation (C) earns a good reputation and defection (D) earns a bad reputation. Standing is very similar to scoring; the only difference is that a good donor can defect against a bad recipient without losing his good reputation. Note that scoring is associated with costly punishment (Sigmund *et al.*, 2001; Fehr and Gaechter, 2002), whereas for standing punishment of bad recipients is cost-free. For judging it is bad to help a bad recipient. Shunning assigns a bad reputation to any donor who interacts with a bad recipient.

defection, D, always leads to a bad reputation, B. Standing (Sugden, 1986) is like scoring, but it is not bad if a good donor defects against a bad recipient. With judging, in addition, it is bad to cooperate with a bad recipient. For another assessment rule, shunning, all donors who meet a bad recipient become bad, regardless of what action they choose. Shunning strikes us as grossly unfair, but it emerges as the winner in a computer tournament if errors in perception are included and if there are only a few rounds in the game (Takahashi and Mashima, 2003).

An action rule for indirect reciprocity prescribes giving or not giving, depending on the scores of both donor and recipient. For example, you may decide to help if the recipient's score is good or your own score is bad. Such an action might increase your own score and therefore increase the chance of receiving help in the future. There are 16 action rules.

If we view a strategy as the combination of an action rule and an assessment rule, we obtain 4096 strategies. In a remarkable calculation, Ohtsuki and Iwasa (2004, 2005) analyzed all 4096 strategies and proved that only eight of them are evolutionarily stable under certain conditions and lead to cooperation (Figure 2.4).

Both standing and judging belong to the leading eight, but scoring and shunning are not. However, we expect that scoring has a similar role in indirect reciprocity to that of TFT in direct reciprocity. Neither strategy is evolutionarily stable, but their simplicity and their ability to catalyze cooperation in adverse situations constitute their strength. In extended versions of indirect reciprocity, in which donors can sometimes deceive others about the reputation of the recipient, scoring is the foolproof concept of 'I believe what I see'. Scoring judges the action and ignores the stories. There is also experimental evidence that humans follow scoring rather than standing (Milinski *et al.*, 2001).

In human evolution, there must have been a tendency to move from the simple cooperation promoted by kin or group selection to the strategic subtleties of direct and indirect reciprocity. Direct reciprocity requires precise recognition of individual people, a memory of the various interactions one had with them in the past, and enough brain

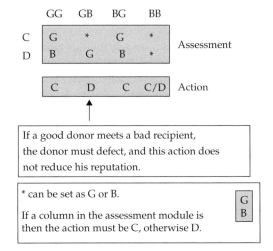

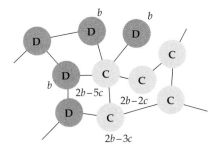

Figure 2.4 Ohtsuki and Iwasa's leading eight. Ohtsuki and Iwasa (2004, 2005) have analyzed the combination of $2^8 = 256$ assessment modules with $2^4 = 16$ action modules. This is a total of 4096 strategies. They have found that eight of these strategies can be evolutionarily stable and lead to cooperation, provided that everybody agrees on each other's reputation. (In general, uncertainty and incomplete information might lead to private lists of the reputation of others.) The three asterisks in the assessment module indicate a free choice between G and B. There are therefore $2^3 = 8$ different assessment rules which make up the leading eight. The action module is built as follows: if the column in the assessment module is G and B, then the corresponding action is C, otherwise the action is D. Note that standing and judging are members of the leading eight, but that scoring and shunning are not.

Figure 2.5 Games on graphs. The members of a population occupy the vertices of a graph (or social network). The edges denote who interacts with whom. Here we consider the specific example of cooperators, C, competing with defectors, D. A cooperator pays a cost, c, for every link. Each neighbor of a cooperator receives a benefit, b. The payoffs of some individuals are indicated in the figure. The fitness of each individual is a constant, denoting the baseline fitness, plus the payoff of the game. For evolutionary dynamics, we assume that in each round a random player is chosen to die, and the neighbors compete for the empty site proportional to their fitness. A simple rule emerges: if $b/c > k$ then selection favors cooperators over defectors. Here k is the average number of neighbors per individual.

power to conduct multiple repeated games simultaneously. Indirect reciprocity, in addition, requires the individual to monitor interactions among other people, possibly judge the intentions that occur in such interactions, and keep up with the ever-changing social network of the group. Reputation of players may not only be determined by their own actions, but also by their associations with others.

We expect that indirect reciprocity has coevolved with human language. On the one hand, it is helpful to have names for other people and to receive information about how a person is perceived by others. On the other hand a complex language is needed, especially if there are intricate social interactions. The possibilities for games of manipulation, deceit, cooperation, and defection are limitless. It is likely that indirect reciprocity has

provided the very selective scenario that led to cerebral expansion in human evolution.

2.4 Graph selection

The traditional model of evolutionary game dynamics assumes that populations are well-mixed (Taylor and Jonker, 1978; Hofbauer and Sigmund, 1998). This means that interactions between any two players are equally likely. More realistically, however, the interactions between individuals are governed by spatial effects or social networks. Let us therefore assume that the individuals of a population occupy the vertices of a graph (Nowak and May, 1992; Nakamaru et al., 1997, 1998; Skyrms and Pemantle, 2000; Abramson and Kuperman, 2001; Ebel and Bornholdt, 2002; Lieberman et al., 2005; Nakamaru and Iwasa, 2005; Santos et al., 2005; Santos and Pacheco, 2005). The edges of the graph determine who interacts with whom (Figure 2.5).

Consider a population of N individuals consisting of cooperators and defectors. A cooperator helps all individuals to whom it is connected, and pays a cost, c. If a cooperator is connected to k other individuals and i of those are cooperators, then its payoff is $bi - ck$. A defector does not provide any help, and therefore has no costs, but it

can receive the benefit from neighboring cooperators. If a defector is connected to k other individuals and j of those are cooperators, then its payoff
is bj. Evolutionary dynamics are described by an
extremely simple stochastic process: at each time
step, a random individual adopts the strategy of
one of its neighbors proportional to their fitness.

We note that stochastic evolutionary game
dynamics in finite populations are sensitive to the
intensity of selection. In general, the reproductive
success (fitness) of an individual is given by a
constant, denoting the baseline fitness, plus the
payoff that arises from the game under consideration. Strong selection means that the payoff
is large compared with the baseline fitness; weak
selection means the payoff is small compared with
the baseline fitness. It turns out that many interesting results can be proven for weak selection,
which is an observation also well known in
population genetics.

The traditional, well-mixed population of evolutionary game theory is represented by the complete graph, where all vertices are connected,
which means that all individuals interact equally
often. In this special situation, cooperators are
always opposed by natural selection. This is the
fundamental intuition of classical evolutionary
game theory. But what happens on other graphs?

We need to calculate the probability, ρ_C, that a
single cooperator starting in a random position
turns the whole population from defectors into
cooperators. If selection neither favors nor opposes
cooperation, then this probability is $1/N$, which is
the fixation probability of a neutral mutant. If the
fixation probability ρ_C is greater than $1/N$, then
selection favors the emergence of cooperation.
Similarly, we can calculate the fixation probability
of defectors, ρ_D. A surprisingly simple rule determines whether selection on graphs favors cooperation. If

$$b/c > k \qquad (2.4)$$

then cooperators have a fixation probability of
greater than $1/N$ and defectors have a fixation
probability of less than $1/N$. Thus, for graph
selection to favor cooperation, the benefit/cost
ratio of the altruistic act must exceed the average

degree, k, which is given by the average number of
links per individual (Ohtsuki *et al.*, 2006). This
relationship can be shown with the method of
pair-approximation for regular graphs, where all
individuals have exactly the same number of
neighbors. Regular graphs include cycles, all kinds
of spatial lattice, and random regular graphs.
Moreover, computer simulations suggest that the
rule $b/c > k$ also holds for non-regular graphs such
as random graphs and scale-free networks. The
rule holds in the limit of weak selection and $k \ll N$.
For the complete graph, $k = N$, we always have
$\rho_D > 1/N > \rho_C$. Preliminary studies suggest that
eqn 2.4 also tends to hold for strong selection. The
basic idea is that natural selection on graphs (in
structured populations) can favor unconditional
cooperation without any need for strategic complexity, reputation, or kin selection.

Games on graphs grew out of the earlier tradi-
tion of spatial evolutionary game theory (Nowak
and May, 1992; Herz, 1994; Killingback and
Doebeli, 1996; Mitteldorf and Wilson, 2000; Hauert
et al., 2002; Le Galliard *et al.*, 2003; Hauert and
Doebeli, 2004; Szabó and Vukov, 2004) and investigations of spatial models in ecology (Durrett and
Levin, 1994a, 1994b; Hassell *et al.*, 1994; Tilman and
Kareiva, 1997; Neuhauser, 2001) and spatial models in population genetics (Wright, 1931; Fisher
and Ford, 1950; Maruyama, 1970; Slatkin, 1981;
Barton, 1993; Pulliam, 1988; Whitlock, 2003).

2.5 Group selection

The enthusiastic approach of early group selec-
tionists to explain all evolution of cooperation
from this one perspective (Wynne-Edwards, 1962)
has met with vigorous criticism (Williams, 1966)
and even a denial of group selection for decades.
Only an embattled minority of scientists continued
to study the approach (Eshel, 1972; Levin and
Kilmer, 1974; Wilson, 1975; Matessi and Jayakar,
1976; Wade, 1976; Uyenoyama and Feldman, 1980;
Slatkin, 1981; Leigh, 1983; Szathmary and Demeter,
1987). Nowadays it seems clear that group selection
can be a powerful mechanism to promote cooperation (Sober and Wilson, 1998; Keller, 1999; Michod,
1999; Swenson *et al.*, 2000; Kerr and Godfrey-Smith, 2002;
Paulsson, 2002; Boyd and Richerson, 2002; Bowles

and Gintis, 2004; Traulsen *et al.*, 2005). We only have to make sure that its basic requirements are fulfilled in a particular situation (Levin and Kilmer, 1974; Maynard Smith, 1976). Exactly what these requirements are can be illustrated with a simple model (Traulsen and Nowak, 2006).

Imagine a population of individuals subdivided into groups. For simplicity, we assume that the number of groups is constant and given by m. Each group contains between 1 and n individuals. The total population size can fluctuate between the bounds m and nm. Again, there are two types of individual, cooperators and defectors. Individuals interact with others in their group and thereby receive a payoff. At each time step a random individual from the entire population is chosen proportional to payoff in order to reproduce. The offspring is added to the same group. If the group size is less than or equal to n then nothing else happens. If the group size, however, exceeds n then with probability q the group splits into two. In this case, a random group is eliminated (in order to maintain a constant number of groups). With probability $1 - q$, the group does not divide, but instead a random individual from that group is eliminated (Figure 2.6)[*].

This minimalist model of multilevel selection has some interesting features. Note that the evolutionary dynamics are entirely driven by individual fitness. Only individuals are assigned payoff values. Only individuals reproduce. Groups can stay together or split (divide) when reaching a certain size. Groups that contain fitter individuals reach the critical size faster and therefore split more often. This concept leads to selection among groups, although only individuals reproduce. The higher level selection emerges from lower level reproduction. Remarkably, the two levels of selection can oppose each other.

As before, we can compute the fixation probabilities, ρ_C and ρ_D, of cooperators and defectors to check whether selection favors one or the other. If we add a single cooperator to a population of defectors, then this cooperator must first take over a group. Subsequently the group of cooperators must take over the entire population. The first step is opposed by selection, the second step is favored by selection. Hence, we need to find out if the overall fixation probability is greater to or less than

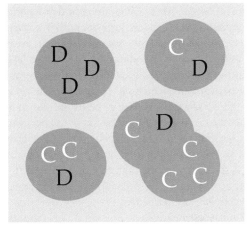

Figure 2.6 A simple model of group selection. A population consists of m groups of maximum size n. Individuals interact with others in their group in the context of an evolutionary game. Here we consider the game between cooperators, C, and defectors, D. For reproduction, individuals are chosen from the entire population with a probability proportional to their payoff. The offspring is added to the same group. If a group reaches the maximum size, n, then it either splits in two or a random individual from that group is eliminated. If a group splits, then a random group dies, in order to keep the total population size constant. This metapopulation structure leads to the emergence of two levels of selection, although only individuals reproduce.

what we would obtain for a neutral mutant. An analytic calculation is possible in the interesting limit $q \ll 1$, where individuals reproduce much more rapidly than groups divide. In this case, most of the groups are at their maximum size and hence the total population size is almost constant and given by $N = nm$. We find that selection favors cooperators and opposes defectors, $\rho_C > 1/N > \rho_D$, if

$$b/c > 1 + n/(m - 2) \qquad (2.5a)$$

This result holds for weak selection. Smaller group sizes and larger numbers of competing groups favor cooperation. We also notice that the number of groups, m, must exceed 2. There is an intuitive reason for this threshold. Consider the case of $m = 2$ groups with $n = 2$ individuals. In a mixed group, the cooperator has payoff $-c$ and the defector has payoff b; the defector/cooperator difference is $b + c$. In a homogeneous group, two cooperators have payoff $b - c$, while two defectors

---

[*] This is *not* the same q as in section 2.3; we have run out of convenient letters.

have a payoff of 0. Thus the disadvantage for cooperators in mixed groups cannot be compensated for by the advantage they have in homogeneous groups. Interestingly, however, for larger splitting probabilities, q, we find that cooperators can be favored even for $m = 2$ groups. The reason is the following: for very small q, the initial cooperator must reach fixation in a mixed group; but for larger q, a homogeneous cooperator group can also emerge if a mixed group splits, giving rise to a daughter group that has only cooperators. Thus, larger splitting probabilities make it easier for cooperation to emerge.

Let us also consider the effect of migration between groups. The average number of migrants accepted by a group during its lifetime is denoted by z. We find that selection favors cooperation provided that

$$b/c > 1 + z + n/m \qquad (2.5b)$$

In order to derive this condition we have assumed weak selection and $q \ll 1$, as before, but also that both the numbers of groups, m, and the maximum group size, n, are much larger than 1. For more information, see Traulsen and Nowak, 2006.

Group selection (or multilevel selection) is a powerful mechanism for the evolution of cooperation if there is a large number of relatively small groups and migration between groups is not too frequent.

2.6 Conclusion

We end by listing the five rules that we mentioned in the beginning. These rules represent laws of nature governing the natural selection of cooperation.

1. Kin selection leads to cooperation if $b/c > 1/r$, where r is the coefficient of genetic relatedness between donor and recipient.
2. Direct reciprocity leads to cooperation if $b/c > 1/w$, where w is the probability of playing another round in the repeated Prisoner's Dilemma.
3. Indirect reciprocity leads to cooperation if $b/c > 1/q$, where q is the probability of knowing the reputation of a recipient.
4. Graph selection (or network reciprocity) leads to cooperation if $b/c > k$, where k is the degree of the graph; that is, the average number of neighbors.
5. Group selection leads to cooperation if $b/c > 1 + z + n/m$, where z is the number of migrants accepted by a group during its lifetime, n is the group size, and m is the number of groups.

In all five theories, b is the benefit for the recipient and c the cost for the donor of an altruistic act.

CHAPTER 3

Single-species dynamics

Tim Coulson and H. Charles J. Godfray

What determines the densities of the different species of plants, animals, and micro-organisms with which we share the planet, why do their numbers fluctuate and extinctions occur, and how do different species interact to determine each other's abundance? These are some of the questions addressed by the science of ecological population dynamics, the subject that underpins all the chapters in this book. In this chapter we introduce some of the basic principles of the subject by concentrating on the dynamics of single-species systems. These are species whose population biology can be studied without also explicitly including the dynamics of other species in the community. The chief justification for this brutal abstraction is that it allows many of the underlying processes to be described simply and more clearly. Moreover, arguments based on the analysis of single-species population dynamics are often surprisingly useful in understanding real populations, especially those in relatively simple environments such as agro-ecosystems.

At the core of population dynamics is a simple truism: the density or numbers of individuals in a closed population is increased by births, and decreased by deaths. If the population is not closed then we need also to include immigration and emigration in our calculation. A population in which births exceed deaths will tend to increase and one where the reverse is true will tend to decrease. But more significant is the mode of change. If birth and death rates remain constant then the consequent increase or decrease in population numbers occurs exponentially—population dynamics occurs on a geometric rather than an arithmetic scale. In the first section of this chapter we describe the calculation of exponential growth rates for different types of population, and explore how such calculations, even though they are based on the simplistic assumption of constant demographic rates, can be very useful for a variety of problems in applied population biology.

The fact that populations persist over appreciable periods of time inescapably means that demographic rates—births and deaths, immigration and emigration—do not remain constant. In fact, population persistence in the long term requires that as populations increase in density the death rate must rise relative to the birth rate and eventually exceed it. In real populations, such demographic rates are what ecologists call density-dependent and mathematicians nonlinear, whereas an engineer might talk about negative feedback. It is this nonlinearity that can give rise to a stable equilibrium, the population density at which birth rates precisely equal death rates. But a far more diverse menagerie of dynamic behaviours is possible; the population may not settle on a stable equilibrium but show persistent cycles. Stranger still, these cycles may not be regular but complex and unpredictable in detail—they may show mathematical chaos. The possibility of chaotic dynamics in simple populations was first appreciated in the 1970s, and ecological problems were very significant in the development of this new field of mathematics. The second section of this chapter explores the consequences of deterministic density-dependent demographic rates, and explores chaos in ecology.

In the last sentence, by deterministic we mean that demographic rates are constant or simply determined by density, and do not also vary by chance. Of course, all real populations are subject to random effects. When the average birth rate is two offspring per year, some individuals will have fewer or more offspring; and in some years, or in some sites within the species' range, the average may be slightly more or slightly less. More radically, the birth rate may be two, year on year, except for the time the meteorite hit and no-one reproduced. The last 10 years has seen notable advances in the study of populations that take into account stochastic effects, and these are the subject of our third section.

3.1 The rate of population growth

All subjects need their founding myths, with appropriate heroes, and while physics has the giants Newton and Einstein, and evolution the peerless Darwin, students of population dynamics are stuck with the far less appetising Thomas Malthus. Malthus was not the first person to appreciate the geometric nature of population growth but he was the first clearly to work through its consequences. In his famous pamphlet *An Essay on the Principle of Population* of 1798, he vividly illustrated the power of geometric growth in terms that mirror the modern clichés that if population growth is unchecked it would take only a few years for the total numbers of aphid/cod/elephant or your favourite animal to weigh more than the Earth (Malthus, 1798). It was the power of this argument that so influenced Darwin—such great potential fecundity must be balanced by great mortality, and any heritable trait that favoured one individual over another would increase in frequency ineluctably. In contrast, the message that Malthus, an upper-class vicar, drew from his own insight was the need to do something about the irresponsibly fecund lower classes (as well as about other problems such as women and the French). Type Malthus into *Google* and you find him a hero to an unpleasant consortium of modern-day social engineers.

But despite its shady origin, the rate of exponential population growth based on current demographic rates is an immensely useful quantity. Consider first a simple, unstructured population; by unstructured we mean that birth and death rates are identical across individuals (clearly an approximation, as a newborn individual cannot

reproduce immediately). Let the rate at which individuals produce female offspring be b (we assume for now that males have no effect on population growth rate, something that is true for most but not all organisms) and the rate at which they die be d. Define the difference between these two rates as $r = b - d$. The population increases if $r > 0$ and decreases if $r < 0$. Moreover, if the current population size is N_0 then the population size t time units into the future is $N_t = N_0 \exp[rt]$. The population increases or decreases at a rate determined by the power of r. Note that in this simple model, to predict population growth rates we do not need to know birth and death rates separately, just their net difference.

Not all species reproduce continuously. Consider a population of an animal or plant with discrete generations that produces λ female offspring before dying. It is straightforward to see that if population size is now N_0 then t generations in the future it will be $N_t = N_0 \lambda^t$, which can be written $N_t = N_0 \exp[\ln(\lambda) \, t]$, the latter expression emphasizing the similarity with the continuous case, with $\ln(\lambda)$ replacing r.

The parameter r (or $\ln(\lambda)$) is the intrinsic growth rate of the population; it allows us to *project* population numbers into the future. Of course we do not believe r will stay constant forever—a projection should not be confused with a forecast—but it tells us something about what will happen in the short term, given the current birth and death rates. This can be a very important management tool. Suppose for example one is trying to assess the potential vulnerability of a series of populations of an endangered species. Calculating their different population growth rates will not give you a complete answer to this question, but it will provide an important clue to their different vulnerabilities. Estimations of population growth rates for more complicated population structures (see below) lie at the heart of population viability analysis, a frequently used tool in conservation biology. Epidemiology provides a rather different example of the importance of population growth rate. Consider a population of susceptible hosts exposed to a small number of infectious individuals. From the point of view of the disease, births consist of new infections and deaths occur when the host either recovers or actually dies. The disease will only spread if $r = b - d > 0$, where b and d are the rates of disease 'births' and 'deaths'. In the epidemiological literature this condition is normally stated as $\exp(r) = R_0 > 1$, which has the simple interpretation that for spread to occur every initial infection must leave at least one secondary infection. As Grenfell and Keeling (Chapter 10 in this volume) discuss in more detail, calculation of R_0, usually called the basic reproductive ratio by ecologists or, more correctly, the basic reproductive number (it is dimensionless) by epidemiologists, lies at the heart of much human and animal health population analysis.

3.1.1 Structured populations

The assumption that all populations are made up of identical individuals with the same demographic rates is clearly a gross oversimplification. How can population growth rates be calculated in more complex structured populations?

We introduce this topic by considering a population that has discrete breeding seasons so that it makes sense to census it once a year. We also suppose that the population is age-structured: demographic rates vary with age but are constant within an age class. To describe population numbers at time t we now need to write down a vector, $\mathbf{n(t)} = \{n_1, n_2, \ldots, n_x\}(t)$, where n_i is the number or density of individuals in their ith year at time t (and x is the oldest age class). To explore how population numbers change over time we need to know the probability that an individual of age i will survive to the next year (p_i) and the number of offspring it produces each year (f_i). Then

$$
\begin{pmatrix} n_1 \\ n_2 \\ n_3 \\ \vdots \\ n_x \end{pmatrix}(t+1) = \begin{pmatrix} f_1 & f_2 & f_3 & \cdots & f_x \\ p_1 & 0 & 0 & \cdots & 0 \\ 0 & p_2 & 0 & \cdots & 0 \\ \vdots & \ddots & \ddots & \cdots & \vdots \\ 0 & 0 & \cdots & p_{x-1} & 0 \end{pmatrix} \begin{pmatrix} n_1 \\ n_2 \\ n_3 \\ \vdots \\ n_x \end{pmatrix}(t)
$$

$$(3.1)$$

which can be written more succinctly $\mathbf{n(t+1)} = \mathbf{A}\,\mathbf{n(t)}$. Note that the numbers in the youngest age class are given by the numbers in each age class the season before multiplied by their fecundity,

while the probabilities of surviving until the new season form the lower subdiagonal. The matrix $\mathbf{A}$ is an example of a population-projection matrix, and this particular form, where the population is structured by age, is called a Leslie matrix (Leslie, 1945).

The simplest way to explore the growth rate of a population described by eqn 3.1 is to iterate it on a computer, an option not available to the originators of these techniques in the 1940s. But there are some important mathematical results that allow much greater insight into the population growth process. We do not have the space to derive these results or explain them in detail, but attempt to give some flavour of their elegance and importance.

The matrix $\mathbf{A}$ includes all the information we need to know about the population's demographic parameters (Caswell, 1989, 2001). From this matrix, a polynomial equation in an arbitrary variable (say η) can be derived. The order of the polynomial is determined by the number of age classes. If there are five age classes than the equation will be of order five (terms up to η^5) and if there are 20 age classes then there will be terms up to η^{20}. Just as the familiar quadratic equation (order two) has two roots (values of η for which the equation equals 0) then these larger polynomials of order x have exactly x roots, though unlike the quadratic they can only be calculated numerically (except for some special cases). A collection of mathematical results called the Perron–Frobenius theorem tells us that for Leslie matrices (with some minor exceptions that we will return to) there will always be one root that is larger than all the others. Moreover, this root, which is a complicated function of the different elements of the matrix $\mathbf{A}$, represents the long-term growth rate of the population. In matrix theory the roots are called *eigenvalues* and calculation of the largest root, the dominant eigenvalue, provides the asymptotic population projection that we require.

This powerful result tells us that whatever the initial distribution of individuals across age classes the population will eventually grow or decline at a rate set by the dominant eigenvalue (this independence of starting conditions is called ergodicity). Another ergodic property of these population models is that the proportions of individuals in different age classes assume constant values (the *stable age distribution*) irrespective of starting values. These values can be calculated directly from the projection matrix: associated with the dominant eigenvalue is a pair of vectors of length x with each element corresponding to an age class. These are the dominant *eigenvectors* and the relative magnitude of the elements of one gives us the stable age distribution (we note in passing that the other dominant eigenvector provides a measure of Fisher's reproductive value for each age class).

Whereas the population growth rate is an important management tool, applied ecologists are often also interested in how births and deaths at different age classes contribute to the overall projection. A conservation biologist may need to know whether to prioritize efforts on old or young individuals, while a game manager might need to know the consequences of allowing animals of different ages to be shot. The marginal effects on the population growth rate of changing the birth or death rate at each age can be calculated, and such relationships, which can be defined in different ways depending on precisely for what they are required, are called *sensitivities* or *elasticities*, terms borrowed from equivalent problems in economics.

The projection matrix can also be used to provide information on the speed at which the asymptotic growth rate and stable age distributions are attained. The key quantity here is called the damping ratio, which is defined as the ratio of the largest eigenvalue of the projection matrix to the second largest eigenvalue (Caswell, 2001). This root may be a complex number, and this provides information about whether there is a smooth or oscillatory approach to the long-term growth rate (Fox and Gurevitch, 2000).

Classifying individuals in populations by their age is perhaps the most common way to relax the assumption that everyone shares the same demographic parameters. But the matrix formulation is much more powerful than this and populations can be classified by size, life-history stage, sex, geographical location, or essentially any other variable. To illustrate this consider a plant whose individuals can be classified in one of three life-history stages: seedlings, small plants, and large

plants, which we shall index as stages 1, 2, and 3 respectively. The projection matrix for such a species might look like this:

$$\begin{pmatrix} 0 & b_2 & b_3 \\ a_{12} & a_{22} & a_{32} \\ a_{13} & a_{23} & a_{33} \end{pmatrix} \qquad (3.2)$$

The top row reflects stage-specific fecundity. Between one time period and the next seedlings do not reproduce but small plants produce b_2 and large plants b_3 offspring that survive to the seedling stage. The subdiagonal a_{12} and a_{23} tell us the probability that seedlings become small plants, and small plants become large plants, just as in the Leslie matrix. But we now have further transitions (or lack of transitions): a_{22} and a_{33} are the probabilities that small and large plants remain the same size (we assume this option is not open to seedlings), while a_{13} allows some plants, perhaps in extremely favourable microhabitats, to transit from seedlings to large plants in one go, and a_{32} allows those unfortunate individuals that encounter a rabbit actually to decrease in size.

All the results that apply to the Leslie matrix transfer to these more complicated structured populations: we can calculate projected population growth rates, and what we should now call the *stable stage distribution*. Analysis of matrix models of stage-structured populations has proven to be extremely valuable in many fields of ecology, but perhaps especially so in plant ecology. However, there can sometimes be difficulties in placing individuals that vary in a continuous variable such as size into the discrete classes that are required of the matrix formulation, and also in choosing the appropriate time step for analysis (Easterling *et al.*, 2000). The decision need not be entirely arbitrary, as there are some theoretical results that suggest optimal choices.

We mentioned that there were a few exceptions to the simple application of the ergodic results of matrix theory, although they are easily dealt with by straightforward extensions. These include populations with post-reproductive age classes or with single reproductive age classes. For example, consider the cicada populations in North America (*Magicicada* sp.) that take precisely 17 years to reach maturity. As long as there is absolutely no mixing of cohorts then each year class will increase or decrease in density as determined by the long-run population growth rate. But the ratio of initial frequencies in year 1, 2,..., 17 remains the same; they do not converge on a stable age distribution. In America adult *Magicicada* are abundant only once in 17 years, and this pattern could be explained by initial conditions (for example all year classes are wiped out except one) plus lack of cohort mixing. In fact this is highly unlikely, the power of even minor cohort mixing to destroy the imprint of starting values is so strong that ecologists have universally rejected this hypothesis and sought active processes to maintain the synchronized cohorts. Because the length of the life cycle is a prime number (and 11- and 13-year cicada populations are also found) the classical explanation is that it is a means of escaping predation, as it is hard for predator populations with life cycles that are not exact divisors of the cicada life cycle to increase in density (Hoppensteadt and Keller, 1976). However, a recent study has suggested that although predator satiation and/or competition among nymphs can explain why the dynamics of *Magicicada* are periodic, they do not explain why the period is a prime number of years (Lehmann-Ziebarth *et al.*, 2005). The authors speculate that a physiological or genetic mechanism or constraint might be responsible.

We have dwelt at length on the matrix formulation partly because it is relatively straightforward to explain, but we finish this section by briefly describing two alternative approaches. Suppose first that we are content to census our population at discrete time intervals (perhaps yearly) but that we are unhappy to shoe-horn individuals into discrete classes of size (or other classifying variable). Instead we want to work with the more natural continuous size distribution. This leads naturally to an integral projection model of the form

$$n(y, t+1) = \int_{x=0}^{\infty} [b(y, x) + p(y, x)] n(x, t) \mathrm{d}x \quad (3.3)$$

Here $n(y, t+1)$ represents the density of individuals of size y at time $t+1$. To calculate this value

we need to know what size classes in the last year might give rise to y-sized individuals this year. This contribution can occur in two ways: first, individuals of size x may give birth to individuals that are size y at the next census point (call this $b(y, x)$); second, individuals of size x may avoid death and grow to become size y (call this $p(y, x)$). The integral on the right hand side of eqn 3.3 simply sums these contributions to the current y class over all possible size classes last year (indexed by x). Analyses of equations such as eqn 3.3 produce very similar results to the matrix formulation; most biologically realistic populations increase or decrease at a growth rate and with an age distribution that is independent of initial starting values (Easterling *et al.*, 2000).

Finally, we can study populations structured by a continuous variable in continuous time using the famous McKendrick–von Förster equation:

$$\frac{\partial n(y, t)}{\partial t} + \frac{\partial n(y, t)}{\partial y} = -\mu(y) n(y, t) \quad (3.4)$$

Again $n(y,t)$ represents the density of individuals of size y at time t. This expression simply states that the numbers in a cohort of individuals born at the same time decline with age and time as mortality ($\mu(y)$, which is likely to be age-specific) inexorably whittles them away. To complete the model we need a birth process which is introduced as a boundary condition:

$$n(0, t) = \int_{0}^{\infty} b(y) n(y, t) \ \mathrm{d}t \quad (3.5)$$

This states that the numbers of individual of age 0 (i.e. newborns) are simply the numbers of current individuals in the population multiplied by their age-specific birth rates ($b(y)$). These equations can be generalized to populations structured by size and by other variables (Wood, 1994).

Again, as you would expect from a change of formalism rather than a change in biology, the behaviour of populations described by this model is similar to those we have discussed above. Most reasonable assumptions give ergodic population growth and age distributions. Although the McKendrick–von Förster equation has been used extensively in many branches of population biology, it is notoriously difficult to work with, both analytically and numerically, and it is generally a less popular approach than the other two described here.

3.2 Density dependence

The value of projections is that they tell us something about current populations; to move from projections to forecasts we need to know not only the current values of demographic parameters, but how they may change in the future. In this section we concentrate on how birth and mortality rates may be affected by changes in population density.

We shall begin to explore this topic using a simple, unstructured population model in discrete time which we sample every generation. We will plot the population density in the next generation as a function of that in the current generation. Figure 3.1 shows the simplest case where population growth rate is independent of density. The dashed line at 45° represents the situation where population density remains the same from generation to generation. The solid line illustrates density-independent growth, the slope of the line being λ, the discrete-time rate of population growth.

Inevitably, as populations grow they will come to exceed their resource base and this will result in

either a decline in birth rates or an increase in death rates. An illustration of this is given in Figure 3.2. At low densities populations increase from generation to generation but as density increases the rate of increase slows and then reverses. At one particular density each individual female precisely replaces herself and the population is at equilibrium. In these simple diagrams the equilibrium, marked by a star, is easily found as the density where the population growth curve crosses the 45° line.

Simple though they are, the diagrams can be used to look at dynamic trajectories as well as equilibria through a geometrical trick called *cobwebbing*. Suppose that the current population density is represented by the point 0 in Figure 3.3. The density in the next generation is read off the population growth curve at point 1. To find the density in the following generation the trick is to draw a line that forms a right angle with the (dashed) 45° line and then intercepts the population growth curve (the line $1 \rightarrow 2$ in Figure 3.3). This process can be repeated indefinitely and by noting the population densities given by the series (1, 2, 3, 4, . . .) we obtain the population trajectory. In the case of a population whose biology is summarized by Figures 3.2 and 3.3, the dynamics are a damped approach to a stable equilibrium.

In Figure 3.3 there is a single equilibrium point. This is also called a *globally stable equilibrium*

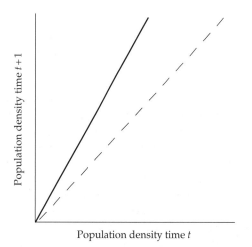

Figure 3.1 Density-independent population growth.

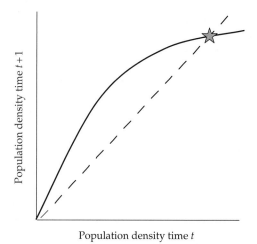

Figure 3.2 Mild density-dependent population growth. The star indicates an equilibrium.

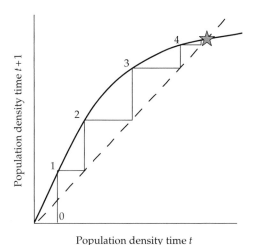

Figure 3.3 Cobwebbing with mild density-dependent population growth.

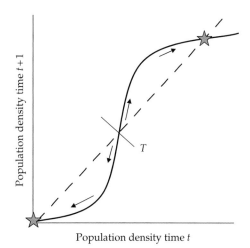

Figure 3.4 The Allee effect: populations that start off above the threshold *T* move towards the upper equilibrium; those that start off or fall below the threshold become extinct (a population size of 0 is a stable equilibrium).

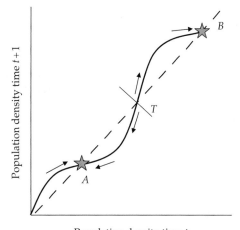

Figure 3.5 Alternative stable states: there are two non-zero equilibria (marked by stars); which equilibrium the population moves to depends on whether it starts at above or below the threshold *T*.

because whatever the initial population density the population trajectory will inevitably home in on the same equilibrium point (try cobwebbing from different starting points). But depending on the natural history of the population in question more than one equilibrium can occur. Consider a population whose growth curve can be represented by Figure 3.4. Below the threshold *T* the population in the next generation is smaller than in the current generation. This might occur if individuals in low-density populations find it difficult to locate mates (or get pollinated), or perhaps in group-hunting or colonial species if cooperation breaks down when there only a few individuals present. Any population that drops below *T* will continue to fall in density until it becomes extinct. Moreover, the population will not be able to increase from low densities. There are thus two *locally stable equilibria,* and which one is attained depends on whether the initial population density is above or below the threshold *T*. The presence of a minimum population density below which the population goes extinct is called an *Allee effect* and for obvious reasons is something that conservation biologists are very concerned about.

A species' population dynamics may include more than one non-zero stable equilibria. Consider Figure 3.5; again we have a threshold (*T*) which determines the final state of the system. If the

population starts above *T* it moves to the local stable equilibrium *B* and if below to *A*. But the lower equilibrium is now not at 0; when very rare the population is still able to increase in numbers. Note that the population in Figure 3.5 actually has four equilibria: at densities of 0, *A*, *T*, and *B*. But while *A* and *B* are stable, 0 and *T* are unstable: a population whose density is precisely 0 or *T* will remain at that density forever, but the slightest

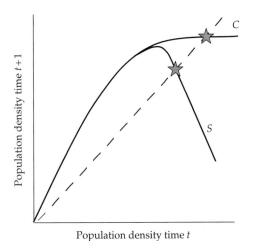

Figure 3.6 Possible population-level consequences of contest (*C*) and scramble (*S*) competition.

perturbation will cause the population to move away to either *A* and *B*.

Are there examples of multiple non-zero equilibria? There are, but to be strictly accurate most involve multispecies interactions that require a more complex approach than the simple, unstructured single-species models we are discussing here. A classic example is the spruce budworm moth, *Choristoneura fumiferana*, which is hypothesized normally to be controlled at a relatively low population density by a suite of specialist invertebrate predators (Peterman *et al.*, 1979). If, however, the population is perturbed such that control breaks down, the moth increases in density to a second equilibrium which is set by bird predation (these predators ignore spruce budworm unless it becomes very common). A further perturbation is needed to switch it back to the low-density equilibrium. The density marked *T* is thus a threshold or tipping point separating two locally stable equilibria.

Competition for food and other resources is without doubt the most important process determining the dynamics of animals and plants whose dynamics can be said to be single-species. The curves in Figures 3.2–3.5 are population-level emergent phenomena based on complex interactions occurring at the level of the individual. For example, ecologists have long distinguished

between *scramble* competition where resources are divided equally among all the individuals in the population, and *contest* competition where some individuals get the resources they require, and others get none. (Of course, many populations will show intermediate patterns.) At the population level, pure contest competition will give growth curves that resemble *C* in Figure 3.6 and scramble competition curves more like *S*. With *C*, as densities get high, a fixed number of individuals survive or reproduce to form the next generation; with *S*, at high densities everyone suffers and population numbers plummet.

We can explore the dynamic consequences of moving from contest to scramble competition using cobwebbing. In Figure 3.3 we had a fairly contest form of competition and this resulted in a smooth approach to a stable equilibrium. In the three panels of Figure 3.7 we progressively increase the scramble component. In the first panel we still have a stable equilibrium but now the approach to the equilibrium is not smooth but through damped oscillations. In the second panel, with more scramble, we no longer get a stable equilibrium but instead a stable two-point cycle. Finally, in the third panel we get curious dynamics: oscillations that never exactly repeat themselves. This is dynamical chaos, which we shall return to again in the following subsection.

We have derived this series of population-dynamic behaviours by discussing the spectrum of types of competition from contest to scramble, but what is important is not the underlying mechanism but the shape of the population growth curve, however this comes about. In particular, the angle of the growth curve at the equilibrium where it intersects the 45° line is informative. If the angle is more than 45° (i.e. the growth curve is below the 45° line to the left of the equilibrium) we have an unstable equilibrium, as at *T* in Figure 3.5. If the angle is between 45 and 0° as in Figure 3.3 we have a smooth approach to a stable equilibrium; between 0 and −45° as in Figure 3.7a we have a damped oscillatory approach to a stable equilibrium; and if less than −45° then the equilibrium is unstable but persistent cycles or chaos may occur. In analysing population models a common practice is to solve for the equilibrium values and

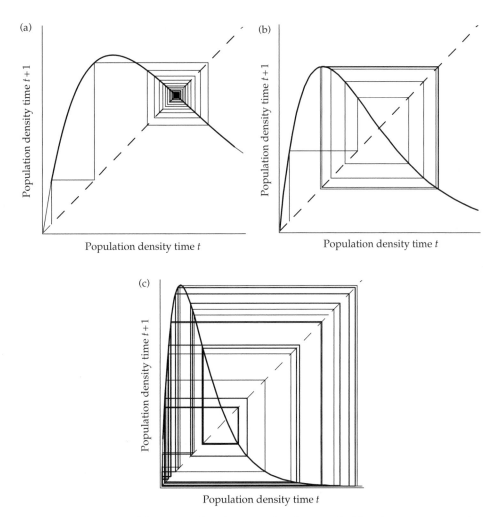

Figure 3.7 Increasing the nonlinearity of the response of population growth to density can lead from the monotonic approach to equilibrium in Figure 3.3 to an oscillatory approach (a) and then a two-point cycle (b) and finally chaos (c). Examples of cobwebbing; the underlying population model is the Ricker equation, $x_{t+1} = x_t \exp(r(1 - x_t))$ where x is population density (scaled to be 1 at carrying capacity) and r is fecundity.

then to conduct a *stability analysis* to determine the equilibrium's properties. In some cases it is possible to show that an equilibrium such as that in Figure 3.3 is globally stable, but such analyses are mathematically challenging and often there is no global equilibrium. Instead, a local stability analysis is performed. For populations described by the type of growth curve shown in Figures 3.2–3.7 we have been able to derive the local stability criteria using intuition helped by a little cobwebbing. Moving to more complex structured or multispecies communities these simple geometrical insights are lost, but algebraic conditions that

are their multidimensional equivalents can be derived (May, 1972).

In this section we have focused on a simple, unstructured population in discrete time. We could alternatively have studied a population in continuous time whose dynamics are described by the equation

$$\frac{dN}{dt} = [b(N) - d(N)]N = r(N)N \qquad (3.6)$$

where birth rate (b), death rate (d), and net population growth (r) are now all functions of population size N. Plotting $r(N)$ against N would similarly tell us much about the dynamics of the system. There is, however, a difference between populations described by eqn 3.6 and those that we studied in Figures 3.1–3.7. In eqn 3.6 the effects of population density act instantaneously on population growth rate. In our discrete-time examples there is an implicit time lag: the level of competition experienced by individuals in the current generation is set by interactions that occurred in the last generation. Such time lags tend to be destabilizing as they delay the onset of a reduction in population growth rates as densities climb, and make it more likely that any equilibrium is overshot. Indeed, it is mathematically impossible for a population governed by eqn 3.6 to show chaos. We should stress that it is not the difference between continuous- and discrete-time formalisms that lies behind the contrasting stability, but the presence of the time lag. Indeed, in continuous time we can get exactly the same dynamics by explicitly making net population growth rates a function of previous population densities,

$$\frac{dN_t}{dt} = r(N_{t-\tau})N_t \qquad (3.7)$$

where τ is a time lag of approximately one generation.

Structured population models with density dependence can be built using the same matrix, integral equation, or partial differential equation approaches discussed in the section on density independence. Naturally they are more complex, and often with a greater potential for destabilizing time lags. Relaxing the assumption that all individuals are equal also leads to the possibility of more complicated types of interaction than are possible for unstructured populations. Competition may be asymmetric, typically with smaller individuals suffering disproportionately at the hands (or roots) of larger individuals. Moreover, cannibalism is much more common in the animal kingdom than often realized, and when it occurs it is nearly always size-related, with older larger individuals consuming their smaller conspecifics. Such age-specific interactions have been studied in detail, particularly in insect systems that can be maintained in the laboratory for multiple

generations. Many of these systems show population cycles with periods shorter than those predicted by unstructured models (e.g. Figure 3.7). The details differ with the natural history of the different systems but a common pattern is for an older cohort of individuals to reduce the numbers in a younger cohort by out-competing them for food or through cannibalism. When the depleted younger cohort grow old enough to be dominant competitors or cannibals themselves there are not enough of them to reduce significantly the next cohort coming through. This means that the next group of individuals to mature into the older cohort are very numerous and decimate the current younger cohort, and the cycle begins again.

3.2.1 Chaos

The pioneers of modern mathematical dynamics, particularly Poincaré at the end of the nineteenth century, realized that the behaviour of highly nonlinear systems could be very odd, but in the absence of computers to help visualize their dynamics, progress on understanding what was happening was very slow. When computers began to become available in the 1960s workers in fields such as meteorology and ecology were able to see the complex dynamics produced by beguilingly straightforward equations, and this led to a burst of interest in both pure and applied mathematics that laid the foundations of the modern field of chaotic dynamics. In population ecology, the classic paper is May's *Simple mathematical models with very complicated dynamics* (May, 1976a), which not only introduced the notion of chaos to the field but showed that lurking underneath the seeming unpredictability of chaotic dynamics was considerable order and pattern. We shall now explore a population model of exactly the type that May analysed.

Chaos has already been encountered in this chapter as the dynamics that emerge in a simple discrete-time population model as the population growth curve (or map) becomes sufficiently nonlinear (the 'humpiness' of the curves in Figure 3.7). Let us now specify a family of curves that can give rise to the maps in Figures 3.3 and 3.7. For reasons that will be explained in a few paragraphs it does not particularly matter which family we chose, and we plump for the Ricker equation as it is commonly used in applied population biology, particularly for fisheries (Ricker, 1954).

$$n_{t+1} = n_t \exp[r(1 - n_t)] \tag{3.8}$$

Here n_t is population density (scaled to equal 1 at equilibrium). When rare the population increases each generation by a factor $\exp[r]$ but as densities approach 1 the increase slows and above 1 it reverses. If r is high there is the potential for the population to overshoot the equilibrium.

We want to picture the dynamics of the whole system for different values of the sole adjustable parameter r. To do this, imagine iterating the equation by cobwebbing as in Figures 3.3 and 3.7 and then throwing away all the transient dynamics, perhaps the first 50 generations. For Figure 3.3 (corresponding to a value of $r = 1$) the non-transient dynamics would not be very interesting: it would simply be a population at stable equilibrium, in this case $n = 1$. In Figure 3.8 we plot r along the x axis and the non-transient dynamics on the y axis; for $r = 1$ there is a single point at $n = 1$. The dynamics of the population described by the first panel in Figure 3.7 ($r = 1.9$) differ only in their transient behaviour and so it too would be represented by a single point at $n = 1$. Indeed, for the Ricker equation a stable equilibrium occurs for all persistent populations with $r < 2$, which gives the straight line at $n = 1$ in the left-hand part of Figure 3.8.

The persistent dynamics depicted by the middle panel of Figure 3.7 ($r = 2.3$) are a two-point limit cycle: the population oscillates for ever between two densities, one greater and one less than the now unstable equilibrium $n = 1$. In Figure 3.8 this appears as two points. The value of this representation now becomes clear because instead of having to try to compare a large number of cobweb diagrams we can see at one glance how the cycles appear at $r = 2$ and then increase in amplitude as r gets bigger. The change of behaviour at $r = 2$ is called for obvious reasons a bifurcation, and the representation itself is a bifurcation diagram. We can also see that at $r = 2.5$ a second bifurcation occurs to give a four-point cycle, and then further bifurcations at increasingly smaller intervals of r until a limit is reached. "*What happens at the point of accumulation [the limit]?*" is what May scrawled on a blackboard in the Theoretical Physics Department at Sydney University in the early 1970s.

May showed that what happens is chaos. As the third panel in Figure 3.7 illustrates, the trajectory never converges on a simple cycle but fluctuates aperiodically around very many values of n, never repeating itself. This is represented in Figure 3.8 by a vertical line containing numerous, in fact an infinite number of, points. Cobwebbing can also be used to demonstrate a cardinal property of chaos: namely sensitivity to initial conditions. Start two trajectories very close together and sooner or later they will diverge. This is not due to a lack of computing power: no matter how close the two initial values they will come to diverge. More accurate estimation of initial values, so that the

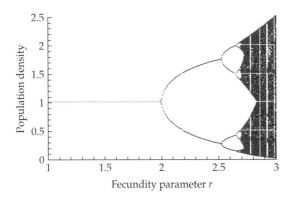

Figure 3.8 The bifurcation diagram for the Ricker population model (see the legend of Figure 3.7).

measured value is close to the 'true' value, can delay the divergence, but not prevent it, and this means that there is an absolute limit to our ability to predict into the future the behaviour of chaotic systems.

The dynamics of a system can be described with a quantity known as the Lyapunov exponent (named after a Russian mathematician whose name is also transliterated Liapunov or Ljapunov). The Lyapunov exponent describes the rate of separation of infinitesimally close trajectories; a positive value means that the trajectories diverge exponentially, and this extreme sensitivity to initial conditions is the hallmark of chaos. Algorithms have been derived to estimate Lyapunov exponents directly from time series (Wolf *et al.*, 1985) and have proved very valuable, especially in the physical sciences where relatively long time series are typically easier to obtain.

Bifurcation diagrams are beautiful objects that contain a wealth of mathematical detail. They have quite literally been the subject of tens or possibly hundreds of mathematics PhD theses. Our focus here is on their relevance to biology and we have space to mention only a very few more technical results. First, May and others showed that the patterns in Figure 3.8 do not just apply to the Ricker equation but to a very broad class of models that all show the same transition through period-doubling from order to chaos (May, 1973a, 1974c; Li and Yorke, 1975). There is a limited number of routes to chaos and one can derive general results that apply to very many systems. For example, the ratio of the interval of *r* values in which two-point cycles are found and in which four-point cycles are found is 4.6692. In fact the same ratio is found for every adjacent interval (four-point/eight-point, etc.) not only for the Ricker equation but for *every* map that shows this type of transition from order to chaos (Feigenbaum, 1978). Second, if you look closely at the bifurcation diagram to the right of the accumulation point you see that the region of chaos contains intervals of simpler dynamics, including period-three cycles that undergo their own transition back into chaos. In fact there is an infinite number of narrow, periodic windows. Finally, the bifurcation diagram has fractal structure: enlarge part of the region of chaos and you

will see a complex pattern of bifurcations, aperiodic and period trajectories; chose part of this picture and enlarge yet again and the same patterns appear in miniature, and so on *ad infinitum*.

The beauty of bifurcation diagrams is fragile: add a little stochastic noise—inescapable in real biological systems—and their more rococo patterns disappear. However, the extreme sensitivity to initial conditions, the signature of chaos, remains. So while it is not mathematically true that the Ricker model predicts chaos for all $r > 2.69$ it might as well be for any biological purposes. Another biologically relevant property of chaos is also shown in Figure 3.8. Although precise prediction is not possible the different population trajectories are bounded, that is they cannot become arbitrarily large or small. A pure random walk would not be bounded (except of course by $n = 0$). Indeed, it is sometimes possible to calculate the probability distribution of different population states. Depending on the system this may be valuable information for ecologists and population managers.

Chaos is not just a property of discrete-time systems and chaos in continuous time systems has also been extensively studied. Consider the non-transient behaviour of a continuous system. If the system is at equilibrium this will be a simple point but if there are persistent cycles or chaos then it will be a continuous line. For single-species populations this line can be plotted in a space where the coordinates are population densities now and at times in the past. For example, on a three-dimensional graph the coordinates might be densities now, 1 month ago, and 2 months ago. In this space a cycle will be a closed loop while a chaotic trajectory will be an object such as that on the left of Figure 3.9. This object looks like a twisted diaphanous sheet and is a fractal: successive magnifications of parts of the sheet show the same self-similar pattern. One point to note is that chaos occurs in simple (ordinary) differential equations only for systems of three or more variables: the dynamics of a two variable-system can be described in a two-dimensional space which does not allow for the twisting and mixing of trajectories that are the hallmarks of chaos.

There is a close link between chaos and fractals. Objects that represent the non-transient behaviour of a dynamic system are called *attractors* (because trajectories originating elsewhere in state space are attracted to them). In continuous time, points (stable equilibria) and closed loops (cycles) are examples of normal attractors whereas fractal objects such as that in Figure 3.9 are termed *strange attractors*. All chaotic systems are governed by strange attractors and, as we shall return to shortly, determining that a system's attractor is fractal is one way of identifying chaos in nature. The attractor in Figure 3.9 also provides an insight into why chaos is always associated with extreme sensitivity to initial conditions. The right-hand panel in Figure 3.9 is a cartoon to illustrate the evolution of a set of initially very similar trajectories: the bundle marked 1, which should be imagined as lying flat on the horizontal surface of the attractor in front of the line X. Flow on the attractor occurs in the counter-clockwise direction and sets of points are first stretched (2, 3) and then folded (4, 5). If you imagine this occurring numerous times it is easy to see how trajectories that start off near each other quickly become separated. The degree of stretching in a system is quantified by the Lyapunov exponent.

Chaos in continuous- and discrete-time systems is intimately related. Consider the section X (called a Poincaré section) through the attractor in Figure 3.9. If the position along the section is treated as a variable, and if the position in the current traverse is plotted against that in the previous, one arrives at a map exactly equivalent to the chaotic Ricker map discussed above. Now, however, the r parameter is not simply a measure of single-species fecundity, but a more complex amalgam of the life histories of all species or development stages that influence the dynamics.

How might one seek to decide whether natural populations are chaotic? Typically this has to be done from time-series data, which at least in comparison with data from the physical sciences are inevitably of relatively short duration. There are two broad approaches. The first is to try to fit a flexible population model to the time-series data and then to determine by iterating the model whether the dynamics are chaotic. The second is to try directly to reconstruct the attractor governing the system and determine whether it is fractal. Both approaches are helped by a very important theorem (Takens, 1981) that states that the attractor of a multi-species or complex single-species interaction can always be reconstructed from single-variable time-series data in a space made up of a sufficient number of time-lagged dimensions (i.e. the coordinates are densities at time t, $t - \tau$, $t - 2\tau \ldots$ where τ is a lag). The major proviso is

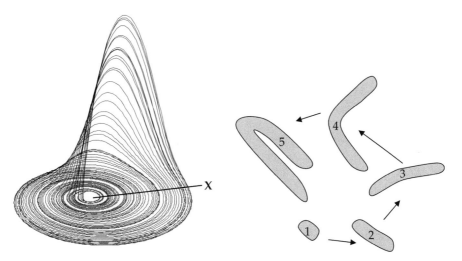

Figure 3.9 Chaos in continuous time. The object on the left is a strange attractor describing the flow of trajectories of a continuous-time system in three-dimensional space (the Rössler attractor). X is a Poincaré section discussed in the text. The cartoon on the right describes how bunches of nearby trajectories become stretched and folded as they move around the attractor. See text for further details.

that you have to have sufficient data, which in practice is usually a very demanding requirement.

The first attempt to fit models to data did not use time series but life-history data on fecundity and density-dependent mortality. Hassell *et al.* (1976) fitted a two-parameter model to data from 24 species of insects with reasonably discrete generations and concluded that the vast majority had stable dynamics, indeed not even showing an oscillatory return to equilibrium. Although the authors were at pains to stress the provisional nature of their conclusions, this paper had a very major impact, and to a certain extent inadvertently licensed ecologists to treat chaos as a theoretical curiosity for the next decade.

The next major attempt to search for chaos used model-free approaches and was spurred by the growth of empirical chaos studies in the physical sciences (Schaffer, 1985; Schaffer and Kot, 1985a, 1985b; Olsen and Schaffer, 1990). The basic idea was to reconstruct the attractor by embedding the time series in time-lagged coordinates and then either to take a Poincaré section and look for a one-dimensional chaotic map, or to estimate the *attractor dimension*. In our daily lives we do not normally need tests to tell us whether an object is one-, two-, or three-dimensional but mathematicians who often work in much higher dimensional space have derived algorithms to estimate arbitrary dimensionality. When these are applied to fractal objects they return a non-integer dimension. A non-integer dimension implies a fractal and a fractal implies chaos. Though clearly worth trying, ultimately this research programme was defeated by the quality of the data available. To quote Schaffer (2000), 'Only in the instance of recurrent outbreaks of measles in human populations, was there sufficient data to justify our initial enthusiasm' and, he added, even here the argument chiefly rested on the comparison of time-series data with the output of epidemiological models.

In the last 15 years, interest has grown again in the challenge of detecting chaos from time series. Sugihara and May (1990) developed a technique called nonlinear forecasting which measures the extent to which predictability decays with time. In chaotic systems this occurs in a characteristic way determined by the magnitude of the Lyapunov exponent. This method has since found wide application beyond biology in econometrics. Model-based approaches have also enjoyed renewed attention. One strand has sought to develop more accurate mechanistic population models, capitalizing on both the more powerful computing tools now available and statistical advances in extracting parameter values from data. A different strand, with similarities to Sugihara and May's approach, fits very flexible non-mechanistic population models to time-series data typically using response surfaces that are optimized either by traditional least-squares methods or more exotic techniques such as thin-plate splines or neural nets (Ellner and Turchin, 1995). The magnitude of the dominant Lyapunov exponent is calculated directly from the fitted model. It is still too early to judge the long-term value of these methods, although they have revealed a number of systems with apparent chaotic dynamics, in particularly involving human–disease and predator–prey interactions.

For single-species interactions, the best examples of possible chaos involve laboratory systems, including Nicholson's famous long-term blow fly experiment. A very nice experimental example is the work of Costantino *et al.* (1997) on the flour beetle, *Tribolium castaneum*. Recall we mentioned above that strong interactions between different life-history stages can give rise to complex dynamics. In *Tribolium*, adults and larvae cannibalize eggs while adults also eat pupae. A population model showed that by varying a single parameter (pupal mortality) the dynamics of the system moved from stability to chaos and then to a three-point cycle. Figure 3.10 shows that experimentally manipulating pupal mortality leads to dynamics that look very like those predicted. It is true that this is a highly artificial system, yet it is an impressive demonstration that the dynamics of these insects have been understood.

3.3 Randomness

3.3.1 Types of random effect

Real animals, plants, and micro-organisms are continually buffeted by the effects of random

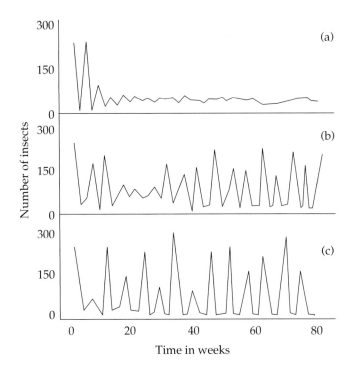

Figure 3.10 Time series of the number of larval beetles in laboratory populations for different rates of pupal mortality which were artificially manipulated. Theoretical models predict that the population in panel a should have a stable equilibrium, panel b should be chaotic, and panel c should have a three-point cycle. The experimental data show good agreement with the predictions (after Costantino et al., 1997).

processes and a critical question in population biology is the extent to which insights gained from the analysis of deterministic models survive the insults thrown at them by stochastic nature.

There are a variety of different ways in which random or stochastic effects can influence population dynamics (May, 1973a). Perhaps the most straightforward is *environmental stochasticity*, where the value of a demographic parameter changes over time. Recall the density-independent, discrete-time model $N_{t+1} = N_t \lambda$ where λ is the annual population growth rate. This model implicitly assumes that the value of λ is constant, but in fact it will almost certainly vary from generation to generation; we might better write the equation $N_{t+1} = N_t \lambda_t$ to emphasize this fact. Note that environmental stochasticity affects the demographic rates of all individuals in a population in the same way, and that this effect is independent of population size (Lande et al., 2003). Much research in identifying factors generating environmental stochasticity has focused on climate (Stenseth et al., 2002), although in principle any other factor with unpredictable effects on population parameters can contribute to this process.

Let us return to the discrete-time model of a population with non-overlapping generations, $N_{t+1} = N_t \lambda$, and for the sake of argument assume that the value of λ is actually constant over time. But this does not mean that every single individual in the population will produce exactly λ female offspring. In the real world there will always be some between-individual variation or *demographic stochasticity*. For example, consider a parasite that searches randomly for hosts into which it lays a single egg; if the average parasite lays λ female eggs then some will by chance discover more hosts and some by chance fewer. This is a Poisson process where the variance is the same as the mean. One can imagine other natural histories where the variance is much less than a Poisson process (vertebrates that normally produce one offspring a year) and others where the variance is much greater (organisms living in a highly heterogeneous environment). Now suppose the population is small: by chance all individuals in one generation may experience low reproduction and so the following year the population size would be significantly less than the expected $N_t \lambda$. Of course, the probability of simultaneous episodes of good or bad luck become progressively more unlikely in larger populations and hence demographic stochasticity is most important in small populations. In many ways, its action is similar to drift in population genetics.

A further random process that is sometimes distinguished is catastrophic stochasticity: random events that destroy the whole population irrespective of its size or current demographic parameters. We shall not discuss this type of randomness further here, although it is particularly relevant to studies of metapopulations (see Chapter 4 in this volume) and also in conservation biology where populations may be wiped out by human action that can at least be approximated as a random process.

3.3.2 Density-independent populations

Let us now see how stochasticity affects population growth rate and population projection. For ease of explanation we shall stick to discrete-time models although the same principles apply to populations that reproduce in continuous time. Return once again to the model $N_{t+1} = N_t \lambda_t$ where the subscript to the population growth rate emphasizes that it varies between generations, specifically with mean $\bar{\lambda}$ and variance σ_λ. This is the way that randomness is most frequently dealt with in population models, and has been referred to as the equilibrium treatment of noise (Coulson *et al.*, 2004). If we take logarithms then we can write

$$\text{Log}[N_t] = \log[N_0] + \sum_{x=0}^{t-1} \log[\lambda_x]. \tag{3.9}$$

If the values of λ vary independently over time, then the right-hand term is the sum of independent random variables, which the Central Limit Theorem tells us is asymptotically normally distributed. This implies that population size itself is lognormally distributed. There are some complexities in calculating long-term population growth rates in this case (Lewontin and Cohen, 1969). An intuitive procedure might be to see how expected population size grows with time. A simple calculation reveals it increases exponentially at a rate determined by $\bar{\lambda}$. But the expected

population size is dominated by very rare, huge population sizes in the upper tail of the distribution. In fact the modal population size, the population size that will actually be observed in the field, grows not at a rate determined by the simple *arithmetic* mean, $\bar{\lambda}$, but the *geometric* mean $(\lambda_0 \cdot \lambda_1 \cdot \lambda_2 \cdots \lambda_{t-1})^{1/t}$.

Several biologically interesting results follow from this. First, as long as there is some variance in λ the geometric mean will always be lower than the arithmetic mean: poor years have a greater negative effect on population growth than the positive effect of good years. Second, a single year with zero net reproduction ($\lambda = 0$) renders the long-term growth rate 0. This makes intuitive sense as the population goes extinct, but note that this is not what a calculation based on the arithmetic mean would suggest. Finally, recall that in the deterministic case persistence was very straightforward: a population would increase if $\lambda > 1$ and decrease if $\lambda < 1$. The situation is now more complicated: populations with geometric mean growth rates less than one will always ultimately go extinct, but some may persist for a long period of time if by luck they experience a chain of propitious years. Similarly, although populations with geometric growth rates greater than 1 will tend to persist, some will by bad luck go extinct. In fact populations which will, on average, grow to infinity also have a probability of extinction of 1 for very long periods of time. This can be seen very simply: if $E(N_t) = T^2$, where t represents time and T is the length of time since the simulation began, the probability of extinction can be written as $1 - 1/T$. When T gets very large, the expected population size tends to infinity and the probability of extinction tends to unity as $1/T$ approaches 0. It is possible to calculate the distribution of persistence times of populations governed by different distributions of growth rates, and this may be helpful in population management.

In many ways the population effects of demographic stochasticity are similar to its environmental counterpart. It will increase the variance in λ and so tend to reduce long-term growth rates, and increase the probability of extinction by bad luck. The major difference is that its effects become very weak as population size increases. Indeed, the total variance in reproductive rates can be thought of as the sum of two components, V_E (environmental stochasticity) and V_D/N (demographic stochasticity divided by population size). A reasonable rule of thumb is that demographic stochasticity can be ignored for populations with more than 50 or so female breeders, though note that the population size of large carnivores, even in extensive nature reserves, can often be below this threshold.

We stated above that we were assuming that stochastic effects were uncorrelated over time. Often this will not be the case, especially for short-lived organisms that might, for example, have several generations in a single summer. Quite frequently there will be a positive correlation between the random component of population growth rates in successive seasons (the term *red noise* is sometimes used for these positively correlated random effects). The most important effect of correlated stochasticity is to increase the severity of poor breeding seasons that now tend to follow one another. We note in passing that correlated red noise may lead to patterns in population dynamics that may be very hard to distinguish from an underlying deterministic cause, especially in structured populations. There can also be correlations between environmental and demographic stochasticity, in particular the effect of demographic stochasticity on population growth may be higher in years when the consequences of environmental stochasticity are most severe, a clear concern in conservation biology.

The arguments above apply also to structured populations, though with some complications. First, there is no longer a simple relationship between arithmetic and geometric population growth rates, but a stochastic equivalent to the deterministic growth rate can be calculated (Tuljapurkar, 1982). As with the unstructured population, adding stochastic effects always reduces long-term growth rates. Second, certain age or stage classes may be much more susceptible to stochastic perturbation than others. Random effects may thus lead to perturbations that disrupt the age-structure of the population (*structural variance*; Coulson *et al.*, 2001; Lande *et al.*, 2002). Here, stochasticity influences the population

dynamics via two routes. First, stochasticity has a direct effect on the size and structure of the current population. Second, these changes influence the future trajectory of the population. This interaction between stochasticity and the deterministic skeleton is sometimes referred to as the active treatment of noise, and is currently an area of considerable interest in population biology research. Such effects always reduce the tendency of the population to reach a stable age distribution and, in anticipation of the next section, can also have important consequences on population regulation if the strength and action of density dependence is also influenced by population structure.

3.3.3 Density-dependent populations

In a real stochastic environment a population is highly unlikely to remain at the exact same equilibrium value from one generation to the next. But it is still reasonable to talk about an equilibrium if populations above a certain value tend to decline in numbers, and those below the same value tend to increase. Conceptually we can think of an equilibrium not as a fixed population density, but as a probability distribution that remains the same over time and which determines the likelihood of observing the population at any particular level of abundance (Turchin, 2003). Of course, we should also consider the possibility that a population, even one that tends to increase when rare, goes extinct through a run of bad breeding seasons.

More generally, stochastic effects can cause a population to shift from one type of dynamic behaviour to another. Figure 3.5 depicted the dynamics of a species with two locally stable equilibria; it is possible that a sufficiently large random perturbation can move the population from the domain of attraction of one equilibrium to that of the other. Similarly, where there is an Allee effect a species is unable to increase in density when rare so zero population density is locally stable; random effects can push a species density below the critical threshold that leads to extinction. It is also possible that a species that for some reason has fallen below the threshold can be rescued by a random set of good breeding seasons. Of course, even when a species can increase when rare, stochastic extinction is permanent if there are no sources of migrants to rescue the population. This treatment of stochasticity in population models has been called the passive treatment of noise.

The shape of the equilibrium probability distribution of abundances will obviously be determined by the magnitude and direction of the stochastic perturbations to the demographic parameters, but also by the dynamic consequences of the perturbations; that is, the interaction of the noise with the deterministic dynamics. Consider unstructured populations with deterministically stable equilibria which are approached either smoothly (Figure 3.3) or by damped oscillations (Figure 3.7a). It is very likely that the first population will tend to return towards the equilibrium faster than the population with damped oscillations, and for the same amount of environmental stochasticity will have a lower variance equilibrium population density. A population with an oscillatory approach to a stable equilibrium can more easily be prevented from reaching that equilibrium and thus appear to the observer to be persistently cyclic. This type of dynamic behaviour has been termed *quasicyclic* (Nisbet and Gurney, 1976) and has been seen in several experimental systems, including the flour beetle study described above as an example of chaos (Costantino *et al.*, 1997).

Consider an unstructured dynamic system that is at the edge of chaos, perhaps showing persistent cycles. If one or more parameters were changed slightly, it would move from persistent cycles into the region of chaos where its dynamics would be governed by a strange attractor. Near this threshold, the transient behaviour of the population before it settles into persistent cycles can be very complex. Although in this region there is not a strange attractor, dynamics may be influenced by an object called a *strange repeller* (Rand and Wilson, 1995), which like a strange attractor is a fractal, but repels rather attracts dynamic trajectories. One can think of the system like the ball in a pinball machine, careering from buffer to buffer, perhaps

for a significant period of time. Indeed, this behaviour may go on for ever if stochastic perturbations are large enough to prevent the system ever from settling on the stable cycles. The time series produced by such a process can be indistinguishable from chaos: it can show exactly the same extreme sensitivity to initial conditions, and attempts to reconstruct the attractor would suggest that it had a non-integer number of dimensions.

In discussing the bifurcation diagram in Figure 3.8 we already noted how random effects would interact with the deterministic component of the dynamics to give chaotic population behaviour throughout the region beyond the 'point of accumulation', even though here there are narrow windows of cyclic behaviour. As with chaotic repellers this is another example of the impossibility of separating the deterministic and stochastic aspects of population dynamics in general and chaos in particular.

Although it may seem unarguable that we should seek to develop models with both stochastic and deterministic components, exactly how to do this is not always obvious. For example, adding one type of noise to a model with a deterministically stable equilibrium and a different type of noise to a model governed by a chaotic attractor can produce dynamics that equally well match the type of data that ecological field studies produce. Also it is often not clear how stochasticity should be introduced into the model, onto which demographic parameters, and with what correlation structure. Nevertheless, we are optimistic about the future. For the analysis of time series and other observational data there are a variety of new statistical methods and techniques that will help identify the major stochastic drivers, and reveal how they interact with the underlying biology of the species (Coulson *et al.*, 2001; Lande *et al.*, 2003; Turchin, 2003; Stenseth *et al.*, 2004). There is also an increasing willingness of ecologists to experiment, both in the laboratory and the field, and to integrate modelling with experimental design and analysis.

CHAPTER 4

Metapopulations and their spatial dynamics

Sean Nee

4.1 Introduction

The study of metapopulation dynamics has had a profound impact on our understanding of how species relate to their habitats. A natural, if naïve, set of assumptions would be that species are to be found wherever there is suitable habitat that they can get to; that species will rarely, if ever, be found in unsuitable habitat; that they will be most abundant in their preferred habitat; that species can be preserved as long as a good-size chunk of suitable habitat is conserved for them; and that destruction of a species' habitat is always detrimental for its abundance. We will see that none of these reasonable-sounding assumptions is necessarily true. Metapopulation biology is a vast field, so to focus this chapter I will be guided partly by questions relevant to conservation biology.

There are two important kinds of metapopulation. The so-called Levins metapopulation idea (Levins, 1970) is illustrated in Figure 4.1. It is imagined that patches of habitat suitable for a species are distributed across a landscape. Over time, there is a dynamical process of colonization and extinction: the colonization of empty patches by occupied patches sending out colonizing propagules and the extinction of local populations on occupied patches. This extinction can occur for a number of reasons. Small populations are prone to extinction just by the chance vagaries of the environment, reproduction, and death—environmental and demographic stochasticity (May, 1974b; Lande et al., 2003). An example of a species for which this is important is the Glanville fritillary butterfly (*Melitaea cinxia*), which has been extensively studied by Hanski and colleagues (Hanski, 1999). This Scandinavian butterfly lives in dry meadows which are small and patchily distributed. Another reason for local population extinction is that the habitat patch itself may be ephemeral. For example, wood-rotting fungi will find that their patch ultimately rots completely away (Siitonen et al., 2005) and epiphytic mosses will ultimately find that their tree falls over (Snall et al., 2005).

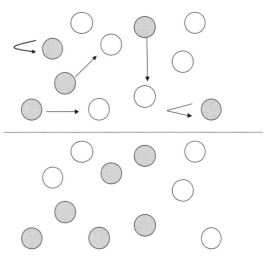

Figure 4.1 The Levins metapopulation. Shaded circles are habitat patches with a local population of a species: empty circles are empty habitat patches. An arrow from a filled circle to an empty circle indicates a colonization event: an arrow turning back on itself onto a filled circle indicates an extinction. Two successive periods in the life of a metapopulation are illutrated. The long-term dynamics have been likened to the asymmetrical blinking on and off of Christmas tree lights (Wilson, 1992). Although extinction is the ultimate fate of any local population, nonetheless the species can persist as a metapopulation.

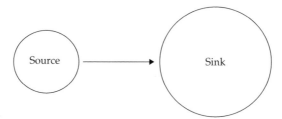

Figure 4.2 Source and sink populations. In the simplest case, there are two habitat patches of different types, only one of which is of sufficient quality to sustain a population (the source population). However, a population can be maintained in the suboptimal habitat (the sink population) if it is topped by immigration from the population in the optimal habitat. Reproduction may occur in the sink habitat, but at a level that is insufficient to maintain the population in the absence of immigration. The relative sizes of the circles are chosen to make the point that the sink population may be much larger than the source.

The second type of metapopulation consists of local populations connected by dispersal, but without the extinction of the local populations. This type has been studied intensively in population genetics, the terminology of which refers to populations and subpopulations rather than populations and metapopulations. (Perhaps ecologists look up, star-wards, whereas population geneticists look down, navel-wards?) As an example of this type of metapopulation we will specifically look at source–sink metapopulation dynamics, illustrated in Figure 4.2. In this model, populations in patches of good habitat sustain populations in poor habitat.

Both types of metapopulation are, of course, abstractions, just like the concept of a population itself. And there are many important related abstractions, like the mainland/island concept of island biogeography (Macarthur and Wilson, 1967), which can be viewed as a hybrid of the above two ideas. For conservation biology, one important way that reality departs from the pure Levins metapopulation is that different local populations may have different extinction probabilities and some patches may be more accessible for dispersal than others (Harrison and Taylor, 1997). Hanski has elaborated metapopulation theory further to incorporate these and other realistic features (Hanski, 1994).

The structure of this chapter is somewhat unusual in that I will split the discussion of the Levins metapopulation in two and sandwich the source–sink section in between. This is done to highlight (1) how much can be done with very simple

models that are essentially nothing more than cartoons and (2) the power of graphical models (see also Chapters 3 and 5).

4.2 The Levins metapopulation model

We will now consider Levins' metapopulation models in more detail. Book-length treatments exist (Hanski and Gilpin, 1997; Hanski, 1999). There are different biological systems that can be studied within the framework of metapopulation theory. The patches and local populations may correspond to their ordinary meanings: local populations of butterflies in a meadow or shrews on different islands, for example. Of course, it must be established that the dynamical processes of colonization and extinction are actually occurring. So, for example, Smith and Green (2005) conducted a meta-analysis to see if the metapopulation paradigm is appropriate for pond-dwelling amphibians (it largely is). But there are other systems that also belong in the metapopulation framework and some of these are listed in Table 4.1. This more abstract view received prominence after the work of Lande in his celebrated study of the Northern spotted owl (Lande, 1987, 1988a, 1988b), a territorial species, although Hastings (1980) may have been the first to adopt it in the study of coral dynamics.

This chapter discusses metapopulations at a general level that can encompass all these different biologies. The disadvantage of this approach is that the devil may be in the detail and a metapopulation analysis that is biology-general

Table 4.1 Some less obvious examples of Levins metapopulations.

Example	Local population	Patch	Colonization	Extinction
Territorial species (Lande, 1987)	Breeding pair	Territory	Occupation of empty territory	Death
Sessile organisms (Hastings, 1980; Stone, 1995)	Single individual	Sufficient, suitable, space for an individual	Establishing in the space	Death
Gut flora	Bacterial population	Individual gut	e.g. Neonatal ingestion of feces	Host death or treatment with antibiotics
Infectious disease organisms (Nee *et al.*, 1997)	Individual viral load, for example	Host individual	Infection	Host death or recovery
Persistent infection, e.g. HIV	Intracellular agent	Suitable cell, e.g. CD4 T-cells	Entry into cell	Cell death

(biology-free?) may lack features essential to understanding a specific problem at hand. The advantage of the approach is that one can gain insight into general properties of metapopulation dynamics that are not specifically tied to any particular biological detail. An additional advantage is that results derived in one context can be seen to be relevant in other contexts. For example, as indicated in Table 4.1, epidemiology is a version of metapopulation biology and, as a vast and mature field, may have results that can be plundered for use in other contexts. We will see examples in this chapter. Another recent example of the virtue of an abstract view of models, if it had been taken, comes from the neutral theory of biodiversity (Hubbell, 2001), where much time and effort was spent re-deriving results that are well known in the neutral theory of population genetics (Nee, 2005).

The basic Levins metapopulation model is illustrated in graphical form in Figure 4.3. The motivation for the colonization curve is simple. If there are no colonized patches, then there are no propagules available to colonize empty patches and therefore no colonization is occurring. Similarly, no colonization of empty patches can occur when *all* the patches are occupied. Colonization of empty patches will occur at the highest rate at intermediate values of patch occupancy, when there are a lot of patches emitting propagules and a lot of empty patches available for colonization. The extinction curve is straightforward: the more patches there are then the more patches in which extinction can occur.

The curves in Figure 4.3 are the simplest ones that satisfy these assumptions—parabolas and a straight line. Various simple, analytical results are derived with these functions below. But we can derive some important results from both the graphical model and with simple reasoning.

4.2.1 Empty habitat

At equilibrium, not all patches are occupied (Figure 4.3). Furthermore, depending on the nature of the curves, a *substantial* fraction of the patches of perfectly suitable habitat may be unoccupied. Hanski (1996) discusses the work of Boycott in the 1920s, who studied the colonization and extinction of snail populations in ponds over a 10-year period. Boycott demonstrated with transplantation experiments that ponds without a population of a particular snail species were nonetheless perfectly suitable habitat. The large differences between people in the species composition of their intestinal flora (Eckburg, 2005) must, at least in part, be a consequence of this fundamental fact about metapopulations. There are many examples of unoccupied—but perfectly suitable—habitat from epidemiology: only a fraction of a population is ever infected by an infectious, endemic disease organism. For example, about 25% of us have the bacteria *Helicobacter pylori* living in our stomachs. A striking demonstration that uninfected people are nonetheless suitable habitat was provided by Barry Marshall, who showed that *Helicobacter* causes stomach ulcers with true Australian directness: he drank a flask of bacterial culture and was successfully colonized and ulcerated. For two millennia, since the Roman physician Galen, medical orthodoxy has held that excess acid in the diet causes stomach ulcers: one Australian drinking a flask of bacteria consigned this orthodoxy to the dustbin of history.

4.2.2 Eradication threshold

Suppose that habitat destruction occurs, either by paving over patches or vaccinating individuals, for example (from the point of view of an infectious disease, vaccination is a wanton act of habitat destruction). As illustrated in Figure 4.3, habitat destruction has the effect of lowering the colonization curve and making it less steep towards the origin. This is because colonization is hampered by the reduction in the number of patches available to actually colonize: for example, visualize seeds landing on tarmac where before there was a patch, or a person sneezing on someone who is vaccinated against a disease. If destruction is so extensive that the slope of the colonization curve at the origin becomes smaller than the slope of the extinction curve, extinction is inevitable and this does not require *all* or even necessarily a substantial fraction of patches to be destroyed. Epidemiology told us long ago that it is not necessary to vaccinate 100% of a population to eradicate an infectious disease (Anderson and May, 1991). So we say there is a threshold level of patch destruction above which extinction will occur, and this is called the extinction or eradication threshold.

Note that metapopulation extinction does not occur *instantly* when destruction exceeds the threshold. Just like a population in which birth rates are always, marginally, less than death rates, the metapopulation may dwindle to extinction on a long time scale. This fact has been called the extinction debt (Tilman *et al.*, 1994) to recognize that the habitat destruction of today may have to be paid for in the future with extinctions.

What is the threshold? This can be computed very simply, using an argument from epidemiology used to estimate R_0 (Anderson and May, 1991),

which is the number of individuals infected by a single infected individual introduced into a wholly susceptible population (see Chapter 10 in this volume). At the equilibrium level of patch occupancy in the pristine world, y^*, which is where the highest colonization curve and extinction curve intersect in Figure 4.3, $1-y^*$ patches are unoccupied and so are available to be colonized. At equilibrium, each patch, over the course of its 'lifetime' colonizes, on average, one empty patch; with fewer than $1-y^*$ empty patches, each occupied patch will colonize less than one other patch. Imagine a world with fewer than $1-y^*$ patches and a metapopulation consisting of a single occupied patch in this world. Over the course of its lifetime, the patch will colonize less than one patch and so the metapopulation will go extinct. Hence, the threshold level of destruction is simply y^*. This argument can be generalized to other ecological relationships (Nee, 1994): for example, it can be used to calculate the minimum prey carrying capacity needed to sustain a predator population.

The above argument does, of course, make assumptions. In epidemiology, these are known as weak homogenous mixing (Anderson and May, 1991) and assume, for example, that all patches are roughly equivalent in their colonization and extinction properties. A particularly important assumption, from the conservation point of view, is that there is no rescue effect.

The rescue effect (Hanski, 1999) is the exact metapopulation equivalent of the Allee effect in population biology (Stephens and Sutherland, 1999). The Allee effect describes a situation in which either birth rates decline or death rates increase at low densities. Many biological features of species can produce an Allee effect. For example, plants may have difficulty getting pollinated at low densities or animals that rely on social defence against enemies may find these breaking down at low densities. For example, smaller colonies of sea birds suffer higher predation (Serrano *et al.*, 2005). The effect is also known as inverse density dependence (Begon *et al.*, 1996a).

In metapopulation biology the rescue effect may result if an input of colonists into a patch lowers the patch extinction rate. In this case, at lower

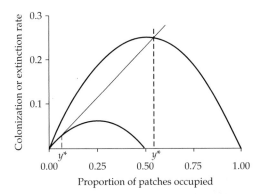

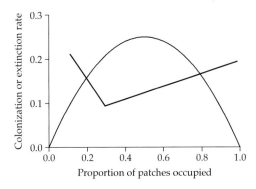

Figure 4.3 The basic Levins metapopulation model. This illustrates how the metapopulation colonization rate (parabolas) and extinction rate (straight line) vary as a function of the overall level of patch occupancy, expressed as a fraction of the total number of habitat patches in a pristine world. Illustrated are two colonization curves corresponding to the situation in a pristine environment (top parabola) and to the situation in which 50% of the patches, relative to the total number of patches in the pristine environment, have been destroyed (bottom parabola). These curves are justified in the text. The equilibrium level of patch occupancy, y^*, is found where the colonization and extinction curves intersect. As explained in the text, habitat destruction lowers, crucially, the slope of the colonization curve near the origin. Marginally more destruction will result in the colonization curve having a lower slope than the extinction curve at the origin and extinction will be inevitable: destruction will have exceeded the eradication threshold.

Figure 4.4 Threshold level of patch occupancy. As in Figure 4.3, the parabola is the colonization curve and here the piece-wise straight line is the extinction curve. The lower, left-hand equilibrium is unstable, and if the metapopulation size falls below it, extinction will result. For a general review of population models with this sort of behaviour see May (1977a).

levels of patch occupancy there will be a lighter rain of colonists over the metapopulation and the patch extinction rate may rise. As illustrated graphically in Figure 4.4, this results in a threshold level of patch occupancy below which the metapopulation will go extinct.

A consequence of this effect is that we may expect to find in nature a bimodal distribution of patch occupancy, with species found in either many patches or none (Hanski, 1982; Hanski *et al.*, 1995). We may also find a minority of species at intermediate levels of occupancy, either because the metapopulation is in transit from one equilibrium to another or because of immigration of the species from high-occupancy metapopulations outside the study area.

An example of this is provided by worms inhabiting mammalian intestines (Arneberg and Nee, unpublished work). Nematode worms have sexual reproduction whereas cestodes are

effectively asexual (they are self-fertilizing). May (1977a) demonstrated that sexual reproduction can produce multiple equilibria as a result of the difficulty of finding a mate at low abundance. This suggests that we should find a bimodal distribution of patch occupancy (the proportion of individuals harbouring worms, i.e. the prevalence) in studies of nematodes, but not cestodes. Data collected and analysed by Per Arneberg show this is, indeed, the case: see Figure 4.5.

4.3 Source–sink metapopulations

Population biologists have long paid attention to heterogeneities in populations—in age, for example—and to the effects of these heterogeneities on population dynamics (see Chapter 3 in this volume). Also, the effects of heterogeneity in time in the environmental factors affecting birth and death rates have also been considered extensively (Lande *et al.*, 2003). It is a somewhat more recent interest to consider the effects of heterogeneity in habitat. But as habitat is modified and destroyed by humans, ecologists are increasingly interested in this heterogeneity as well.

Habitats vary in their suitability for species in terms of the individuals' needs for survival and reproduction. Obviously, penguins would be

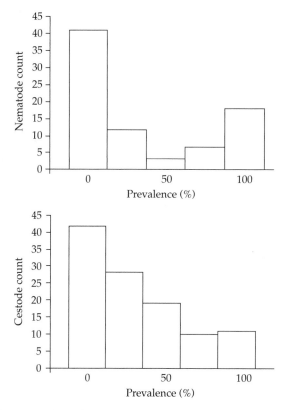

Figure 4.5 Data on worm prevalence were collected from the literature. The counts are of the prevalence of a particular worm species in a particular study. So, for example, one datum is the prevalence of the nematode *Ascaris suum* in a study of 48 feral swine on an island in Georgia, USA. The top histogram shows data for nematodes and the bottom for cestodes. As expected, the sexual nematodes exhibit bimodal prevalences whereas the cestodes do not. This result is unaffected by the choice of bin size.

unsuccesful in rainforest and orchids will not thrive on a glacier, but a full continuum of less extreme examples could be constructed. A very simple model capturing this fact is that of source–sink dynamics (Pulliam, 1988): see Figure 4.2.

Source habitats are those in which there is a net outflow of individuals. In other words, not only are the birth and death rates such that the species can maintain a population in that habitat, but there is a surplus production of individuals. Sink habitats, on the other hand, are those in which birth and death rates could not maintain a population: the population only persists because of immigration into it from source habitats.

Why do organisms end up in sink habitat? In territorial animal species, individuals may be simply making the best of a bad job if they fail to secure territories in prime habitat. An important point is

that sink habitat may not be obviously bad: it may support some reproduction, just not, in the long run, enough to sustain the species and, from an animal's point of view, some reproduction is better than none. Perhaps the most poignant example comes from Arctic ground squirrels (Carl, 1971), discussed with other examples by Pulliam (1996). Prime land for colonies is scarce and squirrels that are forced out burrow in the banks of creeks that are prone to flooding, which drowns them: so this is almost literally sink habitat. Another example of almost literally sink habitat comes from a study of rolled leaf beetles (*Cephaloleia fenestrata*), which suffer lower survivorship in areas prone to flooding (Johnson, 2004).

Plants have little choice about where they end up. Fig seeds are dispersed to Krakatau by birds and grow to healthy trees. Unfortunately, however, there are no fig wasps to pollinate them (Pulliam, 1996). It is possible that the vagaries of seed dispersal are more important for tropical plants than for temperate ones: the latter have the option of waiting in a seed bank until they detect that suitable conditions have arisen for them. Many disease organisms may be like plants, with little say where they end up: humans are sink habitat for Ebola virus, for example.

The presence or absence of predators often determines whether a habitat is source or sink for a species. The presence of trout in beaver ponds turns them into a sink for mayflies (Caudill, 2005) and egg-eating hedgehogs turn Scottish islands into sinks for sea birds (Jackson *et al.*, 2004). Allee effects may also be responsible for sinks. The population of gastropods and bivalves in the deep ocean is maintained by immigration from above: density in the abyss is too low for successful reproduction (Rex *et al.*, 2005).

The source–sink metapopulation idea is an example of a simple model that allows us to immediately see clearly some important possibilities about the world that were not previously part of our mental furniture (Pulliam, 1996). It is entirely conceivable, for example, that most members of a species live in sink habitat. This could be the case if the sink is not too bad, with long-term average birth rates only marginally less than death rates and if source habitat is highly productive. Hence, inferences about a species' preferred habitat based on where we typically find it may be completely wrong. Also, there are clear conservation implications: it is far better to preserve a small amount of source habitat than a large amount of sink, so we need to know the difference.

Unfortunately, this may not be easy: it may be very difficult to actually identify sink habitat even with detailed information on the vital rates of the residents and immigrant numbers. It is entirely possible for what is actually source habitat to be identified as sink habitat: i.e. the death rates of the residents being higher than the birth rates and the population deficiency being made up by immigration. Suppose density-dependent population regulation is acting on birth rates. Since, at equilibrium population size, the total per-capita birth rate must equal the total per-capita death rate then the birth rate of the natives must decline if there is immigration. A graphical model of sinks and pseudo-sinks is given in Box 4.1.

4.4 Two–species Levins metapopulations

Obviously, the simple Levins metapopulation model can be extended in many directions. For example, Hastings (2003) allows patch extinction to be a function of patch age, which might arise as

---

Box 4.1 A graphical model of sinks and pseudo-sinks

For simplicity, we just look at the effects of immigration into a population, so we do not look at reciprocal effects between populations. This would correspond to a situation of asymmetries in migration.

Figure 4.6 illustrates how the per-capita birth rates of residents, b, and death rates, d, vary as a function of local population size, N, in some time interval—for example, a year. The effect of migration, m, is equivalent to an elevation in the birth rate and a stable population, N^*, may exist. In a true sink, as in Figure 4.6, the population cannot exist in the absence of immigration as the b curve is lower than the d curve for all population sizes. Note that no attempt has been made to draw the $b + m$ curve

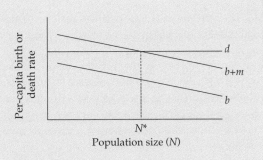

Figure 4.6 Graphical model of a true sink. See box text for details.

realistically: so, for example, if a constant number of migrants arrives each time unit, then the apparent effect on per-capita birth rates will be a function of N.

In a pseudo-sink (Figure 4.7; all symbols are as above) in the absence of immigration a stable population can exist, where the b curve intersects the d curve, so it is not a sink. Immigration is equivalent to an increase in the birth rate, so the population size rises to a new equilibrium, N^*. At the new equilibrium, the resident birth rate alone is insufficient to compensate for deaths.

Further graphical modelling can be done to explore, for example, the effects of migration and symmetry between habitats of different quality.

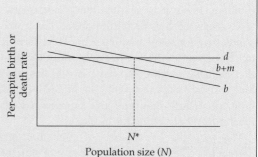

Figure 4.7 Graphical model of a pseudo-sink. See box text for details.

a result of successional processes, for example. Whereas a necessary condition for metapopulation persistence in the Levins model is simply that the colonization rate exceed the extinction rate (see below), the condition for persistence in Hastings' more general model is that the colonization rate be greater than $1/A$, where A is the average age of the patches. Keymer and colleagues (Keymer et al., 2000) study a spatially explicit model of ephemeral habitat patches and find that metapopulation persistence is affected by the rate of patch turnover as well as the amount of habitat destroyed, which resonates with Hastings' result.

Here I discuss the extension of the Levins model to interactions between two species with a variety of ecological relationships. It is now necessary to study simple mathematical models to gain insights. We will start with the one-species Levins model and then extend it. Definitions are as follows.

$x(t)$ The fraction of empty patches, relative to the total number of patches in the pristine world; $y(t)$ the fraction of occupied patches, relative to the total number of patches in the pristine world; h the fraction of habitable patches, relative to the total number of patches in the pristine world; $h=1$ in the original Levins' model; more generally $x+y=h$; c colonization rate parameter; e extinction rate parameter.

As discussed in the text and illutrated in Figure 4.3, the simplest sensible model sees colonization as a multiplicative function of x and y and extinction as a linear function of x. Hence:

$$\frac{dx}{dt} = ey - cxy$$
$$\frac{dy}{dt} = cxy - ey \qquad (4.1)$$

We do not, of course, need both of these equations: we could simply substitute, for example, $x = h - y$ in the second one and work with that alone. This would also make more obvious the fact that colonization is a quadratic function. But this redundancy makes what comes later more transparent. In epidemiological terms, this is a model of an infectious disease with only two host types: susceptible and immune.

At equilibrium there is no overall change: $dx/dt = dy/dt = 0$. There are thus two possible equilibria. One has $y^* = 0$ and, consequently, $x^* = h$. The second equilibrium is:

$$x^* = h - y^* = e/c$$
$$y^* = h - \frac{e}{c} \qquad (4.2)$$

Analysis shows that this equilibrium is stable to perturbations. The first equilibrium is stable only if $h < e/c$. If $h > e/c$ then a small propagule of the species introduced into the empty habitat would grow into a viable metapopulation of size $h - e/c$. In short, if the fraction of patches occupied in the pristine world, $h = 1$, is $y_0^* = 1 - e/c$, then the criterion for the extinction of the metapopulation ($y^* \rightarrow 0$) is to reduce the fraction of habitable patches from $h = 1$ to $h = e/c = 1 - y_0^*$. As discussed in the text that fact is not dependent on this particular mathematical model.

4.4.1 Competing species

Imagine two species coexisting as metapopulations even though they exploit the same habitat patches. This coexistence can come about if there is a competitive asymmetry such that the inferior competitor is either (1) a superior colonizer—for example, a weedy species leading a fugitive existence—or (2) has a much lower patch extinction rate than the superior competitor. With increasing interest in competitive interaction in multiple infections (Read and Taylor, 2001), we expect to soon have a biological example of the latter from epidemiology. For simplicity, we will assume that a superior competitor arriving on a patch occupied by an inferior competitor immediately eliminates it. It is straightforward to extend the previous model. Empty patches are still denoted as x, patches occupied by the superior competitor as y, and patches occupied by the inferior competitor as z. Using subscripts s for superior and i for inferior, the model is:

$$\begin{aligned}
\frac{dx}{dt} &= -c_s xy + e_s y - c_i xz + e_i z \\
\frac{dy}{dt} &= c_s y(x + z) - e_s y \\
\frac{dz}{dt} &= c_i zx - c_s zy - e_i z
\end{aligned} \tag{4.3}$$

Finding the equilibrium solution of the simple Levins model required solving a linear equation. Finding the solution to this simple extension just involves solving a quadratic equation. We will just focus on two interesting features of the solution: a full account can be found in Nee and May (1992) and an extended discussion of related models in Nee *et al.* (1997), including the appearance of models of this form in the study of the evolution of virulence (Nowak and May, 1994).

The equilibrium solution to this model is:

$$\begin{aligned}
x^* &= h - y^* - z^* \\
y^* &= h - \frac{e_s}{c_s} \\
z^* &= \frac{e_s(c_s + c_i)}{c_s c_i} - \frac{e_i}{c_i} - \frac{hc_s}{c_i}
\end{aligned} \tag{4.4}$$

From the expression for z^* we can, with some algebra, derive a necessary condition for coexistence to occur:

$$\frac{c_i}{e_i} > \frac{c_s}{e_s} \tag{4.5}$$

In epidemiological terms, this expression says that the R_0 value of the inferior competitor must be higher than the R_0 value of the superior competitor. The R_0 of a species is simply the number of patches a single patch would colonize, in an empty, pristine world before suffering local extinction.

The expression for z^* has another interesting feature. Notice that increasing habitat destruction—decreasing h—*increases* the abundance of the inferior competitor: this is because the abundance of the superior one declines. This is true only up to a point: once the superior competitor is eliminated entirely, then the abundance of the inferior declines with decreasing h. This result is robust: the same qualitative result has been found in more realistic models (Nee *et al.*, 1997).

4.4.2 Predator–prey metapopulations

We imagine a specialist predator that drives local populations of its prey—subscript v for victim—extinct upon dispersing into them and, consequently, the predator (subscript p) goes extinct itself. A possible biological example of this is the relationship between prickly pear cactus, *Opuntia*, in Australia and its specialist control agent, the *Cactoblastis* moth. Plant–fungal pathogen systems may also belong to this category (Laine, 2004). A remarkable study system combines both this category and the previous one. The Glanville fritillary butterfly, a well-studied metapopulation, has two specialist, competing, species of parasitoids and the inferior competitor is the superior disperser (Lei and Hanski, 1998; van Nouhuys and Hanski, 2002). Huffaker's famous mite-orange experiments belong in this category (Huffaker, 1958).

A simple model of a predator–prey metapopulation is:

$$\begin{aligned}
\frac{dx}{dt} &= e_v y + e_p z - c_v xy \\
\frac{dy}{dt} &= c_v xy - c_p yz - e_v y \\
\frac{dz}{dt} &= c_p yz - e_p z
\end{aligned} \tag{4.6}$$

where y denotes prey-only patches and z denotes patches with both predators and prey.

A full discussion of this model can be found in May (1994) and Nee *et al.* (1997). The equilibrium for this model is:

$$x^* = h - y^* - z^*$$
$$y^* = \frac{e_p}{c_p}$$
$$z^* = \frac{c_v}{c_v + c_p}\left(h - \frac{e_p}{c_p} - \frac{e_v}{c_v}\right)$$
(4.7)

Here we note one interesting feature of this equilibrum—the expression for y^* tells us that habitat destruction has *no* effect on equilibrium prey abundance. (Again, only up to a point: once the predator is extinct, prey abundance declines with increasing destruction.) This is, in fact, true of a very broad class of predator–prey models, not just this metapopulation model. As discussed in Nee *et al.* (1997), as long as predators only affect each other through their consumption of prey (so they are not territorial, for example) then this result holds true.

4.4.3 Mutualism

Think of two species, one of which can survive on a patch for a time, but needs another for colonizing new patches, whereas the second species requires the first for both survival and reproduction: imagine a plant species and its specialist pollinator or disperser. Plant–insect metapopulations are reviewed by Tscharntke and Brandl (2004). Virology provides many examples, some quite curious. Tobacco rattle virus is actually a double act consisting of two viral particles: the long particle carries the replicase and the short particles carries the coat protein, so whereas the long particle can persist in a patch (plant) on its own, a complete cycle of infection requires both particles.

In the following simple model, y refers to patches occupied solely by the plant (subscript p) and z refers to patches with both plant and disperser (subscript d).

$$\frac{dx}{dt} = e_p y + e_d z - c_p z x$$
$$\frac{dy}{dt} = c_p z x - c_d z y - e_p y$$
(4.8)
$$\frac{dz}{dt} = c_d z y - e_d z$$

A complete analysis of this model can be found in Nee *et al.* (1997) and an extended discussion of the viral interpretation in Nee (2000).

The equilibium solution is:

$$x^* = h - y^* - z^*$$
$$y^* = \frac{e_d}{c_d}$$
(4.9)
$$z^* = \frac{1}{2}\left(\alpha \pm \left[\alpha^2 - 4\beta\right]^{1/2}\right)$$

where

$$\alpha = h - \frac{e_d(c_p + c_d)}{c_p c_d}$$
(4.10)
$$\beta = \frac{e_p e_d}{c_p c_d}$$

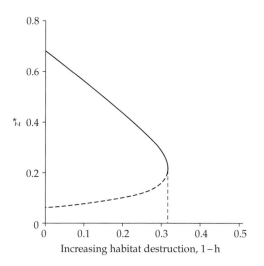

Figure 4.8 Equilibria in mutualistic associations. As can be seen in the expression for z^* in the text, there are two equilibria and local stability analysis (or simulation) shows that the larger one is stable. In the figure, the solid line shows the stable equilibrium for each value of h, and the dashed line is the unstable equilibrium—technically, it is a saddle point. As destruction increases, eventually the two equilibria collide in what is both mathematically and biologically a catastrophe, leaving 0 as the only equilibrium. The parameters used to construct this illustration are: $c_p = c_d = 4$, $e_d = 0.5$, and $e_p = 1.5$.

The equilibrium solution exhibits a new, interesting feature. In the previous models, as habitat destruction proceeds and species approach their eradication thresholds, their equilibrium abundance at each greater level of destruction approaches smoothly to zero, as can be seen in the expressions for the equilibria. In this model, on the other hand, that is not the case: see Figure 4.8. What may appear to be a perfectly healthy mutualistic association, judging by the abundance of the species involved, can be driven to extinction by the addition of one more car park to the landscape.

4.5 Finale

Species are to be found wherever there is suitable habitat that they can get to. Species will rarely, if ever, be found in unsuitable habitat. They will be most abundant in their preferred habitat and species can be preserved as long as a good-size chunk of suitable habitat is conserved for them. Destruction of a species habitat is always a bad thing for the species' abundance. Still sound reasonable? (Lande, 1987; Pulliam, 1996; Nee *et al.*, 1997; Hanski, 1999)

CHAPTER 5

Predator–prey interactions

Michael B. Bonsall and Michael P. Hassell

5.1 Introduction

Predation is a widespread population process that has evolved many times within the metazoa. It can affect the distribution, abundance, and dynamics of species in ecosystems. For instance, the distribution of western tussock moth is known to be affected by a parasitic wasp (Maron and Harrison, 1997; Hastings *et al.*, 1998), the abundance of different competitors can be shaped by the presence or absence of predators (e.g. Paine, 1966), and natural enemies (such as many parasitoids) can shape the dynamics of a number of ecological interactions (Hassell, 1978, 2000). The broad aim of this chapter is to explore the dynamical effects of predators (including the large groupings of insect parasitoids) and show how our understanding of predator–prey interactions scales from knowledge of the behaviour and local patch dynamics to the population and regional (metapopulation) levels. We draw on a number of approaches including behavioural studies, population dynamics, and time-series analysis, and use models to describe the data and dynamics of the interaction between predators and prey.

Predator–prey interactions have an inherent tendency to fluctuate and show oscillatory behaviour. If predators are initially rare, then the size of the prey population can increase. As prey population size increases, the predator populations also begins to increase, which in turn has a detrimental effect on the prey population leading to a decline in prey numbers. As prey become scarce then the predator population size declines and the cycle starts again. These intuitive dynamics can be captured by one of the simplest mathematical descriptions of a predator–prey interaction: the Lotka–Volterra model (Lotka, 1925; Volterra, 1926). Specifically, the Lotka–Volterra model for an

interaction between a predator (P) and its prey (N) is a continuous-time model and has the form:

$$\frac{dN}{dt} = N_t \cdot (r - \alpha \cdot P_t) \tag{5.1}$$

$$\frac{dP}{dt} = P_t \cdot (c \cdot \alpha \cdot N_t - d) \tag{5.2}$$

where r is the prey-population growth rate in the absence of predators, α is the predator attack rate, c is the (positive) impact of prey on predators, and d is the death rate of predators in the absence of their prey resource. Graphical or analytical analyses show that the equilibrium point ($N^* = d/(c \cdot \alpha)$, $P^* = r/\alpha$) for this interaction is neutrally stable. That is, the system has the propensity to oscillate with a period determined by the model parameters and amplitude set by the initial conditions of the predator and prey populations (Figure 5.1a). Although oversimplified, the Lotka–Volterra model is a useful point of departure for understanding further these types of ecological interaction.

A second, equally simple, model framework for understanding predator–prey interactions is the Nicholson–Bailey model (Nicholson and Bailey, 1935). Specifically formulated to explore the dynamics of insect parasitoids (Askew, 1971; Godfray, 1994) and their hosts, the Nicholson–Bailey model is a discrete-time model that takes the form:

$$N_{t+1} = \lambda \cdot N_t \cdot f(N_t, P_t) \tag{5.3}$$

$$P_{t+1} = c \cdot N_t \cdot [1 - f(N_t, P_t)] \tag{5.4}$$

Here, λ is the *per-capita* rate of increase of hosts in the absence of parasitoids, c is the conversion efficiency of hosts to new parasitoids, and $f(N_t, P_t)$ is the function describing the probability of hosts

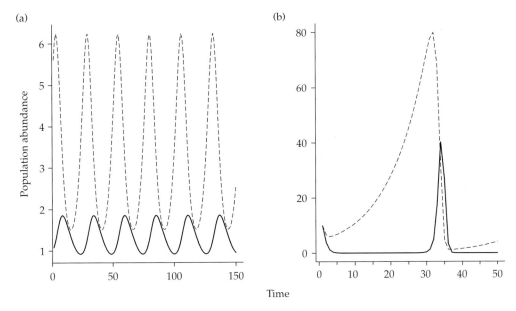

Figure 5.1 Dynamics of predator–prey interactions predicted by (a) the Lotka–Volterra model and (b) the Nicholson–Bailey model. In both models, the dynamics are unstable. In the Lotka–Volterra model, the dynamics are neutral cycles with a period determined by the model parameters and amplitude set by the initial conditions. Predators lag prey by one-quarter of a cycle. In the Nicholson–Bailey model, the dynamics show divergent oscillations with overexploitation by the predator leading to the extinction of both prey and predators. Dashed lines, prey; solid lines, predators.

surviving parasitism. Unlike the neutrally stable Lotka–Volterra model, for simple $f(N_t, P_t)$ this discrete-time model predicts divergent oscillations of hosts and parasitoids with overexploitation by the parasitoid leading to rapid extinction of the populations (Figure 5.1b).

The contrasting dynamics between these two models results from the inclusion of explicit generational time lags, which introduces age structure into the interaction (May, 1974b; Smith and Mead, 1974; Hastings, 1983, 1984; Murdoch *et al.*, 1987, 1997). Thus, a time-delayed version of the Lotka–Volterra predator–prey interaction has the form:

$$\frac{dN}{dt} = r \cdot N_{t-\tau_N} - f(P_t) \cdot N_t \qquad (5.5)$$

$$\frac{dP}{dt} = N_{t-\tau_P} \cdot f(P_{t-\tau_P}) - d \cdot P_t \qquad (5.6)$$

where once again r is the growth rate of prey in the absence of predators and d is the death rate of predators in the absence of prey. $f(P_t)$ is the rate of predation, and the time delays τ_N and τ_P capture aspects of age structure. $N_{t-\tau_N}$ represents the

number of new prey entering the prey population τ_N time steps ago and broadly captures the age structure of an effectively invulnerable juvenile stage in the prey population. Similarly, $N_{t-\tau_P}$ represents the number of prey attacked τ_P time steps ago by a predator population of magnitude $P_{t-\tau_P}$. This lag captures the biology of an explicit developmental lag between predators attacking prey and the production of new predators. Analysis of the equilibria (for particular functions for the rate of predation) reveal that it is the inclusion of a time lag—which is what the transition from continuous to discrete time effectively does—that destabilizes the predator–prey interaction (Figure 5.2). It is this time lag that leads to the difference in the dynamics observed in the Nicholson–Bailey and Lotka–Volterra models (May, 1973b).

5.2 Behaviour and patch dynamics

Predators do not respond instantaneously to changes in prey density, and nor do they immediately convert prey to new predators. It takes time to find, subdue, and consume prey. These

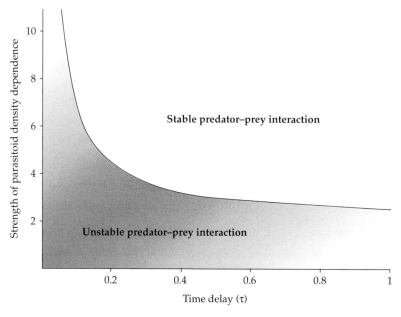

Figure 5.2 Stability conditions of eqns 5.1 and 5.2 in terms of the strength of predator density dependence (*k*) and the time delay $\tau = \tau_N = \tau_P$. The line shows the minimum amount of density dependence need to stabilize the interaction and in the limit of $\tau \to 0$ a version of the continuous-time Lotka–Volterra model (with parasitoid density dependence) is recovered, whereas as $\tau \to 1$ a version of the discrete-time Nicholson–Bailey model is recovered. Differences in the dynamics between the Lotka–Volterra and Nicholson–Bailey models are a result of time lags.

behaviours require the investment of time and energetic resources, and the foraging activities of predators for prey have important consequences for the population dynamics at a variety of different temporal and spatial scales (Sutherland, 1996). For instance, behavioural decisions can lead to the increased tendency for predators to forage in a non-random way: predators may be attracted from long distances to localized patches of high prey density or remain longer in habitats where prey have already been located (Waage, 1979). These behaviours have implications for the way in which predators distribute themselves across patches, move between patches, and exploit prey across patchy environments (Bernstein *et al.*, 1988, 1991).

The original Lotka–Volterra and Nicholson–Bailey models assume that foraging for prey occurs at random. However, predators rarely do this: they encounter prey non-randomly with the result that some prey individuals are more at risk of attack than others. This heterogeneity, the differential susceptibility or risk of predation, can arise through a number of mechanisms. For example, the physiology or genetics of individuals

may predispose some to be found more readily than others, or prey may be segregated into patches of different densities which again may predispose some individuals to be found more readily. The role of this spatial heterogeneity in the distribution of prey across a habitat has attracted considerable interest in the quest to understand how foraging affects the dynamics of predators and parasitoids (Bailey *et al.*, 1962; Hassell and May, 1973, 1974; May, 1978a).

For patchily distributed prey, a number of patterns in the distribution of parasitism by insect parasitoids have been observed (Figure 5.3). The proportion of hosts parasitized can be a positive function, a negative function, or independent of host density. Early theoretical models demonstrated that aggregation by parasitoids to patches of high host density (leading to a positive relationship between the fraction of hosts parasitized and host density) was potentially a significant mechanism leading to the temporal persistence of the interacting populations (Hassell and May, 1973, 1974).

A number of reviews have examined the incidence of the different patterns of parasitism in

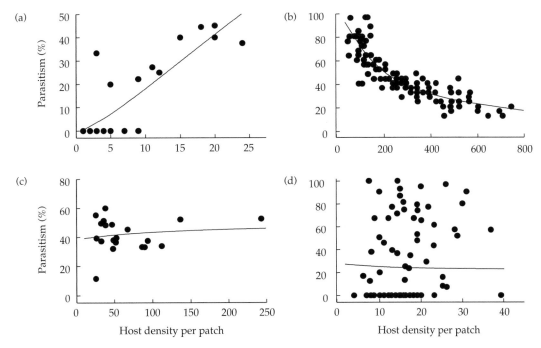

Figure 5.3 Patterns of parasitism from field studies. (a) Positive density-dependent response in the number of hosts (cabbage root fly, *Delia radicum*) parasitized by *Trybliographa rapae* (Jones and Hassell, 1988), (b) negative or inverse density-dependent response to parasitism of gypsy moth (*Lymantria dispar*) by the parasitoid *Ooencyrtus kuwanai* (Brown and Cameron, 1979), (c) independent pattern of parasitism by the aphelinid parasitoid *Coccophagoides utilis* on the olive scale *Parlatoria oleae* (Murdoch *et al.*, 1984), and (d) independent pattern of parasitism of the gall midge *Rhopalomyia califonica* by the torymid parasitoid *Torymus baccaridis* (Ehler, 1987).

Figure 5.3 (Lessells, 1985; Walde and Murdoch, 1989; Hassell and Pacala, 1990). In these, 29% show positive patterns of parasitism, 26% show a negative relationship, and 45% show patterns independent of host density. Explanations for how these patterns might arise have focused on the role of the functional response (Soloman, 1949; Holling, 1959a, 1959b)—that is, the relationship between the number of prey attacked and prey density (Hassell, 1982)—on parasitoid biology and foraging (Lessells, 1985), and on the spatial scale of the interaction (Heads and Lawton, 1983). However, following the work of Chesson and Murdoch (1986), the emphasis has shifted to consider the distribution of parasitism among the prey population as a whole rather than the different patterns in Figure 5.3. So long as parasitism exhibits sufficient variability within the host population, this can stabilize the host–parasitoid interaction (Chesson and Murdoch, 1986; Pacala *et al.*, 1990; Hassell *et al.*, 1991a; Pacala and Hassell, 1991). This heterogeneity

in the risk of parasitism is thought to be an important mechanism promoting the persistence and stability of host–parasitoid interactions.

As originally highlighted by Reeve *et al.* (1989), and more recently explored by Gross and Ives (1999), inferring the stability of the population dynamics from such spatial patterns of parasitism is complicated by a number of biological and statistical difficulties. Variation in the between-generation patterns of parasitism (Redfern *et al.*, 1992), the underlying behavioural responses by the natural enemies (Ives, 1992), or the way parasitism is distributed within and between patches (Ives, 1992; Gross and Ives, 1999) can all lead to overestimates of the contribution of the spatial distribution of parasitism to population stability. One alternative approach to tackling the difficulties in understanding the effects of spatial heterogeneity on predator–prey interactions is to make comparisons between model predictions of the dynamical consequences of heterogeneity and the dynamics of actual populations.

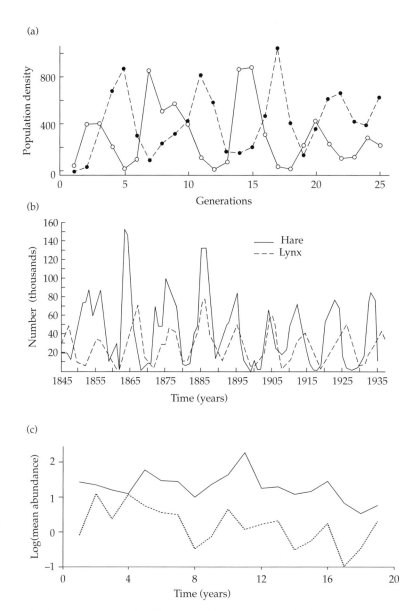

Figure 5.4 Examples of predator–prey interactions from laboratory and field systems. (a) Utida's (1957) host–parasitoid interaction between *Callosobruchus chinensis* (solid line) and the parasitoid *Heterospilus prosopidis* (dashed line) shows cycles consistent with the original theoretical models; (b) lynx–hare interactions (Elton, 1924) fluctuate with a period of 9–11 years (Stenseth *et al.*, 1997); and (c) cabbage root fly (*Delia radicum*, solid line)–parasitoid (*Trybliographa rapae*, dashed line) interaction from Silwood Park shows relatively stable dynamics (Bonsall *et al.*, 2004a).

5.3 Population dynamics

We have seen that theory predicts that predator–prey interactions have an inherent tendency to oscillate. This has indeed been observed in a number

of laboratory and field systems (Figure 5.4). One classic study of a predator–prey interaction is that undertaken by Syunro Utida to explore the dynamics of the interaction between *Callosobruchus chinensis* and its parasitoids, *Heterospilus prosopidis*

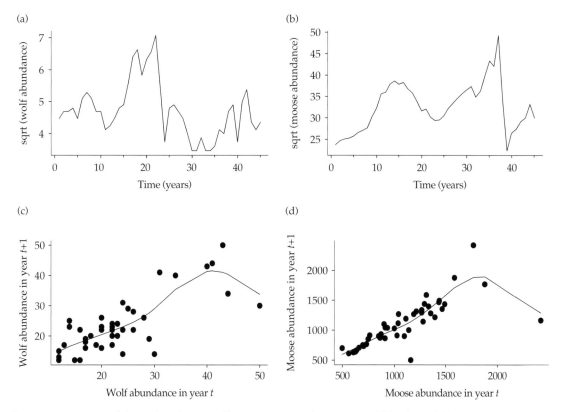

Figure 5.5 Time series and density dependence in wolf–moose interaction. The time series of (a) wolves and (b) moose show multiannual cycles thought to be consistent with theory. Evidence for density dependence in (c) wolf and (d) moose population abundances suggest that additional processes may underpin this predator–prey interaction.

and *Neocatolaccus mamezophagus* (Utida, 1957). Utida was one of the first to show that populations of predators and prey are likely to have this inherent tendency to oscillate (Utida, 1957). Using hosts alone, hosts and parasitoids, and multispecies systems of a host and two parasitoids in laboratory microcosms, Utida showed that the interactions followed the boom-and-bust dynamics predicted by the deterministic models (e.g. Lotka–Volterra) rather than erratic fluctuations driven by changes in abiotic factors (Figure 5.4).

Another, now classic, study of a predator and its prey is the interaction between moose (*Alces alces*) and wolves (*Canis lupis*) on Isle Royale in Lake Superior, USA. This interaction has been studied continuously since 1959 (Peterson, 1999), and it is now one of the longest and most intensively studied predator–prey interactions. The system is

unique in that wolves are the only predators of moose on Isle Royale and moose are overwhelmingly the mainstay of wolf diet. Moose are thought to have colonized Isle Royale around 1910 and wolf invasion occurred in the late 1940s during periods when Lake Superior froze over. Density dependence clearly operates in this predator–prey interaction (Figure 5.5), and the long-term dynamics shown in Figure 5.5 are thought to be cyclic (Peterson, 1999): both wolves and moose show multiple concurrent years of increase and decrease (Peterson, 1999). However, as Vucetich and Peterson (2004a) noted, it is important to distinguish whether these dynamics are aperiodic, multiannual cycles, or population cycles with a constant period, because the underlying set of mechanisms generating these dynamics may be different.

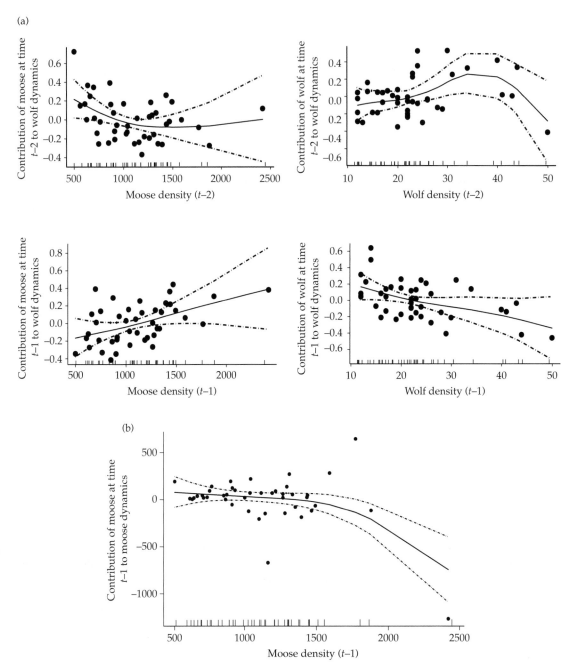

Figure 5.6 Nonlinear time-series analysis for (a) wolf and (b) moose dynamics from Isle Royale reveals asymmetric interaction. (a) Predator abundance is determined by lagged densities of both wolf and moose whereas (b) moose abundance is determined solely by lagged density of moose. Generalized additive models using smoothed splines (with four degrees of freedom) were fitted to three lagged densities of both predators and prey for each of the response variables (wolf abundance, moose abundance) and model simplification was undertaken using χ^2 tests (Hastie and Tibshirani, 1990).

Nonlinear time-series analysis of the wolf–moose interaction highlights that the dynamics might be more complicated than simple predator–prey cycles and reveals interesting differences between the dynamics of predators and prey. As expected, wolf dynamics are determined by both moose and wolf densities (Figure 5.6a). As Vucetich *et al.* (2002) noted, the importance of wolf density as a predictor of wolf dynamics is likely to be important because wolves exhibit group hunting, aggressive interaction between groups, and social interactions. The importance of different time-lagged densities of moose abundance on wolf dynamics is likely to be a consequence of the importance of prey age structure. Juvenile and senescent moose are most susceptible to predation (Vucetich and Peterson, 2004b), and changes in these demographic classes have major effects on the size of the prey population. Such stochastic changes in prey population density lead to a complex set of nonlinear, density-dependent processes acting together to determine wolf abundance. In contrast, moose dynamics are determined solely by moose numbers in the previous year (Figure 5.6b). Recent work suggests that the top-down processes of wolf predation are neither a primary nor a dominant influence of the interannual variation in changes in moose population numbers (Vucetich and Peterson, 2004c). This is corroborated by the nonlinear time-series analysis presented below (Figure 5.6b).

Most interestingly, the time-series analysis of the wolf–moose interaction reveals that this predator–prey interaction is asymmetric: moose dynamics are determined simply by numbers in the previous year, whereas wolf numbers are determined by time-lagged densities of both moose and wolves. A theoretical representation of the time-series analysis presented in Figures 5.5 and 5.6 is:

$$M_t = f(M_{t-1}) \tag{5.7}$$

$$\begin{aligned} W_t = {} & f_1(M_{t-1}) + f_2(M_{t-2}) + g_1(W_{t-1}) \\ & + g_2(W_{t-2}) \end{aligned} \tag{5.8}$$

Here, the functions $f_i(.)$ and $g_i(.)$ are the nonlinear effects of the time-lagged contribution of moose (M) or wolf (W) densities to numbers of wolf and moose in the next generation. Evidence that this (Figure 5.5 and 5.6) and other trophic interactions (Bonsall *et al.*, 2003) are asymmetric, stochastic, and nonlinear emphasizes the need to examine species interactions over considerable lengths of time to explain adequately the type, form, and strength of the species' effects.

The snowshoe hare (*Lepus americanus*) and the Canadian lynx (*Lynx canadensis*) in boreal forests of North America also show cycles in population density. The dynamics of this predator–prey interaction fluctuated regularly with a period of 9–11 years (Figure 5.4b). Originally analysed by Charles Elton (1924) using data from 200 years of fur records traded through the Hudson Bay Company, and explored further by Elton and Nicholson (1942), the interaction been the lynx and hare was interpreted as an example of a classic predator–prey oscillation. Hypotheses for the regularity in the cycles between this predator and its prey have included correlations with sunspot frequency (Elton, 1924), food abundance during winter (Keith, 1983), and food-web structure (Krebs *et al.*, 1995). A thorough analysis of the time series reveals more complexity in the dynamics than those expected from a simple predator–prey interaction (Stenseth *et al.*, 1997). Although the boreal food web within which the lynx and hare are embedded is relatively complex, the dynamics of the lynx and hare are of low dimension. That is, the multiple ecological processes, possibly involving both food limitation and predation, underpin the dynamics of the snowshoe hare population, while in contrast the lynx population appears to be regulated through the single process of food limitation (the availability of hares; Stenseth *et al.*, 1997).

Although predator–prey interactions can show multiannual population cycles (Begon *et al.*, 1996b), these trophic interactions are also capable of generating another type of population-dynamic behaviour not predicted by simple ecological theory. With age structure included (and in the absence of any environmental cues), predator–prey interactions can show cycles in population abundance of roughly one generation period (Godfray and Hassell, 1989; Gordon *et al.*, 1991).

These generation cycles arise as a consequence of the effects of behavioural decisions, density dependence, and age structure. Marked differences in the generation lengths between predators and prey are more likely to lead to generation cycles than situations where predators or prey have more-or-less equal generation lengths (Hassell, 2000). Although generation cycles have been observed in laboratory predator–prey interactions (Begon *et al.*, 1995), Reeve *et al.* (1994) have demonstrated the potential for generation cycles in the interaction between the salt marsh planthopper, *Prokelisia marginata*, and its mymarid parasitoid *Anagrus delicatus*. The parasitoid's life cycle is substantially shorter than the host's, taking 28 days to develop from egg to adult (compared with 45 days for the host). Although host reproduction and attack by parasitoids occur throughout the year, this predator–prey interaction shows distinct generation cycles (Reeve *et al.*, 1994).

The dynamical patterns observed in predator–prey systems, whether they be wolf–moose, lynx–hare, or planthopper–parasitoid interactions, can be explained using quite different approaches, such as manipulative experiments (e.g. Krebs *et al.*, 1995), observational inferences (e.g. Royama, 1992), or fitting ecological models to appropriate population-dynamic data (e.g. Bonsall and Hastings, 2004; Bonsall and Benmayor, 2005). Modern methods of analysis using predator–prey models with different assumptions about the biology of the interaction can be used to test different mechanisms determining the dynamics. For example, heterogeneity, as mentioned above, is thought to have a major effect on the dynamics of predator–prey interactions. It is thus unlikely that any two individuals will experience the same probabilities of predation. Huffaker and colleagues showed how this sort of heterogeneity can affect the long-term temporal dynamics of the predator–prey interaction between the predatory mite *Metaseiulus occidentalis* and its prey *Eotetranychus sexmaculatus* (Huffaker *et al.*, 1963). In simple environments this predator–prey interaction was shown to be unstable and extinction of over-exploited prey occurred rapidly. However, in more complex environments (barriers for dispersal,

refuges for prey) the predator–prey interaction was more persistent. One general way to describe the effects of this sort of spatial heterogeneity (unevenly distributed prey among habitat patches) is to test the population-dynamic consequences of how the functional response varies in the two environments. First, there may be qualitative differences. By assuming that the functional response differs between heterogeneous and homogeneous environments the dynamical effects of patchy environments can be understood. Second, there might be quantitative differences in the magnitude of the same functional response that gives rise to differences in the predator–prey dynamics.

In a homogeneous environment, for example, the population dynamics of a continuous interaction between a predator and its prey might be described by the Lotka–Volterra model (eqns 5.1 and 5.2). Similarly, the population interaction may be driven by a behavioural response of the predator and a type II functional response (Solomon, 1949; Holling, 1959a, 1959b) might be more appropriate:

$$\frac{dN}{dt} = r \cdot N(t) - \frac{\alpha \cdot N(t)}{1 + \beta \cdot N(t)} \cdot P(t) \qquad (5.9)$$

$$\frac{dP}{dt} = c \cdot \frac{\alpha \cdot N(t)}{1 + \beta \cdot N(t)} \cdot P(t) - d_p \cdot P(t). \qquad (5.10)$$

Here, the parameters are as before (eqns 5.1 and 5.2) except that β is a measure of the time taken to handle each prey item. In a heterogeneous environment the dynamics could be described by:

$$\frac{dN}{dt} = r \cdot N(t) - k \cdot \left[1 - \frac{\alpha \cdot P(t)}{k}\right] \cdot N(t) \qquad (5.11)$$

$$\frac{dP}{dt} = c \cdot k \cdot \left[1 - \frac{\alpha \cdot P(t)}{k}\right] \cdot N(t) - d_p \cdot P(t). \qquad (5.12)$$

Here, the parameters are as before (eqns 5.1 and 5.2), except that k is a measure of the degree of clumping or non-random search (see Section 5.2). The dynamical effects of this model are widely known (May, 1978a; Chesson and Murdoch, 1986) and for sufficiently small values of k the predator–prey interaction is stable. Each of these

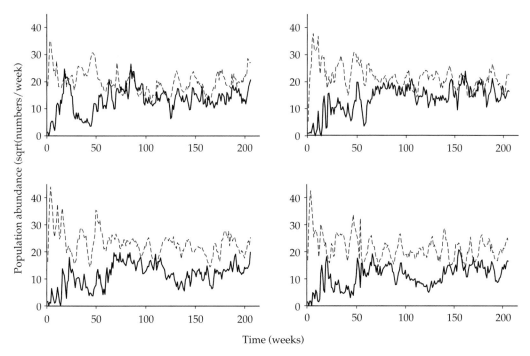

Figure 5.7 Population dynamics of *Drosophila ananassae* and *Leptopilina victoriae* in a patchy (but no refuge) environment. In chronological order, four resource patches (apple-based media) out of 20 were renewed each week. Dynamics show that in the presence of the parasitoid, the abundance of *D. ananassae* is reduced to approximately 490 flies per week. The dynamics of the predator–prey interaction in the absence of a host refuge are expected to be unstable. Model parameter estimates (with 95% confidence interval) for eqns 5.11 and 5.12 were $r = 1.27$ (0.001), $\alpha = 0.008$ (0.001), $k = 4.405$ (0.095), and $d = 0.746$ (0.003). Dashed line, *D. ananasse*; solid line, *L. victoriae*.

predator–prey models provides a different mechanistic interpretation to the underlying population dynamics. And with any of these models significant, quantitative differences in the magnitude of the functional response (e.g. differences in attack rates or degree of clumping), density dependence, or growth rate might be adequate to explain differences in the dynamics.

Time series from experimental population studies can allow us to explore, compare, and contrast these different mechanistic models. One particular case where this is appropriate is the interaction between an insect host, *Drosophila ananassae*, and its parasitoid, *Leptopilina victoriae*, in patchy environments (Bonsall and Hassell, 2005). Time-series experiments were established to explore the interaction in homogeneous (no refuges, and patches equally available) and heterogeneous (refuges on patches) environments. *D. ananassae* is

a cosmopolitan and domestic species found through south-east Asia. In the laboratory, this species has a relatively short generation time (8–10 days) and is attacked by a wide range of natural enemies. *L. victoriae* is a eucolid parasitic wasp attacking the larval instars of a range of drosophilids. This species also has a relatively short generation time. The population-dynamic interactions between these two species in the different (no refuge, refuge) environments are shown in Figures 5.7 and 5.8.

In order to contrast the different mechanistic models, we can use a statistical approach based on maximizing likelihoods (Edwards, 1972). This involves determining the expected numbers of predators and prey at each census point. The expected numbers of hosts and parasitoids are determined by solving the model over the census interval (τ) and comparing observed and predicted

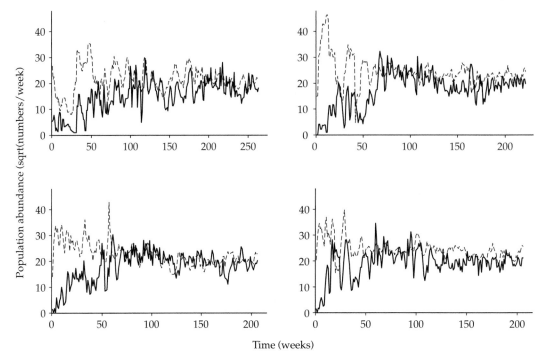

Figure 5.8 Population dynamics of *Drosophila ananassae* and *Leptopilina victoriae* in a patchy environment with a 70% refuge for the host on all patches. In chronological order, four resource patches (apple-based media) out of 20 were renewed each week. Dynamics show that in the presence of the parasitoid, the abundance of *D. ananassae* is reduced to approximately 560 flies per week. The dynamics of this predator–prey interaction in the presence of a host refuge are predicted to show stable dynamics. Model parameter estimates (with 95% confidence interval) for eqns 5.11 and 5.12 were $r = 0.371$ (0.0004), $\alpha = 0.001$ (0.001), $k = 0.622$ (0.0004), and $d = 0.864$ (0.001). dashed line, *D. ananasse*; solid line, *L. victoriae*.

population abundances. So, for the Lotka–Volterra model (eqns 5.1 and 5.2) the expected numbers of hosts and parasitoids would be:

$$E_{N[t+\tau]} = \int_t^{t+\tau} [r \cdot N(s) - \alpha \cdot N(s) \cdot P(s)] \cdot \mathrm{d}s$$

$$(5.13)$$

$$E_{P[t+\tau]} = \int_t^{t+\tau} [c \cdot \alpha \cdot N(s) \cdot P(s) - d_p \cdot P(s)] \cdot \mathrm{d}s$$

$$(5.14)$$

To completely define this problem of comparing different models, we need to give specific and explicit consideration to the role of stochasticity (error or noise) on the dynamics. Given that the abundance of flies and parasitoids are estimates of the flies or parasitoids alive at each census point,

we assume that the variability in population densities could be due to a broad environmental component of noise. Other forms of noise might arise due to the particular demographic process or any bias in the estimates of measurement of population size. Under this environmental noise, changes in population sizes between successive census points occur due to the deterministic processes of birth, predation, and death (e.g. eqns 5.1 and 5.2) and the effects of random noise acting additively on a logarithmic scale (Dennis *et al.*, 1995; Bonsall and Hastings, 2004). The stochastic version of the Lotka–Volterra model under this type of noise is:

$$E_{N[t+\tau]} = \exp(v_1)$$
$$\cdot \int_t^{t+\tau} [r \cdot N(s) - \alpha \cdot N(s) \cdot P(s)] \cdot ds \quad (5.15)$$

$$E_{P[t+\tau]} = \exp(v_2) \cdot$$
$$\int_t^{t+\tau} \left[c \cdot \alpha \cdot N(s) \cdot P(s) - d_p \cdot P(s) \right] \cdot ds \quad (5.16)$$

where v_1, v_2 is a vector of independent, identically distributed random numbers. Under the assumption that the autocovariances (correlation between time points) are weak, the most appropriate description for these dynamics of D. ananassae and L. victoriae in the differing patchy environments is a model incorporating parasitoid density dependence (eqns 5.11 and 5.12; Bonsall and Hassell, 2005). Further, comparing models with different predator functional responses (e.g. type I, type II, heterogeneous distribution in parasitism) revealed no qualitative differences in the mechanisms underpinning the population dynamics between the refuge and no-refuge environments (Bonsall and Hassell, 2005). This mechanism of parasitism introduces density dependence into the predator–prey interaction through a process known as pseudointerference (Free et al., 1977). In the no-refuge treatment, the strength of parasitoid density dependence (k) predicts that the dynamics will be unstable ($k = 4.31$, 95% confidence interval = 0.095) while, in contrast, in the refuge treatments the strength of parasitoid density dependence predicts stable population dynamics ($k = 0.622$, 95% confidence interval = 0.0004; Bonsall and Hassell, 2005).

Other issues in fitting ecological models to ecological data involve measurement (observation) errors that affect model identification and parameter bias (Carpenter et al., 1994; Hilborn and Mangel, 1998). For instance, Carpenter et al. (1994) showed that appropriate estimates for discrete-time versions of the Lotka–Volterra model, for a phytoplankton—cladoceran predator–prey interaction, could only be obtained when measurement error (as well as process error) was incorporated. More recently, de Valpine and Hastings (2002) have shown that it is crucially necessary to separate measurement error and process error in understanding nonlinear ecological interactions, particularly in predator–prey and other trophic interactions.

A similar approach to making predictions about the dynamics of highly nonlinear systems has been used to explore the dynamics of a range of ecological systems, including predator–prey and host–disease interactions (Sugihara and May, 1990; Sugihara et al., 1990). Using a relatively robust conjecture that dynamics can be adequately modelled in time-delay coordinates (e.g. Packard et al., 1980; Takens, 1981; Tong, 1990), it is possible to reconstruct the dynamical attractors of ecological interactions and make short-range forecasts for noisy time series (Sugihara and May, 1990; Casdagli, 1992). Sugihara and May (1990) have shown that the noisy dynamics of measles (a form of predator–prey interaction) from New York between 1928 and 1963 (the prevaccination era) are driven by deterministic chaos. In contrast, the dynamics of measles for all England and Wales between 1948 and 1967 do not show chaotic dynamics but a seasonal (biennial) cyclic pattern influenced by additive, stochastic perturbations (Sugihara et al., 1990). The differences in these epidemic patterns between England and Wales (a regional scale) and New York or individual British cities such as Birmingham (a local scale) are a consequence of averaging out the local epidemic effects associated with towns and cities. That is, the dynamics of this trophic interaction are dependent on spatial scale and spatial structure (see next section).

5.4 Space and noise

Populations are rarely structured simply by births and deaths: the role of space and spatial structure are central to the patterns of distribution and abundance of predators and prey. Although the direct, explicit inclusion of spatial structure is a relatively recent development (Giplin and Hanski, 1991; Hanski and Giplin, 1997; Hanski and Gaggiotti, 2004), a number of early ecological studies argued that space might affect the persistence of different ecological interactions (e.g. Nicholson and Bailey, 1935, Andrewartha and Birch, 1954; Hassell and May, 1973).

Spatial scale can have profound influence on the dynamics, distribution, and abundance of predators and prey. By aggregating patches of predators and prey into increasingly larger units of habitat (e.g. leaf, twig, branch, tree, forest), different ecological processes and mechanisms become important to the dynamics of the interaction. For instance, the statistical patterns associated with the distribution of parasitism are clearly dependent on the scale of the observations (Heads and Lawton, 1983) and what constitutes a patch (Waage, 1979). However, as spatial scale increases, the size of the samples at each scale decline but the range of densities increases. This makes predicting the outcome at different spatial scales difficult, which is clearly highlighted in a comprehensive study of the mortality factors affecting the cynipid gall-former *Andricus quercuscalicis* (Hails and Crawley, 1992). Mortality of this gall-former can be attributed to one of nine different causes, five of which were due to predation. Patterns associated with predator-induced mortalities were shown to vary across spatial scales and between years. For instance, bird predation on galls was positively density dependent across all scales in one year and highly variable (positive and negative) the following year. Similarly, parasitism by *Mesopolobus fuscipies* varied from tree to tree, with some trees showing positive and others showing negative density dependence as spatial scale changed. It is important to note that the persistence of predator–prey interactions does not necessarily require density-dependent processes to operate at all times, in all places, or at all scales (e.g. Taylor, 1988): it is entirely plausible that the persistence of these interactions is masked by population redistributions and stochasticity.

The dynamical implications of mixing of predators and prey populations have been widely explored (Allen, 1975; Reeve, 1990; Hassell *et al.*, 1991b). Assuming implicit space, Hassell and May (1988) show that incomplete mixing of hosts and parasitoids can allow the persistence of an otherwise unstable predator–prey interaction. This is particularly marked if the host completely mixes but the parasitoid is sedentary. Under this extreme scenario, patches of low host density are often unable to support a parasitoid population and are effectively refuges from parasitism leading to a stable host–parasitoid interaction (Hassell and May, 1988). Other spatially implicit representations of predator–prey interactions have explored the stability properties of the overall regional predator–prey interaction compared with the localized interaction (Reeve, 1990; Wilson *et al.*, 1998). Reeve (1990) concluded from his study that the dynamical stability properties at the regional scale were essentially the same as those observed in the localized interaction. If the local population dynamics were unstable then extinction of the predator–prey interaction was expected at the regional scale (Reeve, 1990). However, details about the mechanistic interaction between predator and prey can disrupt this general finding. Rohani *et al.* (1996) showed for a range of ecological models that the broad stability effects did not differ between local and regional scales. However, under high predator (or prey) overdispersion ($k \ll 1$), then the regional-scale dynamics might be unstable even though the local dynamics are stabilized (by the highly nonlinear effects of parasitism; May, 1978a). Similarly, the coupled effects of predation and environmental noise can lead to disparity between the regional and local dynamics (Reeve, 1990). Stochastic variation in host fecundity and parasitoid attack rate between patches can destabilize the local dynamics but prevent extinction of the interaction at the regional scale (due to patch turnover and rescue). Other forms of stochasticity, such as demographic stochasticity (that associated with the inherent fluctuations of birth, death, and dispersal) have similar effects (Wilson *et al.*, 1998). Coupled with restricted dispersal, demographic stochasticity introduces heterogeneity among predator–prey patches, leading to persistent interactions at the regional scale but with local extinction of patches.

More recently, it has been shown that predator–prey metapopulations can be influenced by both stochastic and deterministic processes (Bonsall and Hastings, 2004). Exploring the local and regional dynamics of the interaction between the bruchid beetle *C. chinensis* and its parasitoid, *Anisopteromalus calandrae*, Bonsall and Hastings (2004) showed how demographic stochastic processes dominate at the local scale yet this noise is undetectable at the regional scale. By fitting different population models to the regional predator–prey time series, it was shown that identifying such demographic stochasticity is confounded by noise operating differently in different patches. This leads to noise being mis-identified as environmental rather than demographic stochastic perturbations, and is a consequence of the simple statistical phenomenon of the central limit theorem. By aggregating predator–prey patches which are experiencing demographic stochasticity (often described by a Poisson distribution), we can create the illusion that the regional predator–prey interaction is experiencing 'environmental' stochasticity (often described by a normal distribution). Consequently, the type of noise and its effects operating in predator–prey interactions at different spatial scales can be easily mis-interpreted (Bonsall and Hastings, 2004).

One fundamental way in which space can affect predator–prey interactions is through the processes of limited dispersal linking otherwise local, independent populations. This is the metapopulation paradigm (Levins, 1969, 1970) and it is the central theme in understanding how the dynamics of ecological interactions scale from local to regional levels (e.g. Hanski and Gaggiotti, 2004). Theoretical models of spatially explicit predator–prey interactions reveal that even if the local dynamics are unstable, the regional interaction can persist (Hassell et al., 1991b; Comins et al., 1992; Wilson et al., 1993). This occurs because the local populations tend to fluctuate out of phase, enabling extinct patches to be rescued (through the immigration of prey) and allowing the whole metapopulation to persist. A number of different spatial patterns are associated with such regional dynamics that are generated principally through the process of limited dispersal (Figure 5.9). Under low host and high parasitoid dispersal, crystal lattice patterns may emerge (Figure 5.9c). As host dispersal increases, indeterminate patterns (spatial chaos) are observed (Figure 5.9b), but the predominant type of spatial pattern takes the form of predator–prey spirals (Figure 5.9a). These spirals are characterized by the local population densities forming spiral waves which rotate around relatively fixed focal points. The regional dynamics, however, are relatively complex limit cycles influenced by the position and number of the focal points (which vary through time in a non-repeating way; Hassell et al., 1991b; Comins et al., 1992).

Once the assumption of discrete space or patches is relaxed, we find that the dispersal of predator and prey can lead to a range of dynamical outcomes (Kot, 1992; Neubert et al., 1995). Greater dispersal of the predator (relative to the prey) can lead to a range of dispersal-driven, period-doubling bifurcations resulting in unstable interactions between predators and prey (Kot, 1992). Extending this idea, White et al. (1998) show how wolf-pack territoriality and the spatial interaction between wolves and deer can be described by simple rules

(a)   (b)   (c)

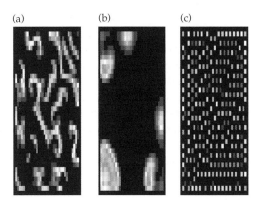

Figure 5.9 Maps showing the spatial distributions of host and parasitoids from a spatially explicit version of the Nicholson–Bailey model (Hassell et al., 1991b, Comins et al., 1992) with parameters $\lambda = 1.3$ and $\alpha = 0.01$. In each case the lattices have absorbing boundaries and interactions are initialized by seeding a single patch with a small number of hosts and parasitoids. The patterns are (a) spiral waves, obtained with host and parasitoid dispersal fractions of 0.5, (b) chaotic (indiscernable) patterns, obtained with host and parasitoid dispersal fractions of 0.2, and (c) crystal lattice patterns, obtained with host dispersal of 0.04 and parasitoid dispersal of 0.9. Lattice sizes were 35×35.

of movement behaviour. For example, movement of wolf packs towards regions of higher prey density leads to spatial segregation of predator packs, reduced competition between predators, and allows the establishment of prey gradients between wolf packs (White *et al.*, 1998). These behaviours give rise to spatial patterns in the distribution and abundance of predators and prey (Gueron and Levin, 1993; Gueron *et al.*, 1996).

Territorial structure, and consequently habitat size, has an influential effect on spatially explicit predator–prey interactions. Ecological interactions in small localized places are often prone to extinction and there is a positive relationship between metapopulation persistence and habitat size (Hanski, 1999; Bonsall *et al.*, 2002). Theoretically, lattices of a small, finite size restrict the possibility of asynchrony in the local dynamics (all patches are in phase) and as such patches can not be rescued from extinction (Hassell *et al.*, 1991b; Comins *et al.*, 1992). As habitat size increases, the possibility of asynchronous dynamics increases and the probability of persistence is greater. However, only rarely has this effect of increased persistence due to spatial processes and habitat structure been observed (Huffaker, 1958; Holyoak and Lawler, 1996; Hanski, 1999; Ellner *et al.*, 2001; Bonsall *et al.*, 2002).

As mentioned above, in an original set of experiments, Huffaker and colleagues (Huffaker and Kennett, 1956; Huffaker, 1958; Huffaker *et al.*, 1963) investigated the effects of spatial structure in a mite predator–prey system. First, Huffaker (1958) showed that in the absence of dispersal, the predator–prey interaction was prone to extinction. Second, by manipulating mite movement he showed that the persistence of the interaction could be increased. Finally, in extending the system to more patches and greater complexity, the long-term persistence of the predator–prey interaction could be attributed to the effects of habitat size and dispersal (Huffaker *et al.*, 1963). More recently, the role of spatial structure on the persistence of predator–prey interactions has been thoroughly explored (Holyoak and Lawler, 1996; Ellner *et al.*, 2001; Bonsall *et al.*, 2002). The overriding consensus from this range of studies on different experimental organisms is that

persistence of predator–prey interactions is critically dependent on the process of dispersal. For example, recent work on the metapopulation dynamics of *C. chinensis* and *A. calandrae* has shown that the persistence of this extinction-prone host–parasitoid interaction is enhanced by metapopulation processes (Bonsall *et al.*, 2002). By controlling for the effects of resource availability affecting the predator–prey interaction, metapopulation persistence was shown to be driven principally by coupling patches through limited dispersal, with larger systems persisting for longer. In a comparable study, Ellner *et al.* (2001) have shown that habitat structure per se has a relatively weak role in the persistence of predator–prey metapopulations and it is the reduced probability of attack by the predator at the patch scale that allows the system to persist. The role of habitat structure and spatial scale has a central role on the dynamics of predator–prey interactions, and these effects have now been observed to operate in more complicated, multispecies predator–prey assemblages (Hassell *et al.*, 1994; Comins and Hassell, 1996; Bonsall and Hassell, 2000; Bonsall *et al.*, 2005). The special case of interactions between microparasitic diseases and their hosts provide some very striking illustrations of the importance of spatial scale and spatial structure. Consider, in particular, the dynamics of measles in England and Wales (Grenfell *et al.*, 1994, 2001, 2002; Grenfell and Bolker, 1998; Bjornstad *et al.*, 2002). Aggregated data from urban and rural regions show the seasonal, biennial dynamical pattern with epidemics in urban places coming ahead of rural ones (Grenfell and Bolker, 1998). This difference in the timing of the epidemics is due to the strength of local coupling between urban and rural places. Patterns of spatial synchrony are related to population size: the number of cases in larger cities (large population size) tend to be negatively correlated, whereas there is no correlation in case reports between rural places with small population size (Grenfell and Bolker, 1998). This regional heterogeneity leads to hierarchical epidemic patterns. In the small rural places, the infection fades out in epidemic troughs, although this is clearly dependent on the degree of coupling to larger urban places (Finkenstadt and Grenfell, 1998).

These processes give rise to a range of spatial dynamical patterns such that larger cities show regular biennial cycles whereas in small towns disease dynamics are strongly influenced by stochasticity (Grenfell *et al.*, 2001). Most recently, the dynamics of measles epidemics have been shown to depend on the balance between nonlinear epidemic forces, demographic noise, environmental forcing, and how these processes scale with host population size (Grenfell *et al.*, 2002).

5.5 Conclusions

Identifying what regulates populations and allows ecological interactions to persist has vexed ecologists for almost 100 years (Howard and Fiske, 1911, Nicholson, 1933, 1957). Nicholson's work on density dependence (Nicholson, 1957) and predation (Nicholson and Bailey, 1935) has been hugely influential in the development of these broader aspects of ecological theory. Building on this, the theory of predator–prey interactions has made significant advances over the past few decades (Hassell, 2000; Murdoch *et al.*, 2003). Nevertheless, the integration of theory and data remains a challenge in ecology. Many of the components of predator–prey interactions have been separately quantified, originally using the classic approach to analysing population dynamics of key-factor analysis (Varley and Gradwell, 1960) or variants on this theme (Sibly and Smith, 1998). However, a number of limitations exist with this sort of approach, including the dynamics being misinterpreted, noise being mis-identified or ignored, and populations being embedded in a complex food web where single key factors are unlikely to operate. Modern methods of analysis of behavioural, population, and metapopulation data are making it much easier to parameterize ecological models with ecological data, and this is opening the way for a closer integration of theoretical and empirical ecology. It seems likely that understanding predator–prey interactions, in all their guises, will continue to provide essential, exciting, and challenging research and career opportunities for ecologists.

CHAPTER 6

Plant population dynamics

Michael J. Crawley

6.1 Introduction

Plants exhibit an extraordinary range of sizes and generation times, from single-celled algae with body sizes of the order of 5 μm and generation times of the order of 1 day, to massive forest trees more than 50 m tall that can live for over 1000 years. Diatoms and trees have the virtue of being easy to count, so it is natural to seek to model the dynamics of changes in numbers. On the other hand, many herbaceous perennials (like clonal herbs or turf-forming grasses) are difficult or impossible to count, and for these plants it is natural to model the dynamics of fluctuation in biomass or proportional space occupancy.

The theory of plant population dynamics is linked to the rest of plant biology through a series of fundamental trade-offs, reflecting the fact that individual plants are constrained in what they can do. There are important trade-offs in reproduction because a plant could produce many small seeds or a few large seeds, but it is not an option to produce many large seeds. Other trade-offs involve investment decisions: for instance a plant can invest in growth or defence and this leads to a trade-off between competitive ability and palatability to herbivores. Alternatively, high growth rate in full sun may trade-off against a high death rate in low light (the cost of shade tolerance). An important set of trade-offs involve competing demands for resource capture. Thus a plant could invest in its root system to forage for phosphorus, or in its shoot system to forage for light, but it cannot maximise investment in competitive ability for light *and* soil nutrients. Finally, there is an important trade-off between competition and colonization because good dispersers tend to be

inferior competitors; this is exemplified by the $r - K$ continuum where colonizers (r strategists) have a set of traits like rapid generation time, small seeds, wind dispersal, and high light requirements, whereas late successional species (K strategists) tend to live longer, produce fewer, larger seeds, and to have more shade-tolerant, slower-growing juveniles.

Underpinning the theory of plant population dynamics is the invasion criterion, which states that all persistent populations must exhibit the tendency to increase when rare. If this were not the case, then successive environmental calamities, like hammer-blows on a nail, would knock population density further and further down towards local extinction. We can state the invasion criterion formally in terms of either a differential equation model ($\mathrm{d}N/\mathrm{d}t > 0$) or a difference equation ($N(t+1) > N(t)$). Because we are dealing with small population sizes (low values of N) we can assume that the plants' vital rates (their birth rates and death rates) are density-independent. Figure 6.1a shows the case where the birth rate exceeds the death rate, as a result of which the population increases exponentially through time (Figure 6.1b). This species passes the invasion criterion. Figure 6.1c shows the contrasting case where the death rate exceeds the birth rate. Under these circumstances, the population declines exponentially (Figure 6.1d) and the species fails the invasion criterion. This simple idea lies at the foundation of the modern definition of a niche. Hutchinson (1957) distinguished between the fundamental niche and the realized niche of a species. He argued that there are combinations of circumstances (weather conditions, substrate, resources, disturbance regime, etc.) where the species would be able to increase when rare, and other combinations of circumstances where they would not be able to increase when rare. You can draw an outline in parameter space defining the margins of the region where dN/dt > 0; this defines the *fundamental niche* of the species. Within the fundamental niche, however, the species is locked into a struggle for existence with its competitors and natural enemies. Hutchinson's *realized niche* is that part of the fundamental niche in which a species' competitors and natural enemies allow it to persist. This great insight gave formal definition to Ellenberg's Rule that you do not find plants in the field

under the conditions that are 'best' for the growth of the species, but rather under the combination of conditions where the plant's competitors and natural enemies allow it to increase when rare (Ellenberg, 1953). Thus, the plant's realized niche is the subset of its fundamental niche defined by 'competitor-free space' and 'enemy-free space'.

Individual plants may quite often be found growing outside the fundamental niche under field conditions, and dispersal continuously blurs the edges of the fundamental niche. Mature plants may appear to be growing perfectly well, but they are expected to leave, on average, less than one

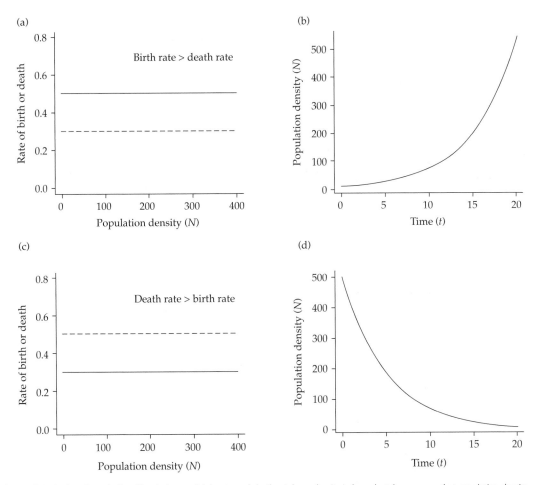

Figure 6.1 The invasion criterion. The vital rates (birth rate and death rate) are density-independent (we assume that population density is sufficiently low that this assumption is realistic). Top row: (a) birth rate (solid line) exceeds death rate (dashed line) so that (b) the population increases exponentially through time (t); the species passes the invasion criterion. Bottom row: (c) death rate exceeds birth rate so that (d) the population declines exponentially through time; the species fails the invasion criterion.

adult plant in the next generation (i.e. $dN/dt < 0$). Unfortunately, this means that you cannot define the fundamental niche by drawing a line around all the places where mature plants are found growing in the field. The job is much more difficult that that. You need to draw the boundary around those individuals for which $dN/dt > 0$, and this is a much more demanding task.

Differences in population dynamics are typically caused by differences in the pattern of density dependence (Figure 6.2). In the typical case, the plant species passes the invasion criterion so that at low densities the birth rate exceeds the death rate and the population increases exponentially in abundance (Figure 6.2a). In both panels of Figure 6.2 there is a stable, high-density equilibrium caused by density dependence in the birth rate (in real systems, of course, the death rate might be density dependent as well). In some plants, however, very low density represents a problem rather than an opportunity. Such 'rare species disadvantage' is called an Allee effect after the American ecologist W.C. Allee, who drew attention to the phenomenon in the 1940s (Allee *et al.*, 1949). In Figure 6.2b the plant population cannot

increase in abundance until its numbers exceed a threshold of 20 individuals. Allee effects in plants may arise because of the breeding system (e.g. obligate out-crossing species may suffer very low rates of pollination when population density is low) or because of herbivore effects (e.g. a threshold number of seeds may be required for granivore satiation). Allee effects are potentially important in plant conservation, because they mean that a local population will go extinct if numbers fall below the threshold.

The approach adopted in this chapter is to introduce the different kinds of population dynamics and the different ways of modelling them, in order of increasing generation times, from diatoms to trees. In parallel, the characteristic mechanisms of density dependence affecting plant dynamics at these different spatial and temporal scales are also introduced in a step-wise manner. Thus, we begin by considering exploitation competition for a single, depletable resource in the homogeneous environment of a chemostat, and end up with spatially explicit interactions between size-structured populations of long-lived woody plants in temporally heterogeneous environments.

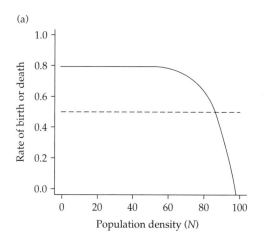 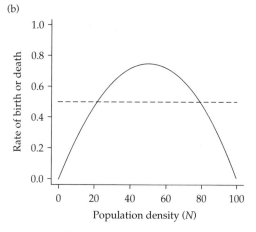

Figure 6.2 Density dependence in plant populations: density-independent death rate (dashed line) and density-dependent seed production (solid line). (a) The usual case, in which the species increases exponentially from low population densities. At high density, however, the birth rate declines and there is a stable equilibrium at approx. 85 plants per unit area. (b) Allee effects (rare species disadvantage) mean that the birth rate is less than the death rate (e.g. because of pollination failure) below a threshold density of approx. 20 plants per unit area. This population also has a stable equilibrium at approx. 80 plants per unit area.

6.2 Population dynamics of diatoms with a single limiting resource

Populations of single-celled algae are amenable to controlled experimentation, and the processes that drive their population dynamics are analytically tractable. Because they have such short generation times (days to weeks) it is appropriate to frame our models as differential equations in continuous time. In the single-species case, there are N individual plankton per unit volume of water, and we model their dynamics by considering the rates of gain and loss of individuals. The change in numbers is written dN/dt and the per-capita rate of change in population size is:

$$\frac{1}{N}\frac{dN}{dt} = \text{birth rate} - \text{death rate} \qquad (6.1)$$

In the simplest case (Figure 6.1), both the birth rate and death rate are density-independent, and this leads to either exponential growth (when the birth rate exceeds the death rate) or exponential decline (when the death rate exceeds the death rate). Numbers change through time as the integral of the differential equation:

$$N(t) = N_0 e^{(\text{birth} - \text{death})t}. \qquad (6.2)$$

where $N(t)$ is the population size at time t and N_0 is the initial population size at time 0. While these patterns of dynamics are important for defining the invasion criterion, they quickly become unrealistic because population growth rate is certain to decline as one or both of the vital rates becomes density-dependent.

In single-celled algae, the most likely cause of density dependence is intraspecific competition for a limiting resource (Tilman, 1982). In some circumstances, phytoplankton might compete for mineral nutrients like nitrogen or phosphorus, or for non-mineral resources like light. In our example, however, we consider that silicate (which is used to construct the ornate cell wall—the frustule—of diatoms) is the single limiting resource. This means that other resources (like light, nitrogen, phosphorus, etc.) are available in non-limiting amounts, reflecting a fundamental rule of plant ecology known as Leibig's Law of the Minimum, which states that the only limiting

factor is the *most limiting factor*. Here it means that because silicate is the most limiting resource, increasing the amount of light, nitrogen, or phosphorus will not increase the algal population growth rate, but decreasing the concentration of silicate *will* decrease the growth rate.

To keep things a simple as possible, we shall assume that the birth rate of the diatoms is a function of silicate but that the death rate is independent of resource supply. When silicate concentration is greater than $4.4\,\mu M$ (say) then the birth rate exceeds the death rate and the population increases exponentially. If the silicate concentration falls below $4.4\,\mu M$ then the death rate will exceed the birth rate and the population will decline exponentially. The silicate concentration, however, is not constant, but depends on the population of plankton. Silicate is removed from the water and tied up in the cell walls of the algae, so as the diatom population grows, the silicate concentration in the water declines. Once the silicate concentration falls below $4.4\,\mu M$ the algal population would begin to decline because its death rate would exceed the birth rate, and silicate would be returned to the water thorough decomposition of algal cell walls. Thus, the silicate concentration of $4.4\,\mu M$ defines an equilibrium amount of resource at which algal births and deaths are equal. The level to which the diatom population reduces the concentration of its most limiting resource was called R^* by Tilman (R means resource and $*$ means equilibrium; see Chapter 7 and Figure 7.2 in this volume). To understand the dynamics of the resource we need to be explicit about the structure of the experimental set-up. For instance, do resources leak out of the system or are they recycled? It turns out that the dynamics of open and closed systems differ in important ways (Daufresne and Hedin, 2005).

6.3 Two or more plant species with a single limiting resource

Now carry out the thought experiment of introducing two different species into the same chemostat, both at low densities. We assume that both species pass the invasion criterion, so both populations increase exponentially to begin with. Given our assumption that there is only a single limiting resource, it is certain that one of the two species will go extinct. The only question concerns the *identity* of the species that survives. Tilman's R^* theory allows us to predict the identity of the winner, because irrespective of its initial population growth rate, the eventual dominant will be the species with the lowest value of R^* (see Chapter 7 in this volume). The species with the highest birth rate will increase fastest and will soon become substantially more abundant than the other species. But does this mean that the species that grows fastest to begin with will persist? It might do, but it might not. The long-term outcome depends on the relative magnitude of the R^* values for the two species and not necessarily on their initial growth rates. Inevitably, one of the species is driven to extinction, and the resource level is reduced to the lower of the two R^* values. This is the process of *competitive exclusion*.

The competitive exclusion principle is one of the central hypotheses underpinning plant population dynamics. The principle was originally introduced in 1934 by the Russian ecologist G.F. Gause, who wrote in *The Struggle for Existence* that 'as a result of competition two species scarcely ever occupy similar niches, but displace each other in such a manner that each takes possession of certain particular [resources] and modes of life in which it has an advantage over its competitions' (Gause, 1934). The principle was developed subsequently by Hardin (1960). The 'advantage' referred to by Gause is what we would now call a lower value of R^* (Tilman, 1982). The current definition of competitive exclusion involves five postulates, as follows.

• Given a set of species, all of which pass the invasion criterion, and all of which are capable of forming self-replacing monocultures, then
• under the prevailing resource regime in which there is a single limiting resource, then
• in a temporally constant environment, and
• in a spatially uniform environment, then
• given long enough,
• one species will persist and all the other species will be excluded.

Obviously there are a lot of caveats here, and we shall relax each of them in due course. We would

certainly not expect to observe competitive exclusion in short-term experiments that are dominated by transient dynamics. The point, however, is that you *can* set up controlled experiments that approximate well to these assumptions (e.g. in chemostats), and, when you do that, you *do* observe competitive exclusion (Tilman, 1982). Moreover, the single species that persists is the species with the lowest value of R^* for the single limiting resource (see Chapter 7 in this volume). Outside chemostats, competitive exclusion has been observed in experiments carried out in spatially uniform environments like arable fields (e.g. where a single cultivar eventually excludes all others from an initially diverse mixture of crop genotypes; Harlan and Martini, 1938) and in well-mixed freshwater systems (Huisman and Weissing, 1994).

6.4 Two or more plant species with two resources

Suppose now that we have a replicated series of separate chemostats in which we can vary the amounts of two different resources, say nitrogen and phosphorus. It is possible that one species has the lowest R^* values for *both* resources, in which case competitive exclusion will occur, just as it did in the case of a single resource. But, at least in principle, species might differ in the resource for which they were the superior competitor. We need to investigate whether coexistence between species can be promoted just by increasing the number of limiting resources from one to two. Tilman investigates this question by drawing a phase plane showing the abundance of two resources, R_1 and R_2 (Tilman, 1982). Now draw on this phase plane the zero growth isoclines for the two species, A and B. A zero growth isocline is just a line across the phase plane separating the region where the species passes the invasion criterion ($dN/dt > 0$) from the region where it fails ($dN/dt < 0$). If species A has a lower value than species B for both R^*_{A1} and R^*_{A2} then competitive exclusion of species B is inevitable. Existence of two resources is clearly not a *sufficient* condition for coexistence. But what if species A has a lower R^*_{A1} for resource number 1 but species B has a lower R^*_{B2} for

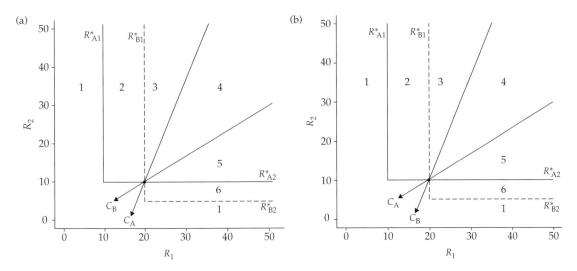

Figure 6.3 Two species with two resources. (a) Stable coexistence is possible in zone 4 because the consumption vector for species A, C_A, is steeper than the vector for species B, C_B. (b) Stable coexistence is not possible in zone 4 because the consumption vector for species B is steeper than that for species A. See the text of this chapter and Chapter 7 in this volume for details.

resource number 2 (Figure 6.3)? Now there is a two-species equilibrium at the point where the two zero growth isoclines cross. The question is whether this is a stable or an unstable equilibrium.

The extra ingredient needed to understand the dynamics of this model concerns the extent to which the growth of the two species is limited by each of the two resources. A particular habitat will provide the two resources at amounts (S_1, S_2) that can be represented by a point in the phase plane; this is called the supply point. A supply point in the top right-hand corner indicates roughly equal supplies of the two resources, whereas a supply point in the top left-hand corner shows a high supply of R_2 and a low supply of R_1. The next thing to understand is that starting close to the supply point, say in the top right-hand corner, population growth of both species will deplete both resources. This will drive both resources downwards, which can be depicted as an arrow pointing roughly towards the bottom left-hand corner (towards the origin of the phase plane). But the key point is that the two species are likely to deplete the two resources at somewhat different rates. One species might use relatively more of R_1 than R_2 during growth, and vice versa. This is reflected in the slope of each species' consumption vector, C. If a species uses both resources at the

same rate then the consumption vector would be at 45°. In an extreme case, if it used none of R_2 the vector would be horizontal, whereas if it used none of R_1 the vector would be vertical. There are two cases of interest: (1) in Figure 6.3a the consumption vector for species A is steeper and (2) in Figure 6.3b the consumption vector for species B is steeper.

The dynamics of the system depend on the location of the resource supply point (S_1, S_2). If it is in zone 1 on Figure 6.3 then neither species can persist; species A is the best competitor for R_1 but its supply is below R^*_{A1}, and species B is the best competitor for R_2 but its supply is below R^*_{B2}. In zone 2 species A will exclude species B, and in zone 6 species B will exclude species A. More subtly, species A will also exclude species B in zone 3, and species B will exclude species A in zone 5, because the supply point lies *outside the consumption vector*. The essential point is that in Figure 6.3a, where the consumption vector for species A is steeper than that of species B, the model predicts that stable coexistence is possible. The equilibrium point is stable because each species consumes relatively more of the resource which limits its growth at equilibrium (and less of the other resource).

In contrast, Figure 6.3b shows the case with an unstable equilibrium, in which the consumption vector for species B is steeper than that of species A. Here, even though the supply point is in zone 4 there is no stable coexistence, and the identity of the winning species is determined by initial conditions (whichever species is initially the most common grows to form a monoculture). The equilibrium point is unstable because each species consumes relatively less of the resource that limits its growth at equilibrium and more of the other resource.

The celebrated Lotka–Volterra coexistence criterion states that coexistence will occur if and only if intraspecific competition is more important than interspecific competition for both species (see Chapter 7 in this volume). As we shall see, all models that exhibit stable coexistence can be translated into essentially this same criterion. Lotka's and Volterra's models have been made more sophisticated and more mechanistic, but their criterion for coexistence remains inviolate. Mechanisms that can be characterized as 'keeping species out of each other's way', promote coexistence, while differences in average fitness, which determine 'how much better one species is than another', favour competitive exclusion (Chesson, 2000). It is beginning to look as if the number of coexisting species is determined by the number of resources. But how many different resources could be limiting for diatom growth? Nitrogen, phosphorus, silicon, iron, light, inorganic carbon, and a few trace metals and vitamins, perhaps. It was this realization that led G.E. Hutchison (1961) to pose his famous paradox of the plankton. How can so many algal species coexist on so few kinds of resource? Hutchinson understood the importance of spatial and temporal heterogeneity, but he thought that competitive exclusion was an all-powerful process; hence the paradox.

6.5 Diatoms in systems with multiple limiting resources

Huisman and Weissing (1999) extended Tilman's resource competition model to deal with multiple diatom species competing for multiple (three or more) resources:

$$\frac{dR_j}{dt} = D(S_j - R_j)$$
$$- \sum_{i=1}^{n} c_{ji}\mu_i(R_1, R_2, \ldots R_k)N_i \quad j = 1, 2, \ldots k$$

(6.3)

$$\frac{dN_i}{dt} = N_i(\mu_i(R_1, R_2, \ldots R_k) - m_i) \quad i = 1, 2, \ldots n \quad (6.4)$$

where the function $\mu_i(R_1, R_2, \ldots R_k)$ is the specific growth rate of species i in relation to the resource availabilities (assumed to be a set of Monod functions, and applying Liebig's Law of the Minimum; see Section 6.2), m_i is the death rate of species i, D is the system's turnover rate, S_j is the supply concentration of resource j, and c_{ij} is the content of resource j in species i. The fascinating new twist in this case is that the process of intraspecific competition itself is capable of producing temporal variability which, in turn, fosters multi-species coexistence. For instance, competitive chaos occurs whenever each species is an intermediate competitor for the resources that most limit its growth rate. With three resources and with five resources, chaotic dynamics result despite almost constant total phytoplankton biomass (caused by the nutrient limitation). This model provides an explanation for plankton biodiversity based on the dynamics of competition itself; the species oscillations generated by the process of competition produced the temporal heterogeneity that prevented competitive exclusion from occurring, and allowed coexistence of many more species than there were limiting resources. 'Once a plankton community is sufficiently complex to generate its own non-equilibrium dynamics, the number of coexisting phytoplankton species may greatly exceed the number of limiting resources, even in a constant and well-mixed environment. In this sense, the paradox of the plankton is essentially solved' (Huisman and Weissing, 1999).

In deep ocean communities rather than chemostats, the dynamics are further influenced by two important sources of spatial heterogeneity: light intensity declines exponentially from the ocean surface to the base of the euphotic zone, while upwelling and diffusing mineral nutrients from the ocean's depths create a concentration gradient in the opposite direction. This spatial heterogeneity interacts with the complex temporal dynamics generated by interspecific competition (above). We do not yet know how much of the observed coexistence can to be attributed to the spatial and temporal heterogeneity, nor within the temporal heterogeneity, how much is generated internally as a result of multi-species competition (as in the models), and how much is externally driven through within-year seasonality and year-to-year climatic differences (Huisman *et al.*, 2006).

6.6 Population dynamics of annual vascular plants

To study the population dynamics of annual plants we adopt a completely different model structure, and work in discrete time (years) using difference equations instead of differential equations. For a spring-germinating annual without a seed bank, the model looks like this:

$$N(t+1) = \lambda \cdot N(t) \qquad (6.5)$$

where the population next year, $N(t+1)$, is calculated by multiplying the population this year, $N(t)$, by the per-capita net multiplication rate, λ (the product of per-capita seed production and survivorship). In the simplest case, all the parameters are density-independent and the population will increase exponentially if $\lambda > 1$ (if the species passes the invasion criterion) and will decline exponentially to extinction if $0 < \lambda < 1$ (having failed the invasion criterion). However, real populations of annual plants will not increase exponentially under field conditions for more than a few years. Sooner or later some form of density dependence will kick in, so that λ declines as the population increases (Watkinson *et al.*, 1989; Watkinson, 1990). There are three contrasting forms that this density dependence might take: competition for access to regeneration niches (microsite limitation), a density-dependent reduction in fecundity (size plasticity), or a density-dependent increase in mortality (self-thinning).

6.6.1 Microsite limitation

Most models assume that recruitment is directly proportional to seed production, but there are two important circumstances when this is not the case (Crawley, 2000; Figure 6.4). In zone III (Figure 6.4), fluctuations in seed production (e.g. caused by herbivore feeding) have no effect on plant recruitment because recruitment is *microsite-limited* and there is sufficient seed to populate all of the available microsites. Note also, that microsite limitation may constrain recruitment to zero for relatively long periods, irrespective of seed production or seed immigration (e.g. in ecosystems where fire or some other relatively extreme form of biomass destruction is necessary to open up the canopy to the point at which seedling recruitment is possible). In zone I variation in seed input is also unrelated to recruitment, but now because of failure of predator satiation. When seeds are scarce, the seed-feeding animals, the granivores, are able to locate and consume essentially all of the seeds (the plants lack enemy-free space). In some systems (like oak, *Quercus robur*, in southern England; Crawley and Long, 1995) recruitment will not occur until seed production exceeds a threshold that allows satiation of the guild of seed-feeding animals and pathogens (these include fungi, insects, birds, rodents, and ungulates). It is only between the zones of granivore-limitation and microsite-limitation (in zone II) that we predict that recruitment will be seed-limited. In this zone, if you carry out the experiment of sowing extra seeds, then you will get extra mature plants (Agren and Fagerstrom, 1984; Schenkeveld and Verkaar, 1984; Shaw and Antonovics, 1986; Fowler, 1986; Turnbull *et al.*, 2000).

The idea of competition for microsites is straightforward. Suppose that there is a fixed number of physical locations per unit area, M, in which a seed could germinate and grow large enough to have a reasonable chance of surviving up to flowering size (this is our operational definition of recruitment for plants). Such a site would need to have the right microclimate, resource availability (especially light and water), and the appropriate physical structure to protect the seed from desiccation and from seed-feeding

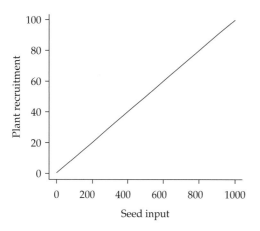
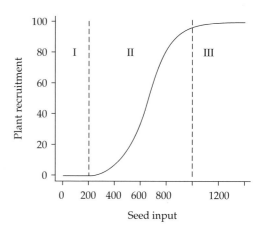

Figure 6.4 Seed limitation of plant recruitment. (a) The idealized model in which recruitment is proportional to seed input. (b) The three-zone model observed in field experiments with seed addition: zone I, granivore limitation (below 200 seeds, all of the seeds are eaten by animals, while above the threshold, predator satiation occurs); zone II, seed-limited recruitment; zone III, microsite limitation (there are only 100 locations in which seedling recruitment can occur) where recruitment is independent of extra seed input.

(granivorous) animals. It might also require the absence of mature plants nearby (i.e. the presence of a canopy gap) and the absence of a deep layer of plant litter. Some species might require physical disturbance of the soil surface so that the seed can become buried to the appropriate depth. In practice, the only way to determine the numerical value of M would be to sow excess seed and to count the number of plant recruits; it is unlikely that you could ever count the microsites directly. But how can a fixed number of microsites cause density dependence in recruitment? A simple numerical example will make it clear. Suppose that there are $M = 500$ microsites. We sow enough seeds that each microsite gets at least one seed, and we assume that only one individual wins the contest to take occupancy of that microsite. We sow 1000 seeds and observe 500 recruits. In another replicate, we sow 5000 seeds and observe 500 recruits. The establishment rate is density-dependent, and has declined from 50 to 10% as seed input increased from 1000 to 5000. This form of density dependence (contest competition) tends to be detected whenever seed-sowing experiments are carried out in relatively undisturbed natural vegetation, because microsites are typically scarce under such conditions (Turnbull *et al.*, 1999). It is observed much less often in greenhouse trials where seeds are typically sown into a competition-free seed bed from which herbivores and pathogens are excluded.

If there has been microsite-limited recruitment, then at maturity the plants may be so widely spaced that classic tests for competition (e.g. removal experiments; Aarssen and Epp, 1990; Goldberg and Barton, 1992) will not show any evidence for plant–plant interactions in the adult stage. It would be a mistake, however, to read this as absence of density dependence. What we are seeing is the Ghost of Competition Past (Connell, 1980), and, as Harper (1977) pointed out, the 'real cause of distribution and abundance will often be missed when mature vegetation is studied.'

6.6.2 Density-dependent fecundity resulting from size plasticity

The modularity of vascular plant growth means that individuals of the same age and genotype can differ from one another by several orders of magnitude in shoot mass at maturity. This environmentally caused variation in plant architecture is often referred to as size plasticity. In an environment where recruitment microsites are not limiting (e.g. a recently cultivated arable field, or a compost tray in a greenhouse), seedling density is likely to be proportional to seed sowing rate, and at high seed inputs very high plant population densities can result. In the absence of density-dependent mortality (discussed below), these increases in plant density are associated with

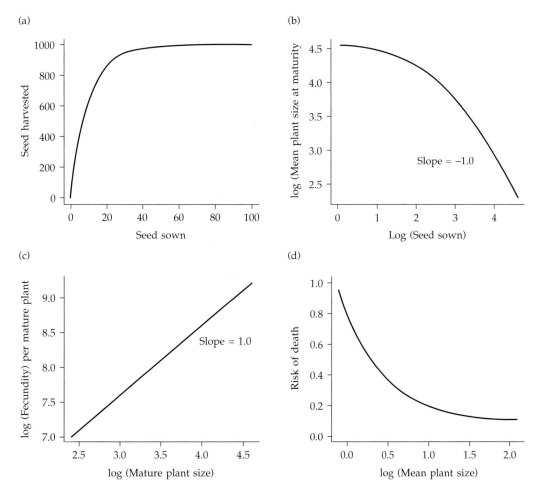

Figure 6.5 The Law of Constant Yield. (a) Above a relatively low threshold, sowing more seeds does not increase the number seeds produced by harvest time. The reason is intraspecific competition, leading to reduced plant size. (b) Once seed yield has reached an asymptote, then mean plant size at maturity declines linearly (on a log–log scale) with increase in sowing density, and the slope of the relationship is − 1.0. (c) For a wide range of plant species, and a wide range of conditions, the number of seeds produced by a plant at maturity is a linear function of shoot biomass, a reflection of the modular construction of the plants. On a log–log scale the slope of the relationship is 1.0. If the per-module fecundity changed with plant size then the log–log slope would depart from 1.0. If bigger plants had more seed per module the slope would be > 1.0 and if bigger plants had fewer seeds per module (e.g. because of disproportionate support costs) then the slope would be < 1.0. Such cases occur but are uncommon. (d) If the intensity of intraspecific competition is sufficiently intense, then the risk of death becomes density-dependent. The risk of death is typically much greater in small plants than in large plants. Note that this relationship is not exponential (the *x* axis is on a log scale) and that all plants, no matter how large, have a non-zero risk of death before flowering (approx. 0.1 in this example).

dramatic reductions in mean plant size. It is typically the *number* of plant modules (like leaves, flowers, or seeds) that varies, rather than the size of individual modules (Harper, 1977). Some authors argue that size plasticity is too vague a concept to be useful. Weiner (2004) proposes an allometric view, where plasticity in allocation is

understood as a change in a plant's allometric trajectory in response to the environment. He distinguishes three degrees of plasticity: (1) allometric growth (apparent plasticity), (2) modular proliferation and local physiological adaptation, and (3) integrated plastic responses. The overall response of a plant to population density is the sum of all modular responses to their local conditions plus all interaction effects that are due to integration (de Kroon *et al.*, 2005).

6.6.2.1 The Law of Constant Yield

This states that, above a low threshold, total seed production per unit area is independent of plant population density; doubling the number of mature plants simply halves the seed production of each plant. If the Law of Constant Yield applies, then the Ricker curve is typically asymptotic rather than humped, leading to very stable population dynamics (see below and Figure 6.5). The implicit assumption is that total biomass is constant and independent of population density, so that mean plant size declines hyperbolically. The law is most likely to apply in even-aged experimental monocultures, and it is important to note that the biomass/seed-density relationship asymptotes well before maximum biomass is achieved (Donald, 1951). In principle, there are three stages to predicting the effect of population density on total seed production:

1 predict the effect of population density on plant size distribution,
2 predict the effect of plant size on fecundity for individual plants,
3 add up seed production for all mature plants in all size classes.

The weight of evidence suggests that for individual plants, seed production (fecundity) is a linear function of shoot dry mass (Samson and Werk, 1986; Rees and Crawley, 1989; Freckleton and Watkinson, 2002), which is what you would predict from a model where shoot size varied because of changes in the *number* of modules, and where the fecundity per module was constant. Thus, stage 2 is often straightforward in practice. If this linearity can be assumed, then plant size variation (stages 1 and 3) is unimportant for predicting total seed production, because we can obtain total seed production simply by multiplying total shoot biomass by the fecundity per unit biomass. This simplicity in predicting fecundity is in marked contrast to the difficulty in predicting the effect of plant size variation on the death rate (as discussed next).

6.6.3 Density-dependent mortality

Death rates in populations of mature plants need not be density dependent, and changes in density might simply cause changes in mean plant size and mean fecundity (as above). However, when pronounced size hierarchies develop, the risk of death is always vastly greater in the stressed, subordinate, suppressed individuals than in the dominant individuals (White and Harper, 1970). There are three broad hypotheses about the origins of size hierarchies:

• non-regular spacing of juveniles in an otherwise spatially uniform environment means that isolated individuals become large and individuals in clumps are small as a result of spatially heterogeneous competition,
• random differences in performance between individual seedlings (germination time, initial growth rate) are amplified during growth, leading to asymmetric competition with a few large plants suppressing the growth of many smaller individuals,
• spatial heterogeneity in growing conditions: big plants develop in the high–quality patches and small plants in the low-quality patches, independent of plant density.

Field evidence suggests that asymmetric competition is a more common cause of size hierarchies than is spatially heterogeneous competition. Regular spatial patterns of surviving individuals should be interpreted as evidence for strong, asymmetric competitive interactions and subsequent density-dependent mortality (Weiner *et al.*, 2001; Stoll and Bergius, 2005). In practice, however, it is very difficult to refute the hypothesis of spatially heterogeneous growing conditions.

Once a certain biomass has been reached, it appears that any further increase in population biomass can only be achieved at the expense of reduced population density (Yoda *et al.*, 1957; White and Harper, 1970; Stoll *et al.*, 2002). There are two generalizations about this process of self-thinning (Sukatschew, 1928): first, density-dependent mortality is greater on more fertile soils, and second, density-dependent mortality occurs at the time of peak plant growth rate.

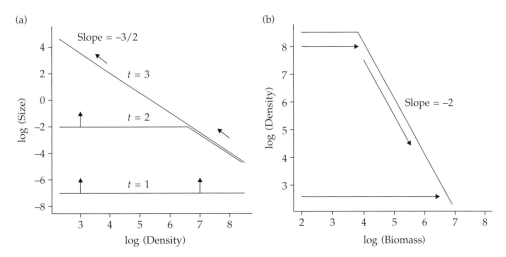

Figure 6.6 Self thinning is density-dependent mortality. (a) The original formulation with the axes muddled up had a slope of $-3/2$. (b) The correct axes with population density declining because of increases in biomass has a slope of -2. Self thinning begins at much lower biomass, when initial population density is very high, and may never begin if initial density is sufficiently low.

Both observations draw attention to the fact that it is the active growth of the dominant individuals in the population that causes the death of the suppressed individuals. The cause of self-thinning is that above a density-dependent threshold value of biomass, increases in biomass are only possible if mortality causes reductions in population density and frees up space for further growth of the survivors. In a plot of log(plant numbers) against log(total biomass) the 'self-thinning line' has a slope of roughly -2 (Figure 6.6).

6.7 Modelling density dependence in annual plants

The simplest way of combining the three density-dependent processes that affect annual plant populations is to construct a Ricker curve: $N(t+1) = N(t) \cdot f(N(t))$ (see Chapter 3 in this volume). This shows next year's population $N(t+1)$ (on the y axis) as a function of this year's population $N(t)$ (on the x axis). A straight line through the origin at a slope of $45°$ is called the replacement line (where next year's population is equal to this year's population). Because the population must pass the invasion criterion, the Ricker curve must lie *above* the replacement line when this year's population density is low (i.e. in the bottom left-hand corner;

Figure 6.7). At the highest densities this year (the right-hand end of the x axis) it is almost certain that next year's population will be lower than this year's as a result of density dependence in growth, fecundity, or mortality; hence the Ricker curve falls below the replacement line. The two extreme cases are called *contest* competition and *scramble* competition. The distinction in the dynamics is due to the operation of overcompensating density dependence in Figure 6.7b. Contest competition produces a constant number of recruits in the next generation, independent of this year's population, and the slope of the Ricker curve at equilibrium is close to 0. Scramble competition means that high populations this year are followed by very small populations next year because of overcompensating density dependence (e.g. mass mortality or complete reproductive failure), and the slope of the Ricker curve at equilibrium is strongly negative.

The consensus from field studies is that annual plant population dynamics are rather tame (i.e. stable, rather than cyclic or chaotic; Samson and Werk, 1986; Rees and Crawley, 1989, 1991; Freckleton and Watkinson, 2002). This is principally because the Law of Constant Yield predicts contest rather than scramble competition (contrast Figure 6.5 and Figure 6.7b). Overcompensating density dependence acting through size plasticity

(a)

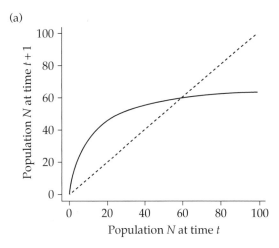

(b)

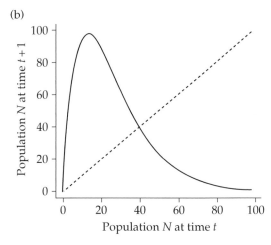

Figure 6.7 Ricker curves illustrating density-dependent recruitment for two contrasting annual plant populations. (a) Contest competition resulting from size plasticity and the Law of Constant Yield produces a Ricker curve with a low positive slope at equilibrium. (b) Over-compensating density dependence resulting from scramble competition produces unstable population dynamics because the slope of the Ricker curve has a large negative value at the intersection with the replacement curve.

and a humped curve of total seed production against seed density (Figure 6.7b) have been reported in some sub-populations of *Erophila verna* (Symonides *et al.*, 1986) but it is not a common phenomenon. Where substantial fluctuations are observed in field populations of annuals (Rees *et al.*, 1996; Hooper and Vitousek, 1997), these tend to be attributed to variations in weather conditions (e.g. rainfall) or the disturbance regime (e.g. fire) leading to fluctuations in seed production or microsite availability (as when drought opens up gaps in a formerly dense perennial cover). Thus, it would not be sensible to infer over-compensating density dependence simply because large fluctuations in population density were observed.

6.8 Annual plants with a seed bank

It is instructive to ask what happens to our annual plant population in the first year that seed production fails completely, so that $\lambda = 0$. Since $N(t+1) = \lambda \cdot N(t)$, the answer, of course, is that the population goes extinct. In the absence of immigration, it would stay extinct. If, however, there is a long-lived bank of seeds in the soil then recruitment could fail for many years in a row, yet the population would bounce back as soon as

conditions favouring recruitment returned. Seeds in the soil bank are 'temporarily opting out of the struggle for existence' (Harper, 1977). Most of the seeds in the bank are not technically dormant, and would germinate as soon as conditions were conducive (typically, when the soil was moist enough, warm enough, and light enough). Genuine dormancy requires some specific process to break it: seeds may be under innate dormancy, enforced dormancy, or induced dormancy (Harper, 1977; Thompson, 2000).

Existence of a seed bank can have a profound stabilizing effect on plant population dynamics as well as reducing the likelihood of local extinction. For instance, systems with parameter values such that they would exhibit chaotic dynamics in the absence of a seed bank show damped oscillations or a stable point equilibrium as the fraction of current seed production entering the seed bank is increased (Rees, 1997). We model the size of the seed bank ($B(t)$ seeds per unit area) and the above-ground population of mature plants $N(t)$ separately, using a pair of coupled difference equations:

$$N(t+1) = B(t) \cdot (1-d) \cdot g \qquad (6.6)$$

where next year's plant population is the germination rate, g, multiplied by the size of the surviving seed bank (survival $= 1 - d$, where d is the death rate of seeds in the bank),

$$B(t+1) = B(t)(1-g)(1-d) + N(t) \cdot f(N(t)) \quad (6.7)$$

and next year's seed bank comprises the survivors from this year's seed bank (seeds that did not germinate or die), topped up by this year's seed production (which is a density-dependent function of this year's mature population size). To act as a stabilizing mechanism on plant dynamics, the essential requirement is that the loss of seeds from the soil (death plus germination) is sufficiently low that, given the characteristic return time of good conditions, u years, and fecundity in a good year F:

$$F(1 - g - d)^u > 1 \quad (6.8)$$

If seed banks are such a good idea, then why don't all annual plant species have a seed bank? The most obvious explanation lies in the trade-off between seed size and seed number. Large seeds give rise to competitive seedlings, but large seeds would suffer very high death rates in a seed bank (they would be too attractive as food for granivores, for instance). But not all small-seeded annuals have seed banks, so this cannot be the whole answer. Perhaps there is a trade-off between strategies of escape in space (effective seed dispersal) and escape in time (effective seed-bank formation). Seed-bank formation would be favoured when fluctuations in reproductive success were positively correlated over large areas (e.g. because of large-scale weather patterns, like regional droughts), because there is no point in trying to disperse if everywhere is likely to be equally bad. Dispersal would be favoured when fluctuations in reproductive success were not spatially correlated, or were negatively correlated. In this case, the benefits of finding a suitable site elsewhere make dispersal advantageous, even though dispersal may be very risky. If these hypotheses are true, then dormancy and dispersal should be negatively correlated traits (Cook, 1980; Klinkhamer *et al.*, 1987).

6.9 Herbaceous perennials

Many herbaceous perennial plants are long-lived and difficult to count because they grow as patch-forming clones through lateral spread of rhizomes or stolons. For these species, it makes sense to model population dynamics in terms of biomass or total cover rather than plant numbers. Lottery models (Yodzis, 1978; Chesson and Warner, 1981) are ideally suited for this. They are spatial models but they are not spatially explicit. The world is divided up into a grid of cells with the size of the cells defined by the size of a typical mature plant module. Each cell can accommodate one and only one module at maturity, but no account is taken of the neighbourhood in which a particular module finds itself (hence the models are not spatially explicit). What makes lottery models so attractive to theorists is the assumption that there is no empty space: all of the cells are occupied all of the time. Interactions between plant species occur only at recruitment. Cells become available for recruitment only through the death of their former occupants, and the total amount of recruitment is exactly equal to the total number of module deaths across all species. The interest hinges entirely on the *fraction of recruitment* that is achieved by each species. Suppose that the arena consists of 1000 cells so that the total population of species A plus species B is always 1000 modules. We begin with 500 modules of each, scattered randomly over the matrix (although the spatial pattern is inconsequential, since this is not a spatially explicit model). We shall assume that both species have the same modular death rate D and differ only in their fecundity, with $F_A > F_B$. In the absence of any density or frequency dependence species A will competitively exclude species B; the two species cannot coexist. The lottery model works like this. We compute the total number of deaths as being $1000 \times D$ (since the two species have identical death rates). Now the only question is how many of these $1000 \times D$ cells are filled with recruits from species A and how many from species B. Let's compute the total seed production by each species. There are 500 individuals of A so there are $500 \times F_A$ seeds of species A. Likewise, there are 500 individuals of B so there are $500 \times F_B$ seeds of species B. The lottery means that each empty cell is colonized by a species with a probability that is simply the *proportion of all seeds that belong to that species*. It is simple to calculate the total number of seeds produced as equal to $500 \times F_A + 500 \times F_B$. So the fraction of empty cells that is captured by species A is $(500 \times F_A)/(500 \times F_A + 500 \times F_B)$ and by species B is $(500 \times F_B)/(500 \times F_A + 500 \times F_B)$. That is all there is to it. In symbols,

$$N_A(t+1) = (1-D)N_A + 1000D\frac{F_A N_A}{\sum_i F_i N_i} \qquad (6.9)$$

The density of species A next year, $N_A(t+1)$, comprises the survivors from this year, $(1-D)$ $N_A(t)$, plus the recruits won in competition with species B for the space opened up by mortality during the year $= 1000D\frac{F_A N_A}{\sum_i F_i N_i}$. The lottery model is put into a loop, and the time series of the species abundances plotted (Figure 6.8a). As predicted, we observe competitive exclusion of species B by species A.

How might an annual plant persist in a community dominated by herbaceous perennials? Consider the case of a rhizomatous grass like *Festuca rubra*, which spreads vegetatively to form a dense carpet, beneath which annual plants are incapable of recruitment from seed. In such a system, the annual plant always loses out when it comes into competition with the perennial. If the perennial is immortal, then the answer is that the annual and perennial species cannot coexist. If, however, the tillers of the grass have a non-zero

death rate, then coexistence becomes a possibility. Death of perennial tillers opens up gaps in the otherwise impenetrable carpet so that at equilibrium, a fraction, E^*, of the ground is unoccupied by the perennial. These ephemeral gaps can be colonized by the annual. So long as the fecundity and dispersal ability of the annual are high enough, it can coexist with the perennial, even though it always loses out in head-to-head competition (Skellam, 1951). In particular, the net multiplication rate of the annuals needs to exceed the reciprocal of the equilibrium proportion of empty gaps:

$$\lambda > \frac{1}{E^*} \qquad (6.10)$$

So, for instance, if E^* is 1% then the net multiplication rate of the annual needs to exceed $1/0.01 = 100$ for persistence (Crawley and May, 1987). This model is based on the assumption that the seeds of the annual are randomly distributed in space, and that seeds falling onto the perennial carpet die (rather than enter a seed bank). If a seed bank can form beneath the perennial, then the invasion criterion for the annual is lower than $\lambda = 1/E^*$, especially if the annual seeds are capable of detecting gaps in the perennial cover (e.g. from the red/far-red ratio of the light) and germinate only in gaps (Rees, 1997).

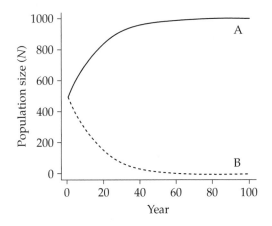
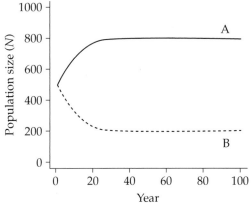

Figure 6.8 Lottery model. (a) In the absence of density dependence the species with the greater fecundity (species A) causes the competitive exclusion of the species with the lower fecundity (species B). (b) Coexistence is possible only if species B obtains a rare species advantage. Here the fecundity of species A declines as its density increases, such that at $N = 800$ it has the same fecundity as species B.

How would a lottery model need to be altered to allow coexistence of two species of herbaceous perennials? Coexistence requires some form of rare species advantage for species B. As configured above in this section the fecundity of species A was greater than the fecundity of species B at all densities. For coexistence, there must be some population densities at which the fecundity of species B is greater than that of species A. The simplest mechanism producing such an effect would be an explicit, global, density-dependent reduction in the fecundity of species A (as might be caused by a build up of fungal pathogens at high population densities; Mills and Bever, 1998; Klironomos, 2002). Suppose that instead of having its fecundity constant at $F_A = 2.5$, species A experiences nonlinear density dependence in fecundity such that $F_A = 2.5 - 2.98 \times 10^{-24} \times N^8$. Once species A has increased to 800 and species B has declined to 200, the fecundity of species B, $F_B = 2.0$, is greater than the fecundity of species A, so species B *increases* in abundance. The equilibrium occurs where both species have the same realized fecundity $N^* = \sqrt[8]{\frac{2.5-2.0}{2.98 \times 10^{-24}}} = 800$ (see Figure 6.8b). As often happens, the mechanism of rare species advantage is actually an abundant species disadvantage. The rare species advantage of species B resulted from direct density dependence acting on the dominant species A (pathogen attack in this example).

6.10 Biomass mixtures

The behaviour of mixtures has fascinated plant ecologists for decades, not least because of the prospect that mixtures might out-yield the highest-yielding monoculture and/or provide a lower variance in yield (the insurance principle). Estimates of the relative abundance of herbaceous perennials in mixed-species communities are often based on sorted biomass from destructive samples. There are two contrasting cases to consider: first, long-established vegetation, where species richness is an emergent property resulting from ecological interactions; second, experimental communities (often only a few years old and established from seed), in which the number of

plant species is manipulated experimentally to create monocultures and polycultures of differing species richness.

A fundamental question concerns the yield of a mixture of species compared with the yields of experimental monocultures of the component species. Can the yield of the mixture ever exceed the yield of the maximum-yielding monoculture? Logic alone suggests that it could not, because this would involve replacing a larger individual of the higher-yielding species with a smaller individual from the lower-yielding species.

The behaviour of biomass mixtures under a range of scenarios is shown in Figure 6.9. The same total number of plants is grown in each case: in the mixtures, half of the individuals are of one species and half are of the other, regularly interspersed. In the simplest case (the building block model) there is no interaction between the species, so the yield of the mixture is exactly half way between the yields of the two monocultures. In the case of extreme contest competition, the species that is the superior competitor makes up all of the final biomass, so the yield of the mixture is the same as the higher of the two monocultures. Over-yielding occurs when the yield of the mixture is greater than the yield of the higher of the two monocultures, and results from some form of niche complementarity (e.g. using soil resources from different depths, or using light at different times of year) or active facilitation (e.g. species A increases the growth rate of species B, as with mixtures including a nitrogen-fixing legume and a grass; Schwinning and Parsons, 1996). Under-yielding occurs when both species inhibit one another's growth (e.g. through reciprocal allelopathy or shared natural enemies; Bais *et al.*, 2003).

For experimental systems, where species richness is manipulated, an intense debate has centred on the relative importance of the *sampling effect* (where more species means a higher probability that the mixture will include the highest-yielding species) and *complementarity* (where niche differences mean that more (or different) resources are exploited by the mixture than by the monoculture). This debate is reviewed by Kinzig *et al.* (2001), but recent data from longer-term experiments provide convincing evidence of over-yielding resulting

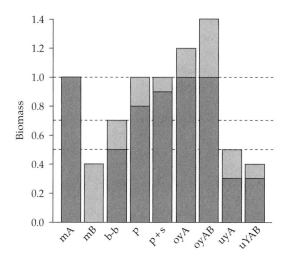

Figure 6.9 Biomass performance in mixtures and monocultures. From left to right, the bars represent the following circumstances: mA, monoculture of species A (dark) with a biomass of 1.0; mB, monoculture of species B (light) with a biomass of 0.4; b-b, the building-blocks null-hypothesis outcome when A and B are grown in a mixture at half the density in which they were grown in monoculture and achieve half their monoculture biomass (i.e. no interactions affecting biomass of either species), so A is 0.5 and B is 0.2; p, plasticity expressed by species A, which expands (0.8) to achieve the same total biomass (A + B) as achieved by A alone in monoculture, while species B attains (0.2) half the biomass as it achieves in monoculture; p + s, plasticity by species A and suppression of species B such that the total biomass is the same as the monoculture of A but species B yields (0.1) less than in monoculture; oyA, over-yielding (total biomass greater than the highest of the monoculture biomasses) caused by species A showing full plasticity (1.0) and species B showing (0.2) its non-plastic biomass; oyAB, over-yielding caused by both species A (1.0) and species B (0.4) showing full plasticity (i.e. total biomass is the sum of the two monoculture biomasses); uyA, under-yielding caused by species A (0.3) producing less than half its monoculture biomass in mixture; uyAB, under-yielding caused by both species A (0.3) and species B (0.1) producing less than half their monoculture biomasses in mixture. It is a genuine statistical challenge to distinguish between these competing hypotheses from field data.

from complementarity in experiments that are run for a sufficiently long time (Tilman *et al.*, 2001a).

6.11 Trees

Trees have the great advantage of being easy to count, and it is straightforward to measure their growth rates, recruitment from seed, and time of death. There is now a global network of permanent quadrats established for the long-term study of tree population dynamics (Wills *et al.*, 2006). In considering the population dynamics of trees we need to think about the relative importance of competition between seedlings and seedlings (recruitment limitation), between seedlings and adults (microsite limitation and herbivore effects), and between adults and adults (resource competition). It is also essential that we consider the neighbourhood in which a tree finds itself.

All of the models so far have been underpinned by the mean-field assumption that organisms encounter one another in proportion to their spatial average densities, ignoring all spatial information on the aggregation and segregation of plants (Murrell, Dieckmann and Law, 2004). But individual plants do not experience mean population density; they interact with a small number of neighbours (Mack and Harper, 1977; Weiner, 1982). Local dispersal causes clustering, but this is within-species clustering. Local competition pushes individuals apart, whatever species they belong to. Thus, collectively, these forces lead to intraspecific spatial aggregation and interspecific spatial segregation (Pacala, 1997). The class of spatially explicit models takes account of these neighbourhood effects, typically through the construction of individually based models. The performance of each individual plant is modelled throughout its life as a function of the number, size, spacing, and identity of its immediate neighbours. The aim is to understand the long-term and large-scale consequences of local inter-actions and finite dispersal. There are four essential components to an individually based model: the growth kernel, the mortality kernel, the fecundity kernel, and the dispersal kernel.

The word kernel is used to refer to the distance-dependence of the different processes from a focal individual. For instance, the dispersal kernel describes the probability density with which new-born offspring are displaced at different distances from their parents. The size of the kernel can be estimated by maximum likelihood from field data, having chosen a particular functional form to describe the shape of the distance-dependence of the function (e.g. Gaussian, hyperbolic, inverse square, negative exponential). All four processes might be affected by the size, spacing, orientation, and specific identity of the plants growing in the immediate neighbourhood of each individual.

Pacala *et al.* (1996) present a spatial and mechanistic model for the dynamics of transition oak/northern hardwood forests in north-eastern North America. The purpose of the model was to extrapolate from measurable fine-scale and short-term interactions among individual trees to large-scale and long-term dynamics of forest communities. The model makes population-dynamic forecasts by predicting the fate of every individual tree throughout its life. Species-specific functions predict each tree's dispersal, establishment, growth, mortality, and fecundity. Trees occupy unique spatial positions, and individual performance is affected by the availability of resources in their local neighbourhood. Competition is mechanistic; resources available to each tree are reduced by neighbours. The complex version of the model included light, water, and nitrogen, but the simplified version includes only competition for light (shading and light-dependent performance) because the field data provide little evidence of competition for nitrogen and water over the range of sites examined. The model predicts succession from early dominance by species such as *Quercus rubra* and *Prunus serotina*, to late dominance by *Fagus grandifolia* and *Tsuga canadensis*, with *Betula*

alleghaniensis present as a gap-phase species in old-growth stands. The model also predicts that old-growth communities will exhibit intraspecifically clumped and interspecifically segregated spatial distributions. Coexistence between tree species involves a variety of strategic trade-offs. For example, species that grow quickly under high light tend to cast relatively little shade, have low survivorship under low light, and have high dispersal. In contrast, species that grow slowly under high light tend to cast relatively dark shade, and to have high survivorship under low light and low dispersal. These trade-offs define one of two dominant axes of strategic variation, providing a simple explanation of community-level pattern in terms of individual-level processes (Pacala *et al.*, 1996).

What makes this study so important is that it represents a classic example of model simplification. The data collected from the field were extremely detailed. There were measurements of several resources (light, water, nitrogen, phosphorus, etc.) throughout the year in multiple locations. The canopies of all the individuals of all the different trees species were measured and described, along with their timber increments. Seed production, dispersal, seedling recruitment, and sapling survival were mapped in detail. The first model was a very complex spatially explicit simulation model, making use of most of the field data. Subsequent models were stripped-down versions which aimed to retain as much explanatory power as possible, but with many fewer parameters. It turned out that a spatially explicit model with only one resource (light) and greatly simplified descriptions of canopy architecture captured most of the details of botanical composition, relative abundance, and successional dynamics. The next simplification, however, in which the spatially explicit model was replaced by the mean-field approximation, proved to be a step too far. The spatially explicit model performed dramatically better than the mean field model: the extra detail in the spatially explicit model was fully justified by its greatly increased explanatory power.

It is a real challenge to capture the essential dynamic behaviour of a spatially explicit, individually based model in a simple, analytically tractable model, but the mathematical problems of doing this are formidable. The mean-field models make life easy by assuming that spatial pattern is irrelevant. To account for the effects of differences in spatial pattern requires that attention be paid to the details of the spacing of pairs of plants, triplets of plants, and so on. In an aggregated pattern, pairs of individuals are closer to one another, on average, than they would be in a random pattern. Likewise, in a regular pattern, pairs of individuals are further apart (more spaced out) than they would be in a random pattern. If we had a two-species model, then in statistical jargon we need to be concerned with spatial auto-covariance of each species and the spatial cross-covariance of the two species. The problem is that the analytical solution for pairs of individuals depends on the distribution of triplets of individuals, and the solution for triplets depends on the distribution of fours, and so on. A line needs to be drawn, and the mathematical name for the drawing of that line is moment closure (Murrell *et al.*, 2004). For the mean field models used so far, the moment closure was based on the assumption that individuals encountered one another at random, so the interaction term simply involved the product of the two mean densities. In a ground-breaking paper, Bolker and Pacala (1999) used moment closure to address the question of how limited seed dispersal affects coexistence in a multi-species plant competition model; the idea is that limited seed dispersal by the competitive dominant provides a refuge for the inferior competitor. Limited dispersal and local competition create spatial pattern by causing intraspecific clumping and interspecific segregation. Bolker and Pacala showed that three fundamental spatial strategies of plants— colonization (favouring wide dispersal), rapid exploitation, and tolerance (both favouring short dispersal)—emerge from a simple, continuous-space, stochastic model of spatial competition and local dispersal. Each of the three strategies exploits a different part of the spatial covariance structure of the community and, between them, the three strategies partition all of the spatial variability in the environment (Bolker and Pacala, 1999). Note, however, that there is still considerable debate about the best approximation that allows moment closure (Murrell *et al.*, 2004).

6.12 Herbivores and plant population dynamics

Every aspect of plant performance can be influenced by herbivorous animals and plant pathogens (germination, growth, seedling survival, flowering, seed-set, and seed mortality both pre- and post-dispersal; Crawley, 1983, 1997, 2000). This does not mean, however, that herbivore feeding has any impact on plant population dynamics under field conditions. For instance, plants may be able to compensate for herbivore feeding, there may be density dependence in mortality or fecundity, recruitment may not be seed-limited, and so on (Crawley, 1983). In many cases, it appears that plant–plant interactions (interspecific competition) are a more important determinant of plant community dynamics than plant–herbivore interactions (Crawley, 1997). It is important to note, however, that when we say that herbivory has no effect on plant population dynamics, this is *not* the same as saying that herbivory has no effect on plant fitness. For instance, the seeds that take possession of limited microsites are disproportionately likely to come from parents that were more resistant to (or tolerant of) herbivory.

6.12.1 Herbivory and plant productivity

The simplest kind of plant–herbivore interaction is exemplified by a sown grassland grazed by domestic livestock. The number of herbivores is determined by the farmer, and the farmer looks after the animals during the unfavourable season. This is important, because it means that the animals do not need to be supported by grassland productivity through the winter. Since the number of animals is fixed, the only dynamic variable is grass biomass (typically measured by leaf area index). In spring the grass begins to grow, and leaf area index increases exponentially. By early summer, self-shading of leaves means that increase in leaf area index slows and eventually stops (Figure 6.10). Once the animals are introduced to the pasture, they feed at a constant rate, C. So long as the animals are introduced to the pasture late enough that plant production exceeds herbivore consumption, then the system is stable. However, if the animals are introduced too early, then consumption exceeds growth rate and leaf area index is driven inexorably down to zero (to the left of the open symbol in Figure 6.10). By judicious choice of animal numbers and timing and duration of grazing, the farmer can harvest substantially more biomass from the grazed grassland than from the ungrazed system cut once at the end of the growing season (Black, 1964). In this very restricted sense, herbivory can increase plant productivity. This is not the same as saying that herbivory increases the Darwinian fitness of the plants. On the contrary; defoliation by herbivores prevents flowering of the grasses and reduces fecundity to zero. Note also that grazing increases the food *quality* for the herbivores (by stimulating regrowth, hence reducing mean leaf age and increasing mean nitrogen content) while at the same time reducing food *quantity*. In the absence of grazers, the grasses would flower, run to seed and then die back, dramatically reducing the food quality for herbivores.

To understand the dynamics of a free-ranging wild herbivore population we need to model the causes of fluctuations in herbivore numbers. In the absence of predators and diseases, wild herbivore populations are likely to be limited by food availability, as exemplified by the Soay sheep on the island of St Kilda off the coast of Scotland (Crawley *et al.*, 2004). Here, plant production is strongly seasonal, and herbivore numbers are limited by minimum (winter) levels of food availability. As a consequence, grazing pressure is relatively low during the plants' rapid growth phase in the summer, and plants can recover their condition (i.e. a system like this would not be said to be over-grazed). The equation for plant dynamics contains gains through density-dependent growth (i.e. plant biomass has a maximum in the absence of herbivores) and losses through grazing (the functional response of the herbivores):

$$\frac{dV}{dt} = \frac{rV(K - V)}{K} - aN(1 - e^{-bV}) \tag{6.11}$$

where V is vegetation biomass, r is the maximum per capita plant growth rate, K is the herbivore-free equilibrium plant biomass, N is the number of herbivores, and $a(1 - \exp(-bV))$ is the functional response (the per-animal feeding rate as a function of plant availability, V). The herbivore equation contains a term for the numerical response (which is related to per-capita feeding success) and a term

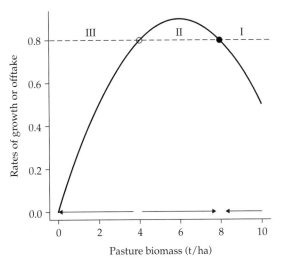

Figure 6.10 Grazing a sown pasture with domestic animals. So long as the animals are introduced after pasture biomass exceeds 4 t/ha the rate of production (solid curve) exceeds the offtake rate by the animals (dashed line) and pasture biomass increases (zone II) to a stable equilibrium (solid circle). Above a biomass of 8 t/ha, grazing causes a decline in plant biomass (zone I). If plant biomass falls below 4 t/ha (open circle) grazing eventually removes all of the above-ground plant mass (zone III).

for herbivore deaths. The behaviour of the system is critically dependent on the linearity of this herbivore equation. Here is the simplest case:

$$\frac{dN}{dt} = fN(1 - e^{-bV}) - dN \qquad (6.12)$$

where f is the numerical response and d is the per-capita herbivore death rate. At equilibrium we have $fN(1 - e^{-bV}) = dN$, so dividing both sides by N allows a solution for the equilibrium plant abundance V^*:

$$V^* = \frac{-\log(1 - (d/f))}{b} \qquad (6.13)$$

As in many versions of the Lotka–Volterra predator–prey model (see Chapter 5 in this volume), we observe the strong, counter-intuitive property that equilibrium plant abundance *does not* depend on the biology of the plants: if you double the plant growth rate, r, you have no effect at all on equilibrium plant abundance. Plant abundance is determined by the herbivore's death rate, d, the numerical response parameter, f, and the functional response, b. Increasing plant growth rate makes the *herbivore* population more abundant, but has no effect on equilibrium plant abundance. Alternatively, the herbivore equation might be nonlinear: here is a case where the herbivore population experiences direct density dependence in its death rate (as might be caused by a viral disease, for instance):

$$\frac{dN}{dt} = fN(1 - e^{-bV}) - dN^2 \qquad (6.14)$$

Now, at equilibrium, the Ns do not cancel out, so

$$V^* = \frac{-\log(1 - (d \cdot N^*/f))}{b} \qquad (6.15)$$

Since V^* now depends on N^*, increasing plant growth rate *does* increase equilibrium plant abundance. In both models, however, the plant–herbivore interaction is highly stable (for details see Crawley, 1983). In field systems like the Soay sheep, fluctuations in herbivore numbers are driven by severe winter weather and by over-compensating density dependence in herbivore mortality (mass starvation) in years when winter

food supply is low. The fluctuations are not the result of over-exploitation of the plant resource, and the plant–herbivore interaction is consequently highly resilient (Crawley et al., 2004).

Pacala and Crawley (1992) investigated the relationship between herbivory and plant species richness. They used a lottery model of herbivory and assumed that partial defoliation affects a plant species' relative competitive ability (Cottam et al., 1986) rather than (as in predator–prey models) increasing its death rate. They modelled the dynamic effects of herbivore feeding on plant abundance rather than the detailed dynamics of the herbivore population. Their key finding was that herbivory can enhance plant diversity in two qualitatively different ways: first, through global density dependence, when the level of herbivory increases monotonically as the plant increases in abundance, and second, when spatial variation in herbivory creates ephemeral local refuges from herbivory for competing plant species.

Both mechanisms require that there is a positive correlation between palatability and competitive ability and that the herbivores exhibit strong dietary preferences (or, conversely, that the plants show strong differences in grazing tolerance). This accords with the common field observation that the plant species that grow most quickly to dominance (i.e. the most competitive species) when herbivores are excluded (e.g. by fencing) are the first to decline (i.e. are most palatable) once herbivores are readmitted (e.g. by removing fences; Crawley, 1990). If there is no correlation between palatability and plant competitive ability, then herbivory will only enhance plant species richness if the animals are relatively monophagous on the most competitive species.

Many plant species benefit from the presence of herbivores in the sense that their population densities go up (grazing managers refer to such species as increasers). Increasers are species that are avoided by the grazing animals because they are unpalatable (either toxic or of low food quality) and selective grazing on the palatable species (the decreasers) provides competitor-release for the unpalatable species (Crawley, 1990, 1997). When herbivores are excluded, the competitive balance is reversed, and the palatable species (with their high growth rates but low investment in defence) out-compete the unpalatable species. Note, however, that the traits associated with unpalatability (toughness, low nitrogen content, etc.) are confounded with other adaptations like drought-tolerance (narrow, drought-adapted (xerophyllous) leaves), nutrient deficiency, tolerance of ultraviolet radiation, and leaf surfaces that afford protection from fungal pathogens and other micro-organisms, and this means that unequivocal interpretation of herbivore effects is difficult without manipulative experiments.

6.13 Conclusion

The patterns of density dependence exhibited by plants lead to a predominance of contest competition and a consequent stability of population dynamics. This applies both to resource competition (e.g. for limiting nutrients or for light) and to interactions with herbivores and pathogens (e.g. Janzen–Connell effects; Janzen, 1970; Connell,

1971). Even in a system as apparently simple as diatoms in the ocean phytoplankton, spatial heterogeneity can play a key role in determining population dynamics (e.g. opposing vertical gradients in light intensity and nutrient supply), and competitive interactions between the species may be capable of generating population fluctuations strong enough to play a role in promoting coexistence. For rooted terrestrial plants, density dependence manifests itself as interactions with a small number of neighbours, and the size and specific identity of neighbours are the main determinants of plant performance. Under these conditions, spatial pattern is often as important as global density, and spatially explicit models tend to out-perform their mean-field counterparts. The theory of plant population dynamics is now reasonably well developed, and there are many cases where we require data from long-term manipulative field experiments to distinguish between alternative theoretical models.

CHAPTER 7

Interspecific competition and multispecies coexistence

David Tilman

7.1 Introduction

Interspecific competition is an interaction in which species inhibit each other such that increased abundance of one species leads to lower growth rates of the other species. Numerous field studies have shown that interspecific competition is a major force determining species abundances for a wide variety of taxa in many different ecosystems (Harper, 1977; Tilman, 1982; Connell, 1983; Schoener, 1983; Aarssen and Epp, 1990; Goldberg and Barton, 1992; Casper and Jackson, 1997; Miller *et al.*, 2005). Predator–prey interactions can also be of simultaneous importance in determining the abundances and dynamics of species (e.g. Sih *et al.*, 1986), as can host–pathogen interactions (e.g. Hassell and Anderson, 1989; Hochberg *et al.*, 1990; Dobson and Crawley, 1995; Mitchell and Powers, 2003) and mutualistic interactions (e.g. Kawanabe *et al.*, 1993; Richardson *et al.*, 2000; Stachowicz, 2001). Although this chapter focuses on competition, all types of interaction operate simultaneously in nature.

Much of the early and continuing interest in competition has centered on how so many competing species coexist. G.E. Hutchinson (1959, 1961) posed the paradox of the plankton, asking how 30 or more species of algae could coexist in a few milliliters of lake or ocean water when there were only one, two, or three limiting resources and when the open waters of lakes and oceans were so homogeneous because of wind-driven mixing. Theory predicted that no more species could coexist than there were limiting factors or resources (e.g. MacArthur and Levins, 1964; Levin, 1970; Armstrong and McGehee, 1980). The same paradox occurred for terrestrial plants and animals. The

Earth's 250 000 species of vascular plants compete for a few limiting factors (usually a subset of nitrogen, phosphorus, potassium, calcium, water, and light). A large part of their diversity can, of course, be explained by the heterogeneity seen along major continental-scale and smaller-scale spatial gradients (Tilman, 1988). Expressed another way, these 250 000 vascular species are spread among perhaps 50 different biomes that occur in each of the five major biogeographic realms of Earth. One might expect different species in different biomes because of their differing climates. Because of their often separate histories of speciation, one might expect different species to occupy a given biome on each continent. Still, given about five separate instances of each of about 50 different biomes, each biome contains a rough average of about 1000 plant species that seemingly compete with each other for only a few limiting factors and yet still coexist. Animal diversity is about 20 times higher. The peak of life's diversity occurs in the taxa of smaller-sized organisms (May, 1986) which, for animals, are insects, of which there are thought to be roughly 5 million species, or about 20 000 species per biome, although the total is still far from certain (e.g. May, 1988, 1990a, 1992; Ødegaard, 2000).

As a theoretical challenge, Hutchinson's paradox of diversity has now been resolved by a large number of theories showing that, given certain assumptions, numerous competing species can coexist within local habitats. For instance, high diversity of coexisting competitors can result from spatial heterogeneity (e.g. MacArthur, 1972, pp. 46–58; Tilman, 1982), temporal variability (e.g. Levins, 1979; Armstrong and McGehee, 1980; Chesson, 1986;

Huisman and Weissing, 1999), spatial structure and interactions between competition and colonization (e.g. Horn and MacArthur, 1972, pp. 46–58; Hastings, 1980; Levin et al., 1984; Gaines and Roughgarden, 1985; Gilpin and Hanski, 1991; Tilman, 1994), and trophic complexity (e.g. Paine, 1966; Levin et al., 1977; Lubchenco, 1978; Menge and Sutherland, 1987; Tilman, 1982; Tilman and Pacala, 1993). In essence, these studies show that, in theory, an almost unlimited number of species can coexist if the simple classical models of competition are made one step more realistic by adding spatial heterogeneity, or non-equilibrial dynamics, or implicit or explicit space, or another trophic level (Tilman, 1982; Tilman and Pacala, 1993). In addition, a high diversity of competitors could result from the interplay of speciation and extinction processes in systems in which all competing species were functionally identical; that is, neutral (Hubbell and Foster, 1986; Hubbell, 2001).

Although no longer paradoxical, biodiversity remains a mystery. We now know that it is plausible for many competing species to coexist in local habitats, but do not know, for any ecosystem, what actually explains the coexistence that is observed. The solutions to this mystery are of both academic and societal importance. A variety of analyses suggest that the Earth may be on the verge of a major, human-caused extinction event (Terborgh, 1974; Ehrlich and Ehrlich, 1981; Nitecki, 1984; Wilson, 1988; Tilman et al., 1994; Pimm et al., 1995; May et al., 1995; Manne et al., 1999; Pimm and Raven, 2000). Habitat destruction and fragmentation, nutrient loading (especially nitrogen and phosphorus), introduction of exotic species, climate change, and other aspects of human-driven environmental change each have the potential to cause extinctions, perhaps massive extinctions, during the coming century. However, forecasts of extinctions will remain speculative and limited until we know the underlying mechanisms whereby the species were coexisting in nature. Academically, we do not yet understand many of the fundamental processes that caused Earth to become so amazingly rich with life. For instance, although it is empirically clear that, for 3 billion years, speciation rates have tended to exceed extinction rates, causing global diversity to rise, the reasons for this are unclear. Similarly, latitu-dinal gradients in species diversity are much better described and understood, in an empirical sense, than in the era of Wallace and Darwin, but their underlying cause remains an open question. Finally, although we now understand much more fully the effects of the diversity of competing species on stability (e.g. May, 1973a; McNaughton, 1993; Tilman et al., 2006) and productivity (Darwin, 1859; Loreau et al., 2001; Tilman et al., 2001a; Hooper et al., 2005), the effects of productivity on diversity have not yet yielded to a single general theoretical explanation.

This chapter will explore competition theory, including how it is that so many competing species can coexist, and thus why one or a few species have not been able to displace all their competitors and dominate portions of the Earth's surface. Several alternative mechanisms of competition, all of which may be able to explain observed multi-species coexistence, will be considered.

From its earliest beginnings, theoretical ecology has explored the effects of competition. The logistic growth equation, in its continuous form (Verhulst, 1838) and especially its discrete form (May, 1974c, 1976b), abstracts the essence of the effects of intraspecific competition on population dynamics. Lotka (1925) and Volterra (1931), two founders of theoretical ecology, proposed a theory of inter-specific competition that was a logical extension of Verhulst's continuous logistic growth equation. Hassell and Comins (1976) proposed a similar discrete-time model. These models have provided fundamental insights into the forces determining the coexistence, abundances, and dynamics of competing species. They have been followed by more mechanistic models of competition and of the interplay between competition and disturbance, dispersal, predation, or other factors. Here we will consider several different models of interspecific competition, the conditions that each requires for stable coexistence of multiple species, and some of their broader implications.

7.2 The Lotka–Volterra competition model

Let us first consider the classic continuous Lotka–Volterra competiton model, which describes the phenomenology of competition—how the rate of change of a species (dN_i/dt) depends on its own population density (N_i) and on the densities of its competitors (N_j):

$$\frac{dN_i}{dt} = r_i N_i \left(\frac{K_i - N_i - \sum\limits_{j=1}^{n} \alpha_{ij} N_j}{K_i} \right) \quad (7.1)$$

where subscripts i and j refer to species i and j, r_i is the intrinsic rate of increase of species i, K_i is its carrying capacity (the equilibrium number of individuals for the species in the absence of interspecific competition), α_{ij} is the competition coefficient specifying how many individuals of species i are displaced by the presence of each individual of species j, and n is the number of competing species.

In this model, an increase or decrease in the abundance of one species causes decreases or increases, respectively, in the growth rates of its competitors. The conditions for stable equilibrial coexistence of two competing species is easily derived by examining the isoclines (conditions for which are $dN_i/dt = dN_j/d_t = 0$) of the two species (Figure 7.1). Simply put, the two species stably coexist when each species inhibits itself more than it inhibits the other species. Mathematically, two species stably coexist when $K_1 < K_2/\alpha_{21}$ and $K_2 < K_1/\alpha_{12}$. To understand the basis of the simple generalization offered above, consider the ecological meaning of the four terms in the two inequalities that give the necessary conditions for two-species coexistence. K_1, which is the carrying capacity of species 1, is the number of individuals, N_1, required for species 1 to fully inhibit itself; that is, to drive dN_1/dt to 0 in the absence of species 2. K_2/α_{21}, which is the point at which the isocline for species 2 intersects the N_1 axis, is the number of individuals of species 1 that are required for species 1 to drive dN_2/dt to 0, also when species 2 is vanishingly rare. Thus, the inequality illustrated by the two points of intersection on the N_1 axis simply states that it takes more individuals of species 1 to fully inhibit species 2 than for it to fully inhibit itself; that is, that species 1 inhibits itself more than it inhibits species 2. Similarly, the two points of interaction of isoclines on the N_2 axis correspond with species 2 inhibiting itself more than it inhibits species 1. These two inequalities give the simple rule that two species can stably coexist at equilibrium when each inhibits itself more than it inhibits the other species, when inhibition is defined as above.

The dynamics of competition among two species competing according to the Lotka–Volterra model are simple. Sustained oscillations do not occur, and

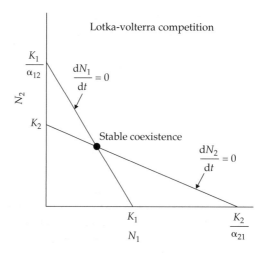

Figure 7.1 Lotka–Volterra competition isoclines for two species. Each isocline shows the densities of species 1 and 2 at which $dN/dt = 0$ for a species. The two-species equilibrium point is stable because, as discussed in the text, each species effectively inhibits itself more than it inhibits the other species, as indicated by the points of intersection of the isoclines with the axes.

there can be no limit cycle. A limit cycle may occur for two species if the model is modified to give nonlinear isoclines (Case, 2000). As is often true when the dimensionality of models is increased, the Lotka–Volterra model can have much more complex dynamics when three or more species are considered (Strobeck, 1973; May & Leonard, 1975; Case, 2000). For instance, given the right combinations of r_i, k_i (the resource level at which the species grows at a rate equal to $r_i/2$; also called the half-saturation constant), and α_{ij}, it is possible for three species to coexist via a stable limit cycle, although, according to Case (2000), 'it is quite difficult to find situations that yield a limit cycle' for three competing species. It is trivially easy to find conditions for which three species can stably coexist at equilibrium—at a three-species equilibrium point. Coexistence in this case can occur if each species inhibits itself, as defined above, more than it inhibits each of the other species, with each pair of species able to stably coexist in the absence of the third species and the three-species equilibrium point being above the plane defined by the carrying capacities of the three species (Case, 2000).

Much larger numbers of species can also stably coexist. Consider, first, a case in which species have identical values for r_i and k_i and the competition coefficients, α_{ij}, are directly derived from a model of niche overlap along a resource-utilization gradient (May, 1973a). In May's niche overlap model, species are uniformly spaced along an environmental gradient and thus have clear interspecific trade-offs in their competitive abilities, each species having its highest competitive ability at a different point on the gradient. Species are numbered in order along the gradient. Letting α be the competition coefficient for adjacent species, May (1973a) showed that

$$\alpha_{ij} = \alpha^{(i-j)^2} \tag{7.2}$$

and that $\alpha < 1$. As discussed more fully in Chapter 9 in this volume (see Fig 9.2), this model predicts stable coexistence of any number of competing species, but also shows that the degree of stability, as determined by the dominant eigenvalue of the Jacobian matrix (i.e. local stability) becomes vanishingly small as diversity becomes large and species are packed tightly on the gradient (May, 1973a). A second version of the Lotka–Volterra competition model that also predicts stable coexistence of any number of species is also based on an assumption of strict interspecific trade-offs. In this case, species also have identical values for r_i and k_i, have all $\alpha_{ii} = 1$, and have all α_{ij} (for $i \neq j$) equal to each other and less than 1. This is just a highly symmetric extension of the conditions for two species (Figure 7.1) or three species to stably coexist, at equilibrium, that we have already discussed. In this case, each species inhibits itself more than it inhibits any other species. This case has been misinterpreted as being one in which all species are identical or equivalent, belying the underlying biology that is needed to make all α_{ij} values be equal and less than 1. Rather, these parameters mean that species traits are highly and precisely constrained in a manner consistent with strong and consistent interspecific niche differentiation.

During the past two decades, the Lotka–Volterra competition model has faded from the literature and been replaced with a variety of more intuitive and mechanistically realistic models with parameters that are more clearly dependent on observable traits of species and thus are more easily measured. The Lotka–Volterra model provides a description of competition that is a near-equilibrium approximation to these more mechanistic models. Its assumptions of constant α values and linear isoclines have been widely criticized. Although changes in the density of one species are likely to lead to changes in the growth rates of other species, as the Lotka–Volterra model assumes, this effect is rarely a direct one in nature. Rather, an increase in the density of one species often leads to increased consumption of limiting resources and thus to lower levels of the resources. These lower resource levels, in turn, influence the growth rates of other species. Surprisingly, there need not be a complexity cost to such mechanistic detail. If there are a relatively small number of limiting resources or limiting factors (e.g. two or three, as is often the case), multispecies models that directly include simple mechanisms of resource competition require estimation of fewer parameters than do Lotka–Volterra models (Tilman, 1982). Moreover, these parameters can be determined via observations of species traits (Tilman, 1976, 1982; Grover, 1997) rather than via pairwise competition experiments among all possible pairs of species (e.g. Vandermeer, 1969).

7.3 Resource competition

7.3.1 A single limiting resource

All plants require the same suite of essential resources to survive and reproduce, including water, light, and biologically available forms of nitrogen, phosphorus, carbon, potassium, calcium, magnesium, sulphur, and about 15 additional elements. Resource-addition field experiments have shown that between one and four of these resources may limit growth in a given habitat. At higher trophic levels, each plant may function as several relatively independent resources for herbivores (leaves, roots, xylem, phloem, seeds) or for mutualists (nectar and pollen for pollinators; root area and sugars for mycorrhizal fungi or nitrogen-fixing microorganisms). Each herbivorous species may be a resource for its predators, parasites, parasitoids, pathogens, etc. These consumer–resource linkages, in total, define a large part of the topology and mechanistic dynamics of food-webs, with much of the rest provided by decomposition, which itself has elements that are consumer–resource interactions. Underlying these linkages are a few basic mechanisms of consumer–resource interactions that can be abstracted in simple theory. Let us start by considering the simplest mechanism—competition for a single limiting resource, such as competition between two herbivorous insect species for a single plant species, or between two plant species for a single limiting resource.

A factor, R, is defined as being a resource for species i if increases and decreases in R lead to increases and decreases, respectively, in the specific growth rate, $f_i(R)$, of the species and if the species consumes the factor (i.e. $\partial R/\partial B_i < 0$). Note that $f_i(R)$ has units of $dB_i/B_i dt$, with B_i being the abundance, or biomass, of species i per area. Let us assume that the consumer species experiences a resource-independent loss rate, m_i, from mortality,

loss of tissue to senescence, etc. Then, the dynamics of the consumer would be:

$$\frac{dB_i}{dt} = f_i(R)B_i - m_i B_i \qquad (7.3)$$

where $f_i(R)$ describes the resource dependence of the specific net growth rate of the species. A variety of experiments have shown that $f_i(R)$ is an increasing but saturating function of R. A commonly used form for $f_i(R)$ is the Michaelis–Menten or Monod formulation, for which $f_i(R) = r_i R/(R + k_i)$ so that $(1/B_i)dB_i/dt = r_i R/(R + k_i) - m_i$, where r_i is the maximal specific growth rate of species i, and k_i (the half-saturation constant) is the resource level at which the species grows at a rate equal to $r_i/2$. The dynamics of the resource would be:

$$\frac{dR}{dt} = h(R) - \sum_{i=1}^{n} Q_i f_i(R) B_i \qquad (7.4)$$

where $h(R)$ defines the habitat's rate of supply of the resource and Q_i is resource consumed to make a unit of biomass. A simple and commonly used form is $h(R) = a(S - R)$ where S is the supply point—that is, the total amount of all forms of the resource in the habitat—and a is the rate at which unavailable forms of the resource are converted into available forms.

There is a resource level, R_i^*, required for the growth of species i to exactly balance its sources of loss. At equilibrium, for which $dB_i/dt = 0$, growth equals loss, giving $f_i(R) = m_i$. This occurs at the resource level R_i^*, at which the loss curve intersects the resource-dependent growth curve (Figure 7.2a). By inverting $f_i(R)$, it can be seen that $R_i^* = f_i^{-1}(m_i)$. A species will be able to maintain a population and survive in a habitat only if the habitat has a resource level greater than or equal to its R_i^* value. A given species i would be able to continue growing ($dB_i/dt > 0$) as long as $R > R_i^*$, and would go locally extinct if $R < R_i^*$. If a single limiting resource were to be reduced to and held at a level less than R_i^* by a competitor, species i would be driven locally extinct by this competitor. For instance, in Figure 7.2b, species 1 has a lower R^* value than species 2. Species 1 will grow until it reduces R to R_1^*. At this level of the limiting

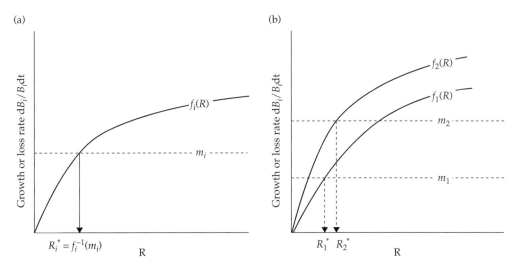

Figure 7.2 (A) Resource-dependent growth, $f_i(R)$ and loss, m_i balance each other at a resource concentration of R_i^*, which is the concentration to which this species reduces R at equilibrium. (B) When two species compete for a single resource, the species with the lower R_i^*, which is species 1, wins in competition. At R_1^*, the growth rate of species 2 is less than its rate of loss, causing it to have a negative exponential decline toward local extinction.

resource, the growth rate of species 2, which is $f_2(R_1^*)$, is less than m_2, causing it to be competitively displaced.

This result highlights an essential assumption of resource competition theory: species interact only through their effects on shared resources. At equilibrium in a habitat in which there is a single limiting resource, the best competitor is the species that has the lowest R_i^*. It will be able to keep growing and reducing R down to its R_i^* value. At this value of R, the resource-dependent growth rates of all other species would be less than their loss rates, and all of the other species would experience a negative exponential decline toward an equilibrial abundance of 0 (Tilman, 1976, 1977; Hsu et al., 1977).

This deceptively simple R^* rule of competition has proven to be surprisingly robust in controlled competition experiments studying freshwater and marine plankton (Tilman, 1976, 1977; Tilman et al., 1982; Sommer, 1986, 1990; Rothhaupt, 1988), bacteria (Levin et al., 1977; Hansen and Hubbell, 1980), and terrestrial plants (Tilman and Wedin, 1991a, 1991b, Wedin and Tilman, 1993), although many additional tests will be needed to determine its generality and predictive power, especially for higher plants and animals (Miller et al., 2005).

Because resource competition is mediated through effects of each species on resource levels, it is not surprising that a single 'trait' of each species, its R^* value, predicts the outcome of competition for a single limiting resource. R^* summarizes and abstracts the effects of many underlying physiological and morphological processes. Let us consider plants, though similar issues arise for any type of organism. The resource-dependent growth rate of a species depends on its ability to capture the limiting resource. For plants, this would depend on root mass, root surface area, and root nutrient kinetics, assuming that the limiting resource was a soil nutrient. It also would depend on its ability to retain the resource, which would be increased by having long-lived tissues and by translocating the limiting nutrient from tissues before they senesced. Another aspect of resource retention would come from decreasing loss of tissues to herbivores, such as via secondary compounds in tissues and tissue toughness. Just such underlying plant traits were included in several complex differential equation models of plant growth on, and competition for, a single limiting soil nutrient (Tilman, 1990). Just as did the simple version of the resource competition model (eqn 7.3), these models predicted that the superior competitor would have the lowest R^* value. Interestingly, this R^* depended on every plant trait included in the models. For example, one of these models predicted that

$$R^* = \frac{rhk(c+sq)}{v(r-c-s)-rh(c+sq)} \qquad (7.5)$$

where, for this equation, r and v are maximal growth and uptake rates, and h and k are saturation constants for Droop growth Michaelis-Menten uptake models; q is proportion of tissue nutrient lost to senescence, s is senescence rate; c is rate of tissue loss to all other causes. For realistic values of these parameters, a species is a better competitor (lower R^*) by having greater tissue longevity (lower s and c), a lower minimal tissue nutrient level (h), greater nutrient translocation (lower q), and greater nutrient uptake (higher v and lower k). The central point is that equilibrial competitive ability for a single resource is ultimately determined by numerous plant traits, the effects of which are summarized in R^* precisely because, under conditions of exploitative competition for resources, the effect of one species on another comes solely from its effect on the level of the limiting resource.

7.3.2 Multiple limiting resources or factors

What, then, can allow numerous species to coexist, at equilibrium, when competing for resources? Major ways for this to occur are for there to be a limiting environmental factor, such as temperature, salinity, or soil pH factor in addition to a single resource, or for there to be two or more limiting resources, or for both of these conditions to occur. Spatial heterogeneity in the limiting environmental factor and/or in the rates of supply of the limiting resources could then allow numerous species to coexist.

Let us first consider how spatial heterogeneity in the supply rates of two or more limiting resources can lead to the coexistence of many species. Consider, for instance, a case in which there are two interactive-essential resources, as illustrated with resource-dependent zero net-growth isoclines in Figure 7.3a. Each of these isoclines shows the levels of the two resources for which $dB_i/dt = 0$ for species i. The abundance of a species would

increase if the levels of R_1 and R_2 fell outside the isocline, would decrease if levels fell inside the isocline, and would be constant for levels on the isocline. When growing by itself in a habitat with a resource supply point (S_1, S_2) that fell outside its isocline, a species would increase in abundance, consuming resources, and eventually reducing resource levels down to a point on its isocline.

When two species compete for the two resources, the isoclines may cross, which would indicate that those species have a trade-off in their competitive abilities for the resources. If species do have such a trade-off, and if each forages optimally for the two resources, the two-species equilibrium point is one of stable coexistence. Each stable two-species equilibrium point has associated with it a region of resource supply points that lead to the equilibrium point. If multiple species all have interspecific trade-offs such that a species that is better at competing for one resource is necessarily inferior at competing for the other resource, then there will be regions in which various pairs of species stably coexist (Figure 7.3a). For examples with other types of resources, such as substitutable resources, switching resources and perfectly essential resources, and for a full treatment of the underlying mathematics, see Tilman (1980a, 1982) and Grover (1997).

These trade-offs cause the competitors to become competitively separated along resource-supply gradients. The gradient from point y to point y' illustrates sites with progressively lower rates of supply of R_1 and progressively higher rates of supply of R_2. At equilibrium in a habitat in which there is no dispersal limitation this would lead first to dominance by species A, then to coexistence of species A and B, then to dominance by species B, and so on, as illustrated in Figure 7.3b, along the gradient from y to y'. In total, all five species could stably persist in the heterogeneous habitats represented by the gradient from y to y'. These same species could also coexist in a more intermingled manner if sites within a habitat differed in their resource supply rates, such as by having the range of supply points encompassed by the ellipse of Figure 7.3a. A simple extension of the logic behind Figure 7.3a shows that there is no obvious simple limit to the number of species that

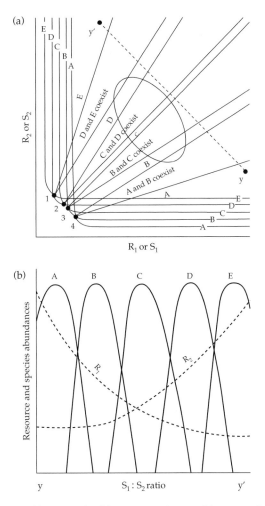

Figure 7.3 (a) The equilibrial requirements of five species (A–E) for two interactive-essential resources, R_1 and R_2, are summarized in their zero net growth isoclines. Each isocline is a right-angle-like curve with a rounded corner. The isocline for species A shows that it has the lowest requirement for R_2 of the five species, but the highest requirement for R_1. There is an interspecific trade-off such that species B–E are each progressively better competitors for R_1 and progressively poorer competitors for R_2. The four dots indicate two-species equilibria. The habitat conditions, as defined by their resource supply points, (S_1, S_2), that are drawn into each equilibrium point via resource consumption, are indicated as regions in which various pairs of species coexist. The line from y to y' represents a gradient in resource supply points, and the ellipse represents a habitat that has spatial heterogeneity in supply point as encompassed within the ellipse. (b) Resource competition causes the five competing species to become competitively separated along the y to y' gradient in a manner determined by their resource-requirement trade-offs.

can stably coexist if all species are constrained by the same interspecific trade-off in their competitive abilities for the two limiting resources. To see this, note that any species that has a trade-off intermediate between that of any two adjacent species can invade and coexist with each of them. If a single species did not experience this trade-off, but rather was a superior competitor for both resources (isocline inside that of all other species), it

would dominate all habitats, displacing all competitors.

Let us now consider a case of competition for a single limiting resource in a habitat in which temperature varies along a gradient, such as from low to high elevations on a mountain side, or from the north to south side of hills at higher latitudes, or from tropical to more poleward sites along a continental gradient. The growth rate of each species would be a function of both R and temperature, x (Tilman, 1999; Lehman and Tilman, 2000). With Michaelis–Menten growth, we could have

$$\frac{dB_i}{dt} = \left[r_i g_i(x) \frac{R}{R + K_i} - m_i \right] B_i$$

$$\frac{dR}{dt} = \alpha(S - R) - \sum_{i=1}^{n} Q_i g_i(x) \frac{R}{R + K_i} B_i \qquad (7.6)$$

$$g_i(x) = \exp\left[-\frac{1}{2}\left(\frac{x - \tau_i}{w} \right)^2 \right]$$

where growth, $g(x)$, is a Gaussian function of temperature, x, with τ_i being the optimal temperature of species i. Let us assume, for simplicity,

that species have identical parameters except for τ_i. Each species then has its lowest R_i^* value in habitats that have its optimal temperature, τ_i (Figure 7.4a). At equilibrium in a habitat in which there is no dispersal limitation on abundances, each species is the superior competitor for the resource for the range of temperatures for which it has the lowest R^* value (Figure 7.4b).

If a habitat has spatial heterogeneity in temperature, it is possible for a large number of species to stably coexist within this heterogeneous habitat. Each species would be dominant in those sites along the gradient for which it had the lowest

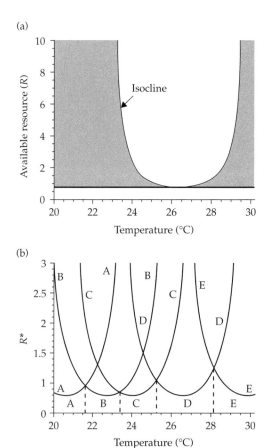

(a)

(b)

Figure 7.4 (a) Gaussian dependence of growth on temperature causes the R^* of a consumer species to depend on temperature, as shown by its temperature-dependent resource zero net-growth isocline. In habitats with a temperature and resource level for which the species can survive (unshaded region), the species would reduce the resource level down to the point on the isocline associated with that temperature. (b) When many species have trade-offs in their temperature optima, their isoclines define temperature ranges for which each species is the superior competitor because of its lower R^* value. For instance, the region between the two vertical dashed lines labeled B shows the habitat temperatures for which species B is the superior competitor. Modified from Tilman (2004).

R^* for the single limiting resource. Just as was so for competitive separation along a spatial gradient in two limiting resources, the competitive separation predicted when species compete along a temperature gradient requires interspecific trade-offs. In this case, the trade-off is simply that a species that is a superior competitor at one temperature must be an inferior competitor at other temperatures. There would be no multispecies coexistence, and no separation along a temperature gradient if a single species had a lower R^* than all other species at all temperatures. That species would dominate all habitats, displacing all competitors.

Numerous studies of the temperature dependence of physiology have shown that species have temperature optima and also differ in the resource ratios (R_1:R_2 ratios) at which each has its greatest competitive ability (Tilman, 1982; Grover, 1997). More importantly, separation of species along habitat gradients in temperature and resource-supply rates, which is the predicted result of these interspecific trade-offs, is a long-standing and recurring theme in ecology based both on observations in a wide variety of habitats (e.g. Beard, 1944, 1955, 1983; Olson, 1958; Whittaker, 1975; Mooney, 1977; Tilman, 1982; Pastor *et al.*, 1982, 1984; Whitney, 1986) and related theory (e.g. MacArthur, 1969, 1972, pp. 46–58; May and MacArthur, 1972; May, 1973a; Tilman, 1982, 1988).

There are other types of resource for which competition may occur (Tilman, 1982). In all cases, multispecies stable equilibrial coexistence requires interspecific trade-offs. If a species arose that was able to avoid such trade-offs and be a superior competitor relative to all other species, it would eliminate its competitors. Indeed, except for the diversity explanation based on neutrality (Hubbell, 2001), all other explanations for the high diversity of life on Earth require trade-offs. The trade-offs need not be purely competitive, as illustrated by consideration of multispecies coexistence resulting from a colonization/competition trade-off or a competition/predation trade-off.

7.3.3 Non-equilibrium coexistence

The examples discussed so far have explored stable coexistence under conditions in which spe-cies attain equilibrium. Species may also stably persist without ever reaching equilibrium or having a stable multispecies equilibrium point. Indeed, Armstrong and McGehee (1976a, 1976b, 1980), Levins (1979), and Huisman and Weissing (1999) have shown that many more competing species can persist in spatially homogeneous habitats than there are limiting resources. The species persist with fluctuating population densities and fluctuating resource levels, including potentially chaotic dynamics (Huisman and Weissing, 1999). The coexistence of more competitors than there are limiting resources occurs because nonlinearities in the resource-dependent growth functions of the species interact with resource fluctuations to create, in effect, new resources or new limiting factors (Levins, 1979). As in all the previous cases, this coexistence requires interspecific trade-offs. In these cases, the trade-offs stem from the degree and type of non-linearities in the dependence of specific growth rates on resource levels and their fluctuations. The results of these nonlinearities are that species both create resource fluctuations and respond differently to them such that each species tends to inhibit itself more than it inhibits the other species.

Chesson (1994) and Chesson and Huntly (1997) have provided an alternative way to understand a broad class of cases of coexistence in varying habitats. They show that multiple competing species can coexist in varying environments (e.g. temperature variation) only if they have appropriate nonlinearities in their growth responses. Coexistence requires that each species has a growth advantage when at low density and that this growth advantage must exceed the costs of interspecific competition.

7.4 Mixed models

7.4.1 Competition/colonization trade-offs

Individuals of a species are spatially discrete entities that interact most strongly with their closer neighbors (Pacala, 1986, 1987). Such neighborhood interactions can be treated at a variety of levels of detail, such as by partial differential equation models (e.g. the models of the spread of novel species by Lewis, 1997), in models that use moment closure to approximate effects of space (Pacala and Levin, 1997), and cellular automata models (e.g. Hassell *et al.*, 1991b). At the simpler end of the spectrum, space can be treated implicitly, as it is in the metapopulation and metacommunity models of Levins and Culver (1971), Horn and Macarthur (1972), Levin and Paine (1974), Hastings (1980), Tilman (1994), and Chase *et al.* (2005). Let us consider a spatially implicit model in which there is an interspecific trade-off between competitive ability for a single limiting resource and colonization ability.

Consider a habitat that is divided into sites, each the size occupied by an individual adult. To consider space implicitly, rather than explicitly, let us (as in Chapter 3 in this volume) make the simplifying assumption that a species occupies a portion, p, of these sites. Where c is its colonization rate and m is its mortality rate, a simple model of the dynamics of this species is (see also Chapter 2):

$$\frac{dp}{dt} = cp(1-p) - mp \tag{7.7}$$

Because $1 - p$ is the proportion of open sites, then $cp(1 - p)$ is the rate at which open sites are filled by the species. The term mp is the rate at which they are made empty by the death of the species. At equilibrium, this species would occupy a proportion p^* of the sites, where $p^* = 1 - m/c$. This shows that, in a spatial habitat, a species cannot (on average) occupy all available sites.

This model is easily generalized to any number of species that have a trade-off between competitive ability and colonization ability. To do this, let species 1 be the best competitor, species 2 the next best competitor, and so on. Let colonization rates vary inversely with competitive ability, with species 1 being the poorest colonist, species 2 a better colonist, species 3 a still better colonist, and so on. This then gives

$$\frac{dp_i}{dt} = c_i p_i \left(1 - \sum_{j=1}^{i} p_j\right) - m_i p_i - \sum_{j=1}^{i-1} c_j p_i p_j \tag{7.8}$$

where the first term on the right-hand side of the equation states that species i can only invade sites not occupied by itself or any superior competitor (species 1 to i). The second term is its mortality rate. The third term is competitive displacement caused by invasion by a superior species; that is, by species 1 to $i - 1$.

This model has the interesting feature that there is no simple limit to the number of competing species that can stably coexist at equilibrium in a habitat that is physically homogeneous, as long as the species have the appropriate competition/colonization trade-off (May & Nowak, 1994; Tilman, 1994). Coexistence occurs because the best competitor necessarily leaves a portion m_1/c_1 of sites empty. An inferior competitor, species 2, that is a sufficiently superior colonist to species 1 can invade these empty sites and maintain a stable population on them despite being periodically displaced as species 1, the superior competitor, invades the sites species 2 occupies. This second species, however, cannot fill all such sites, and the sites that it and the first species leave open can support a third species, and so on, with no simple limit to diversity (Tilman, 1994). This coexistence hinges on the trade-off being structured such that each successively poorer competitor overcomes a limit to similarity in colonization ability (Tilman, 1994). For instance, assuming that species 1 and 2 both have mortality rates of m, then species 2 must have $c_2 > c_1^2/m$ to invade and persist. Merely having $c_2 > c_1$ is insufficient to assure coexistence.

A variety of studies suggest that such competition/colonization trade-offs occur and may explain some of the diversity observed in nature. For instance, species that produce larger seeds may have competitive advantages because of the greater ability of their seedlings to acquire deeper soil resources and/or the higher light levels that occur at greater heights (Westoby *et al.*, 1997; Schwinning and Weiner, 1998; Hewitt, 1998). However, plants that produce larger seeds are likely to produce fewer of them, and to have them dispersed shorter distances (Shipley and Dion, 1992; Augspurger and Franson, 1987), giving them a colonization disadvantage. Another allocation trade-off that can cause a competition/dispersal trade-off occurs because species that allocate more biomass to a competitive structure, such as roots or stems, have less biomass to allocate to seeds (Gleeson and Tilman, 1990; Tilman, 1990, 1994).

Another variant on this is the 'successional niche' of Pacala *et al.* (1996), in which shade-intolerant species and shade-tolerant species coexist because the former are better colonists of open sites (tree-fall gaps) and the latter are better long-term competitors.

7.4.2 Competition and predation—trophic-level trade-offs

Paine's classic starfish-removal experiment (Paine, 1966, 1969) demonstrated that a trade-off between competitive ability and susceptibility to predation was allowing many species to coexist in the presence of starfish that did not coexist when they were removed. Levin *et al.* (1977) modeled such trophic-level trade-offs, showing that they could explain the coexistence of a large number of competitors and predators on a single limiting resource. They also tested their model using various combinations of bacterial prey and phage virus predators. Their tests confirmed the role of trade-offs in the coexistence of more prey species than there were resources limiting the competing prey species. Another tropic-level trade-off is

encompassed in the hypothesis of Janzen (1970) and Connell (1971) for the existence of high tropical and coral-reef diversity. Their hypothesis assumes that species-specific predators, herbivores, or diseases act most strongly against high-density patches of a given species, and thus prevent otherwise superior competitors from dominating a site. Lubchenco (1978) found, in an experiment in tide pools, that a trade-off between plant competitive ability and susceptibility to herbivory allowed multispecies coexistence.

7.5 Discussion

All of the models summarized above predict, as a first approximation, that there is no simple limit to diversity. All thus provide potential solutions to Hutchinson's paradox, as do many variants on these and related models, but they also raise the question of why diversity is not higher. It has often been suggested that diversity is limited because species are less likely to persist in the same habitat when they have closely similar competitors. As discussed more fully in Chapter 9 in this volume, May (1973a) showed in his

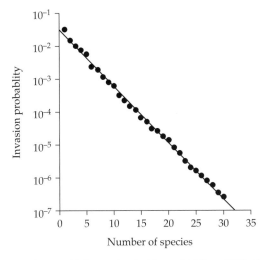

Figure 7.5 Species diversity during community assembly in a stochastic niche model (Tilman 2004). The probability that a propagule of an invading species would survive to become an adult and that the adult would successfully reproduce declined as a log function of the number of coexisting species. The decline occurred because of the lower resource levels that resulted from more complete use of limiting resources at higher diversity.

niche-overlap model (eqn 7.2), the decline in stability with increasing diversity (and thus, also, with increasing competitive overlap among species adjacent to each other on the resource gradient) meant that environmental fluctuations (noise) would impose a limit to similarity and thus to diversity.

In a model in which species competed for a limiting resource along a spatial temperature gradient (as in Figure 7.4), Tilman (2004) showed that, despite a constant rate of arrival of propagules of novel species, the chance of successful invasion and establishment of new species declined dramatically as diversity increased (Figure 7.5). In this stochastic niche theory, propagules had to survive stochastic mortality while growing to maturity on the resources left unconsumed by established species. The theory predicted that the species that became established each dominated an approximately equally wide 'slice' of the habitat's spatial heterogeneity, suggesting a limit to similarity. This stochastic limit to similarity resulted not from diversity per se, but from the uniformly lower levels of resources left unconsumed by the assembled higher diversity communities and from the strongly inhibitory effects of these lower and more uniform resource levels on propagule growth and survival.

Relative to Lotka–Volterra models, which can also predict multispecies coexistence, a major benefit of these more mechanistic models is that each model has its own signature that lays bare the requisite environmental conditions and species traits and trade-offs required for that mechanism of coexistence. If, for instance, spatial heterogeneity in a soil nutrient and light caused high plant diversity, then nutrient addition and shading should impact species abundances and diversity. Also, species should have trade-offs in tissue nutrient chemistry, allocation to different tissues (fine roots versus stems), and resultant soil-nutrient concentrations (R^* values) consistent with patterns in their abundances on natural and experiment nutrient:light gradients. If, though, coexistence came from competition/colonization trade-offs, the trade-off should be between structures important for resource competition (e.g. greater fine root mass, lower tissue nutrient levels,

and lower R^*-like values on nutrient-poor soil) and dispersal (higher seed number, viability, and dispersal distances). Experimentally, abundances of better competitors should be increased relatively more by seed addition than by disturbances, and abundances of better dispersers should be increased relatively more by disturbances than by seed addition. Studies such as those suggested by these brief sketches are needed in a wide variety of habitats if we are to gain a deeper understanding of the factors allowing and maintaining biodiversity. In pursuing this issue, it should be remembered that there is no *a priori* reason to suppose that trade-offs in nature are limited to one or two axes. Rather, trade-offs may reflect selective forces acting on many constraints and limiting factors that influence fitness, raising the possibility that species that seem highly similar when viewed along one or two trade-off axes might have little competitive overlap when additional limiting factors were considered (Pianka, 1976). However, it is also plausible that a few major trade-off axes may explain diversity, but that different axes operate in different ecosystems.

Deeper knowledge of the mechanisms of coexistence is of central importance to efforts to preserve biodiversity in the face of the ever-increasing impacts of humans (e.g. Vitousek *et al.*, 1997b). Let us briefly consider a few possibilities. If, for instance, plant species coexist because of trade-offs that involve competition for nitrogen, elevated rates of atmospheric nitrogen deposition could cause substantial loss of biodiversity, as recent analyses suggest may be occurring (Vitousek *et al.*, 1997a). If a competition/colonization trade-off is important, then habitat destruction and fragmentation may lead to an 'extinction debt' characterized by the time-delayed extinction of the best competitors (Tilman *et al.*, 1994). These extinctions would be in addition to those predicted to occur, via species-area effects, if the destroyed habitat represented the only remaining sites of occurrence of some species. If diversity were to result from multi-trophic-level interactions, then shifts in trophic structure, such as occur in response to human-caused loss of top predators or to invasions by exotic predators, parasites, parasitoids, and pathogens, could similarly lead to the time-delayed extinction of prey species. Their loss could lead to further time-delayed extinctions that cascaded down the food chain. In each of these cases, the path toward preservation of biodiversity hinges on knowing the mechanisms allowing coexistence and the ways that human-driven environmental change impact these coexistence mechanisms.

CHAPTER 8

Diversity and stability in ecological communities

Anthony R. Ives

In this lake, where competitions are fierce and continuous beyond any parallel in the worst periods of human history; where they take hold, not on goods of life merely, but always upon life itself; where mercy and charity and sympathy and magnanimity and all the virtues are utterly unknown; where robbery and murder and the deadly tyranny of strength over weakness are the unvarying rule; where what we call wrong-doing is always triumphant, and what we call goodness would be immediately fatal to its possessor–even here, out of these hard conditions, . . . an equilibrium has been reached and is steadily maintained that actually accomplishes for all the parties involved the greatest good which the circumstances will at all permit.

Stephen A. Forbes, *The Lake as a Microcosm* (1887)

8.1 Introduction

How the diversity of an ecological community affects its stability is an old and important question (Forbes, 1887; Elton, 1927; Nicholson, 1933). The science of ecology grew out of the study of natural history in the nineteenth century, when nature was viewed as wondrous, mysterious, complex, and largely in balance (even if murderous to experience from an individual's point of view; Forbes, 1887). Whereas our current scientific view is more textured and guarded, the 'balance of nature' still permeates the popular press. Some vestiges also remain in the scientific literature.

Over the last 100 years, conclusions about the relationship between ecological diversity and stability have varied wildly (May, 2001; Ives, 2005). The goal of this chapter is to show that these wildly varying conclusions are due largely to wildly varying definitions of both stability and

diversity. To do this, I will take two tacks, one for stability and the other for diversity. For stability, I will give an abbreviated history of the changing definitions of stability, merging both empirical and theoretical studies. I make no pretence of being comprehensive, but will instead pick highlights that show how the definition of stability often changes from one study to the next. For diversity[1], I will present a theoretical model to illustrate how different 'diversity effects' on stability can be parsed out. This model shows in a concrete way how any theoretical study (and, for that matter, empirical study) necessarily makes a long list of assumptions to derive any conclusion about diversity and stability. The multiple definitions of stability, and the multiple roles of diversity, argue against any general relationship between stability and diversity.

In the final section of the chapter, I will argue that understanding the relationship between diversity and stability requires the integration of theory and experiment. Theory is needed to define in unambiguous terms the meanings of stability and diversity. Experiments are needed to ground theory in reality. Unfortunately, rarely is this done.

8.2 History of stability

To present an abbreviated history of the changing definitions of stability, I will discuss theoretical and empirical studies side by side. The empirical

---

[1] Throughout this chapter, I use diversity in a broad sense, capturing many of the properties that have been used in the literature to characterize community complexity. Thus, I am using diversity and complexity synonymously.

studies generally represent case studies in which stability is assessed in a way that is most meaningful and/or practical for a specific ecological system. The same is true for theoretical studies. Although theoreticians often claim that their results are general, when addressing an issue as multifaceted as diversity and stability, the theory quickly becomes case-specific; the case is circumscribed by the specific assumptions made about diversity, stability, and the structure of the hypothetical community under study. Thus, theoretical studies have little more claim to generality that empirical studies.

In the first half of the twentieth century, views on community structure fell into two camps. The first was represented by Frederic Clements, who viewed plant communities as super-organisms made up of discrete assemblages that eventually achieved a stable climax: 'stabilization is the universal tendency of all vegetation, [and] climaxes are characterized by a high degree of stability when reckoned in thousands or even millions of years' (Clements, 1936, p. 256). In contrast, Henry Gleason viewed communities as largely coincidental associations, 'the resultant of two factors, the fluctuating and fortuitous immigration of plants and the equally fluctuating and variable environment' (Gleason, 1926, p. 23). From the Gleasonian perspective, communities did not necessarily have stabilizing properties beyond those conferred by the characteristics of individual species. In terms of determinants of community composition, the Gleasonian perspective prevailed over the Clementsian perspective by the 1950s, with unequivocal demonstrations that communities do not categorize into discrete assemblages (Whittaker, 1951, 1956; see Kingsland, 1991). Despite the death of the Clementsian organismic view of community composition, the powerful image of an integrated and inter-digitated community is still alive today. Communities are still often viewed as somehow greater than the sum of their parts, with emergent properties dictated by the complex web of interactions among species. Diversity is often discussed as a trait of a community, and stability as a community function. Thus, the Clementsian perspective is still common in ecology, with communities treated as integrated

units having their own properties above those of the individual species found within. There is a second parallel between today and yesteryears; Clements largely ignored animals, as is the case today in the majority of studies on ecosystem function.

In the 1950s, Charles Elton presented an argument that more complex communities were more stable, where he defined stable as the absence of 'destructive oscillations' of the type found in predator–prey dynamics (Elton, 1958). His argument was based on six lines of reasoning, all of which have since been largely overturned. For example, he argued that both simple models and simple laboratory experiments with just one prey and one predator lead to unstable oscillations; therefore, he conjectured, to explain the stability of natural communities, systems with more prey and more predators must be more stable. Unfortunately, there is—at least mathematically—a general expectation that the opposite is true[2]. Elton was rather uncritical of the idea that diversity begets stability, as he writes 'it is a question for future research, but an urgent one, how far one has to carry complexity in order to achieve any sort of equilibrium' (Elton, 1958, p. 153).

A quasi-mathematical concept of stability was proposed by Robert MacArthur: a stable community is one that, if any single species becomes abnormally common or rare, it nonetheless continues to convey energy from the bottom to the top of its trophic levels (MacArthur, 1955). MacArthur used the analogy of a food-web as a network in which species served as nodes to energy flow. Almost by definition, communities with greater numbers of species were more stable, because any gain or loss of energy flow caused by an 'abnormal' species would be compensated by flow through the remaining species. Species diversity provided redundancy guaranteeing that no single species controlled the flow of energy through the food-web. Although given mathematical gloss with an equation to measure the complexity of food-webs, MacArthur's view of diversity and

---

[2] Of course, it is possible to construct examples in which theoretical communities with more predators are more stable, but these have to be constructed carefully; this is addressed for a specific model in the next section of this chapter.

stability was nonetheless qualitative, simply asserting that species serve as conduits, shunting energy to where it is needed.

In the 1960s, a set of experimental studies emphasized the importance of trophic structure in the relationship between diversity and stability. Pimentel (1961) studied insect communities on cabbages, comparing cabbage monocultures with cabbages planted in polycultures with 300 other species. The herbivores of cabbage occurred at higher aggregate density in the monocultures. Pimentel provided three hypotheses for this effect of plant diversity on herbivore suppression. First, prey (herbivore) species react differently to fluctuations in environmental conditions, so that even if some prey species were depressed by 'bad' conditions, others would be experiencing 'good' conditions. Therefore, the diversity of prey species attacking the variety of plants in polyculture ensured that at least some prey were available to predators at all times, and this dampened the oscillatory tendency of predator–prey interactions. Second, the diversity of predators that attacked a given prey in polycultures increased the chances that there was a particularly effective predator to control the prey. Third, the polycultures had a greater number of highly polyphagous predator taxa (like spiders), and these ensured that the dynamics of the predators were less tightly coupled, and therefore less sensitive, to the dynamics of any one prey species. The conclusion that diversity creates stability might seem at odds with Pimentel's data: in fact, a greater number of prey taxa feeding on cabbage and a greater number of predator taxa attacking these prey were found in cabbage monocultures! However, Pimentel's measure of diversity was not the number of prey and predator taxa, but instead the variation among prey and predators in key traits. The diverse prey in polycultures responded to the environment in different ways (hypothesis 1), and the diverse predators in polycultures were more variable in their efficiencies when attacking different prey (hypothesis 2), and were more polyphagous (hypothesis 3). Nonetheless, the numbers of both herbivorous and predatory species on cabbage in monocultures were greater than in polycultures. That Pimentel designated monocultures as the

less-diverse community leaves one wondering how much his interpretation of the data was influenced by the established view that diversity begets stability.

In contrast to Pimentel's field system, Hairston *et al.* (1968) studied a laboratory microbial system with bacteria, three species of *Paramecium*, and two protozoan predators, measuring stability as the persistence and evenness in abundance of species. While increasing bacterial diversity increased the stability of protozoans, increasing the number of *Paramecium* species from two to three had differing effects depending on the species; these differences appeared to be due mainly to a single species being outcompeted by the other two. Adding the predatory protozoans separately or together caused the system to collapse, as they consumed the *Paramecium*. In their conclusions, Hairston *et al.* emphasized the importance of the identity of species and the lack of power of any general prediction about diversity and stability. A similar message was given by Bob Paine from his studies on species diversity in rocky intertidal communities (Paine, 1966). At his study site in Washington State, USA, removal of the dominant predatory starfish, *Pisaster ochraceus*, caused the community of invertebrates to drop from 15 to eight species. The role of *Pisaster* in maintaining the stable coexistence of 15 species was to selectively consume the dominant competitor, a species of mussel, that otherwise would outcompete other species for space. Like Hairston *et al.* (1968), this underscores how the peculiarities of a single species can be key to community persistence, and led Paine to coin a term for this: keystone species (Paine, 1969).

To summarize so far, definitions of stability have included suppression of outbreaks (Elton and Pimentel), maintenance of the number of species in the community (Hairston and Paine), and uninterrupted flow of energy through a food-web (MacArthur). While all contenders have generally agreed that greater diversity begets greater stability, the proposed explanations are all different. Elton accepted the belief that simple systems are not stable, and so concluded that it must be greater complexity that makes real systems stable. MacArthur emphasized that greater diversity creates greater redundancy, with many species doing pretty much the same thing—shunting energy through a food-web. Pimentel, Hairston, and Paine all emphasized the need to understand the diversity of what species actually do, emphasizing differences among species rather than species redundancies. This is especially true for Pimentel, as he designated the community with fewest predator species (polycultures) as having greatest predator diversity. While everybody upheld the common belief that greater diversity leads to greater stability, their arguments are so different they cannot really be viewed as mutually supportive. And a dissenter to the dogma that diversity begets stability was about to arrive.

In 1972 Bob May published a theoretical argument showing that more-diverse model communities were less likely to be stable (May, 1972; see also May, 1974b). He employed the ideas and mathematical techniques from dynamical systems, approaching ecology with a physicist's eye. He constructed communities by creating species and randomly selecting interaction strengths between them. For these communities, the probability that all species will return to their equilibrium densities decreases with increasing numbers of species, the proportion of possible interactions among them that are actually realized (connectance, C), and increasing average strengths of interaction (α). Specifically, he showed that such a randomly assembled community with a large number of species ($S \gg 1$) would almost certainly persist with all its members if $\alpha(SC)^{1/2} < 1$, but not otherwise. These results were presented with appropriate caveats, especially emphasizing that real ecosystems are not assembled randomly, and subsequent analyses refined the results (Pimm, 1991). Nonetheless, May dispelled absolutely the idea that diversity necessarily begets stability.

Sam McNaughton responded that regardless of mathematical results, the ultimate court in the diversity/stability debate had to be experimental, and the jury was still out (McNaughton, 1977). McNaughton presented experimental data from the grasslands of the Serengeti, in which species-rich and species-poor plant communities were subjected to fertilization, grazing, and drought. In all cases, green biomass in the diverse community was more stable than in the simple community,

with stability assessed by less growth in the fertilization experiments and more rapid recovery following grazing and drought. Although more stable, the diverse communities nonetheless experienced greater changes in the relative abundance of species. Thus, 'diversity stabilizes function, not diversity' (McNaughton, 1977, p. 522). McNaughton argued that it was in fact the compensatory responses of species in species-rich communities that caused both the stability of their function and the change in relative abundances of their species. Finally, reflecting the views of Hairston and Paine, McNaughton emphasized the importance of specific species for the behaviour of entire communities, explicitly stating that 'the stability of community functional properties is a consequence of adaptive properties of constituent species' (McNaughton, 1977, p. 522), not emergent properties.

Stuart Pimm emphasized that the multiple definitions of diversity and stability made any attempt to synthesize theoretical and empirical studies impossible (Pimm, 1984). Therefore, he proposed a set of four types of stability: resistance to the initial impact of a disturbance, resilience measured by the time it takes to recover from disturbance, persistence of species in the community, and variability in species densities. He also categorized different types of complexity (by which he meant diversity): the number of species, the strength of interactions between them (α), connectance (C), and species evenness. By mixing and matching different combinations of these definitions of stability and diversity, many relationships between diversity and stability could be obtained. While this certainly helped to make sense of confusion, at least among theoretical results, it still left major gaps between theoretical and empirical approaches, since applying these definitions of stability and diversity in a mechanistic sense is very difficult.

As a final example of a definition of stability, Dave Tilman made a compelling empirical case for the separation of stability as measured by the response of individual components of a community from stability as measured by the response of the entire system (Tilman, 1996). This harks back to McNaughton's emphasis that diverse communities may preserve function when disturbed, but they do so through more dramatic changes in species abundances. Tilman's idea was framed through a stochastic view of the world, in which continuous, repeated disturbances batter a community, and stability is measured by the size of the fluctuations in the response of species or the response of the entire community. In more-diverse communities, individual species might fluctuate more, yet the community as a whole fluctuates less. For theoretical support, Tilman turned to a result derived by May (1974a, 1974b, 2001)-not the result discussed above, but another result derived from stochastic theory: with increasing numbers of competitors in a community, variance in individual species densities increases while the variance in combined species densities remains the same. By contrasting stability measured for individual species with stability measured on the aggregate community, Tilman's argument attempts to reconcile the competing claims about the relationship between diversity and stability.

Unfortunately, I don't think it really does, because it does not overcome the issue of the numerous definitions of stability that are commonly used. There has been an accelerating number of empirical studies on diversity and stability (Cottingham *et al.*, 2001; Hooper *et al.*, 2005), and with each new study there is often a new definition of stability. Almost 10 years ago, Grimm and Wissel (1997) catalogued 167 definitions of stability. This plethora of definitions makes it difficult to draw any general conclusions from the collection of studies. I am not critical of this proliferation of definitions; it seems almost unavoidable as people study a wide variety of real or mathematical systems. While all definitions somehow relate to communities 'staying the same' for a long time or in the face of some disturbance, it is only reasonable that staying the same will mean different things for different communities. But this does bring up some awkwardness. A given community might be stable according to one definition of stability but unstable according to another (Ives, 2005). For example, a diverse plant community might preserve biomass very effectively in response to drought, even though many species are lost, as found by McNaughton (1977).

Furthermore, a community might be very stable to one type of disturbance but not to another (Lehmann-Ziebarth and Ives, 2006). For example, a diverse plant community might be more stable against burning than a simple community, while diversity has no effect on invasibility by other species (except after burning), as found by Mac-Dougall (2005). Although there is general support for the prevalent belief that diversity begets stability, the evidence is surprisingly mixed and muddy (Hooper *et al.*, 2005).

8.3 Effects of diversity

The numerous definitions of stability, and the trouble this causes any synthetic view of the relationship between diversity and stability, has been well recognized in the literature (Pimm, 1984; Hooper *et al.*, 2005). Multiple definitions of diversity are equally problematic. The most simple definition of diversity is the number of species in a community, or species richness. For realistic communities, this needs to be extended to somehow include functional groups or guilds, such as the number of species in different trophic levels. Diversity can include additional information about the strength of interactions among species, or the number of possible direct interactions among species that are realized (connectance). Extending things further, diversity could include phylogenetic information, with evolutionarily divergent species contributing more to diversity than evolutionarily related species (Webb *et al.*, 2002). The goal of this section is to address how multiple facets of diversity may affect stability in different ways.

My argument is predicated on a truism: a community of 100 species that are identical in every way is no different from a community of only one species. Therefore, determining how diversity affects stability (regardless of the definition) requires asking how species differ (Hooper *et al.*, 2005), and a good definition of diversity should measure the differences among species in a way that predicts stability. The question is not how the number of species in a community affects its stability, but instead how the differences among multiple species affect stability.

To address this question, I will use a simple model of community dynamics. I confess at the outset that the results are no more general than is the simple model. Generality is not the point. Instead, I want to illustrate two things. First, different types of diversity may have different consequences for stability, even when using a single definition of stability. Second, any conclusion about diversity and stability is inherently comparative. When concluding that one community is more stable than another, it is only possible to attribute this to diversity if the only attribute that differs between communities is diversity. The comparison must be made keeping all else equal. This leads to a fundamental dilemma: there is no absolute, universally accepted rule about what to keep equal. Therefore, any conclusion about diversity and stability is contingent on the decisions made to compare communities. This is an issue for both empirical and theoretical studies, but it is more transparent in theoretical studies because decisions about what to keep equal must be made explicit.

Consider a model community created by coupling predator–prey pairs (Ives *et al.*, 2000). Let α denote the strength of prey–prey coupling, and β denote the strength of predator–prey coupling. Then for a system with n predator–prey pairs, let the dynamics be governed by Lotka–Volterra-like equations:

$$x_i(t+1) = x_i(t)\exp[r(1 - X_i(t)/K) - aY_i(t) + \varepsilon_i(t)]$$
$$y_i(t+1) = y_i(t)\exp[caX_j(t) - d + \phi_i(t)]. \qquad (8.1)$$

Here, $x_i(t)$ and $y_i(t)$ give the prey and predator densities of the ith predator–prey pair. The per-capita population growth rate of prey i depends on its carrying capacity, K, and the per-capita attack rate from predators, a, both of which (at least for now) are the same for all prey species. Competition depends on the combined prey density, $X_i(t) = x_i(t) + \alpha \sum_{j \neq i}^{n} x_j(t)$, in which the density of interspecific competitors is diminished by the strength of coupling between prey, α. Similarly, predation depends on the combined predator density, $Y_i(t) = y_i(t) + \beta \sum_{j \neq i}^{n} y_j(t)$, in which the density of predators is diminished by the strength

of coupling β. The per-capita population growth rate of the predators depends on the number of prey killed, $aX_j(t)$ where $X_j(t) = x_j(t) + \beta \sum_{j \neq j}^{n} x_i(t)$, the conversion of prey into predator reproduction, c, and the predator death rate, d. Finally, environmental variation is modeled by including the normal random variables $\varepsilon(t)$ and $\phi(t)$ that have variances σ_ε^2 and σ_ϕ^2; for simplicity, I assume that the correlation between $\varepsilon(t)$ and $\phi(t)$ is 0. To let environmental fluctuations affect prey species similarly, the correlation between $\varepsilon_i(t)$ and $\varepsilon_j(t)$ is ρ_ε, and correspondingly for predators, the correlation between $\phi_i(t)$ and $\phi_j(t)$ is ρ_ϕ. To simplify the analyses below, I will set $\rho = \rho_\varepsilon = \rho_\phi$.

The central mathematical result for this discussion was first derived by May (1974a) for communities containing only competitors, and re-derived for systems with multiple trophic levels by Ives *et al.* (2000)[3]. The result is easy to intuit. Suppose first that, in a community containing n predator–prey pairs, each pair is dynamically independent ($\alpha = \beta = 0$) but in all other respects the species are dynamically identical. In this case, adding replicate identical predator–prey pairs (increasing n) should have no effect on the dynamics of the combined densities of prey or predators. Now suppose that all prey species are identical, so that individual prey are indistinguishable regardless of what species they are assigned to. This gives $\alpha = 1$, and a similar assumption about predators gives $\beta = 1$. In this case, adding more predator–prey pairs (increasing n) should have no effect on the dynamics, because all individual prey are identical to each other, and similarly for predators. Thus, for both extremes ($\alpha = \beta = 0$ and $\alpha = \beta = 1$), the number n of species in the community has no effect on dynamics—and hence no effect on stability—of the combined densities of all prey and all predators. The mathematical proof demonstrates that, at least to a first-order approximation, this is true not only for extremes, but also for all values of α and β in between, provided $\alpha = \beta$.

---

[3] The derivation in May (1974a) differed substantially from that in Ives *et al.* (2000), in that the former used continuous-time models and stochastic calculus, whereas the latter used discrete-time equations and results from autoregressive stochastic processes. However, the underlying explanations of the results are fundamentally the same.

The mathematical proof involves community matrices. Community matrices are linear approximations that describe population dynamics around equilibrium in deterministic systems (May, 1974b), and describe the variance of stochastic systems that are not too variable (Ives, 1995). For a single predator–prey pair in isolation, the community matrix $\mathbf{C}$ is derived by taking the derivatives of eqns 8.1 with respect to x and y, and evaluating them at equilibrium:

$$\mathbf{C} = \begin{bmatrix} \frac{\partial x(t+1)}{\partial x(t)} & \frac{\partial x(t+1)}{\partial y(t)} \\ \frac{\partial y(t+1)}{\partial x(t)} & \frac{\partial y(t+1)}{\partial y(t)} \end{bmatrix} = \begin{bmatrix} \left(1 - \frac{rd}{acK}\right) & -\frac{d}{c} \\ r\left(c - \frac{d}{aK}\right) & 1 \end{bmatrix}. \quad (8.2)$$

The mathematical proof shows that for the system with n predator–prey pairs, the dynamics of the combined prey and predator densities, $\sum_{i=1}^{n} x_i(t)$ and $\sum_{i=1}^{n} y_i(t)$, are governed by the two-dimensional community matrix

$$\mathbf{C} = \begin{bmatrix} \left(1 - \frac{rd}{acK}\right)\frac{1}{\Delta} & -\frac{d}{c}\Delta \\ r\left(c - \frac{d}{aK}\right) & 1 \end{bmatrix}, \text{ where}$$
$$\Delta = \left(\frac{1 + (n-1)\beta}{1 + (n-1)\alpha}\right). \quad (8.3)$$

Thus, to a first-order approximation, the dynamics of combined prey and predator densities are identical to the dynamics of a single predator–prey pair whenever $\alpha = \beta$ ($\Delta = 1$). This mathematical result is quite general (given, of course, the structure of the model). For communities constructed by replicating modules of interacting species (such as the replicated predator–prey pairs here), the dynamics of the entire community are dictated by the dynamics of the isolated modules. Increasing the number of identical species has no effect on the dynamics, because more of the same does not increase diversity.

This mathematical result refocuses the question of how diversity affects stability. The question is not how increasing numbers of species affect stability, but instead how differences among species affect stability. Of course, if there are differences among species, then having more species will affect stability. But this is really an effect of the differences among species, not the number of species per se. This mathematical result meshes well with the empirically motivated conclusions of Pimentel, Hairston, and McNaughton, that we need to know how species differ to understand any effect of diversity on stability.

To illustrate how various types of differences among species affect stability, I will consider two types of difference: (1) differences in species–environment interactions, in which species respond to the environment differently, and (2) differences in species–species interactions, in which different predator–prey pairs are governed by different demographic parameters (specifically, K and a). For these comparisons, I will assume $\alpha = \beta$, so that predator–prey pairs are connected with equal strength through prey and predator interactions. To complete the comparison, I will also consider (3) the case in which $\alpha = 0$ and $\beta > 0$, so prey have no effect on each other whereas predators attack all prey species (e.g. Thebault and Loreau, 2005). The reason for exploring this case is to demonstrate how assumptions about what to keep equal determine the conclusions. Figure 8.1 gives example trajectories from eqns 8.1 when there is a single predator–prey pair (Figure 8.1a) and when there are four predator–prey pairs combined under the assumptions of cases 1–3 (Figure 8.1b–d). This figure, while boring, is simply to demonstrate that the model does give output that looks reasonable.

For each of these three cases, I will use two closely related measures of stability. First, I will calculate the dominant eigenvalue of the $2n$-dimensional community matrix for the entire community, λ^* (see Chapter 3 in this volume for a detailed discussion of eigenvalues). Since it is calculated solely from the pattern and strength of species interactions, the magnitude of λ^* gives a measure of how the stability of the community depends on interactions among species. This is the measure that May used to obtain the result that greater diversity begets lower stability (May, 1974b). Second, I will calculate the coefficient of variation (CV) for combined prey densities, which depends not only on the interactions among

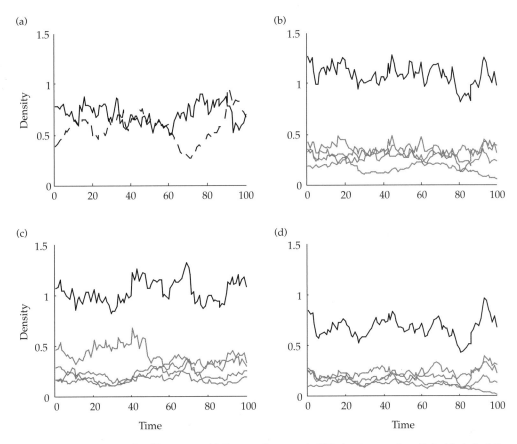

Figure 8.1 Population dynamics produced by eqns 8.1. (a) The case of one prey (solid line) and one predator (dashed line). (b–d) Cases 1–3 (see text) with $n = 4$ predator–prey pairs; the combined prey densities are given with heavy lines. Shared parameter values are $r = 0.5$, $K = 1$, $a = 0.3$, $d = 0.2$, $c = 1$, $\alpha = 0.5$, $\beta = 0.5$, $\sigma_\varepsilon^2 = 0.1$, $\sigma_\phi^2 = 0.1$, and $\rho = 0$. For (c), values of K for the four prey are 0.9107, 0.9163, 1.0277, and 0.9038, and values of a are 0.2784, 0.3081, 0.3352, and 0.3252. For (d), $\alpha = 0$.

species, but also interactions between species and the environment. The CV (the standard deviation divided by the mean) is a convenient measure of population variability, with higher values of CV corresponding to greater variability and hence lower stability for a given level of environmental variability. I present only the CV for combined prey densities, rather than also for combined predator densities, because the two show very similar patterns. The measures of stability, λ^* and CV, are mathematically closely related, and it is possible to use the community matrix, along with information about the pattern of environmental variability, to predict the CV of combined species densities (Ives *et*

al., 2003). The advantage of using both measures is that λ^* depends only on species–species interactions, whereas CV depends on species–species and species–environment interactions.

Consider case 1, in which predator–prey pairs share the same parameter values but may differ in how they respond to environmental variability. Setting $\rho = \rho_\varepsilon = \rho_\phi = 1$ makes all prey and all predators from different pairs dynamically identical, because they all respond to the environment in the same way. When all species respond to the environment in the same way, increasing the number of species n has no effect on the stability of combined species densities (Figure 8.2a and b), as intuition

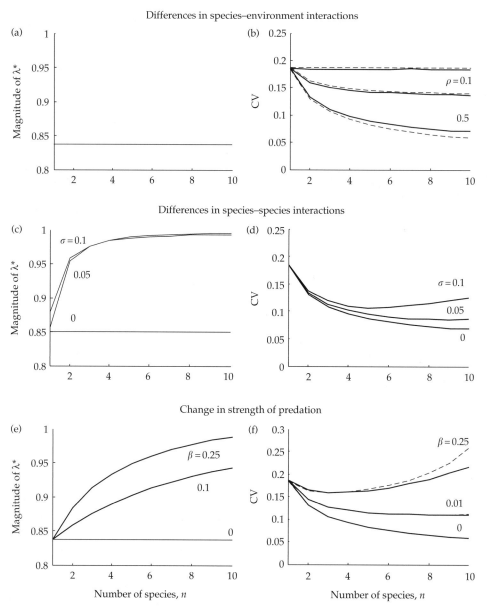

Figure 8.2 Stability versus diversity for cases 1–3. For each case, the panels on the left give the magnitude of the dominant eigenvalue λ^* of communities matrices and the panels on the right give variability in combined prey density (measured by the coefficient of variation, CV) for communities with n predator–prey pairs. For (a) and (b), the model given by eqns 8.1 has identical species interactions between each predator–prey pair. In (b), different lines correspond to different values of ρ, and to the correlation between $\varepsilon_i(t)$ and $\varepsilon_j(t)$, and between $\phi_i(t)$ and $\phi_j(t)$. As the differences among species increase (ρ decreases), the CV in combined species densities decreases, implying greater stability. Dashed lines give the CV predicted from the model for a single predator–prey pair, for which the CV declines according to $(1 + (n-1)\rho)/n$. For (c) and (d), the values of a and K vary among predator–prey pairs; lines are the averages of 10^4 simulated communities with values of a and K selected from normal distributions with means $\bar{a}$ and $\overline{K}$, and standard deviations given in the panels. For (e) and (f), communities with n predator–prey pairs had no interactions between prey ($\alpha = 0$), and different lines correspond to differing levels of β. As β increases, the predator attack rate on the combined prey species increases relative to the strength of competition. As a result, the magnitude of the dominant eigenvalue of the system of combined prey and predator densities increases with n (eqn 8.3) and the CV of combined prey densities is higher for larger values of β (f, solid line). Nonetheless, the CV is still predicted by the dynamics of the single predator–prey pair when accounting for the increasing attack rate (f, dashed lines).

would suggest. However, if species from different pairs differ in their responses to environmental fluctuations ($\rho<1$), then increasing the number of species n in the community decreases the CV of combined species densities (Figure 8.2b). The decrease in CV is due to a statistical averaging effect (Doak *et al.*, 1998); when the environment is bad for one species, it is not necessarily bad for another (since $\rho<1$), and therefore the combined density of both species does not fluctuate as much as either species separately. There is still no effect on the magnitude of λ^* (Figure 8.2a), which is not surprising because species–species interactions remain identical among predator–prey pairs. Also, using the community matrix for a single predator–prey pair (eqn 8.3) as a linear approximation of eqns 8.1 gives fairly good predictions of the CV calculated by simulating eqns 8.1 (Figure 8.2b, dashed lines). This indicates that the role of species–species interactions in the stability of the model with n predator–prey pairs is predicted by the predator–prey interactions of just a single predator–prey pair[4].

In case 2, predator–prey pairs differ in demographic parameters. To create these differences, suppose values of a and K for a given predator–prey pair are drawn from random variables with means $\bar{a}$ and $\bar{K}$, and standard deviations σ scaled relative to the mean. Values of $\bar{a}, \bar{K}$, and the other parameters are set so that when $\sigma=0$, case 2 gives exactly the same model as in case 1 under the assumption that $\rho=0$ (Figure 8.2a and b). Differences in a and K among predator–prey pairs decrease the endogenous stability of the community; the dominant eigenvalue λ^* of the n-species community matrix increases with even small differences ($\sigma=0.05$) among predator–prey pairs (Figure 8.2c). The effect of this is apparent in the CV in combined prey densities, which decreases but then increases with increasing n when there are large differences in values of a and K among pairs (Figure 8.2d). The net CV of combined prey density is driven by two forces. First, increasing n

causes a decrease in CV due to differences among species in species–environment interactions. This causes the CV to decline in the absence of differences among pairs ($\sigma=0$). Counteracting the effect of differences in species–environment interaction, differences in species–species interactions ($\alpha>0$) raise the CV. Thus, these two types of diversity act in opposition.

I included the final case, case 3, to address the issue of making comparisons. Making comparisons requires assumptions about what to keep equal. In both cases 1 and 2, predator–prey pairs were added while keeping the strength of coupling among pairs the same ($\alpha=\beta$). This has the effect of maintaining the overall strength of competition and the overall strength of predation in the system as n increases; formally, when $\alpha=\beta$, the sum of effects of competitors and predators on the per-capita prey population growth rates at equilibrium, $\sum_{j=1}^{n}\frac{\partial(x_i(t+1)/x_i(t))}{\partial x_j(t)}$ and $\sum_{j=1}^{n}\frac{\partial(x_i(t+1)/x_i(t))}{\partial y_j(t)}$, are independent of n and $\alpha=\beta$; the same is true for the per-capita population growth rate of predators. For the predation rate, this means that the same proportion of the total prey population is eaten regardless of either the number of predator–prey pairs or the strength of coupling between them. One could envision alternative scenarios, with for example increasing numbers of pairs leading to greater impacts of predation relative to competition. This might occur if prey occupy different habitats and therefore do not compete, whereas predators move among habitats and feed on all prey. In mathematical terms, this scenario would mean $\alpha=0$ and $\beta>0$.

When the coupling between predator–prey pairs only occurs through predation ($\alpha=0$ and $\beta>0$), increasing the number of pairs n is strongly destabilizing, causing increases in both the magnitude of λ^* and the CV in combined prey density (Figure 8.2e and f). There is a simple explanation for this. Consider eqn 8.3, which gives the dynamics of the combined prey and predator densities in the $2n$-species community. The element in the top left-hand corner of matrix **C**, $\partial x(t+1)/\partial x(t)$, gives the effect of prey density on the prey population growth rate. When $\alpha=0$,

---

[4] The deviation between the predicted and observed CV is due to nonlinearities in eqns 8.1, rather than the inability of the dynamics of the single predator–prey pair to predict the dynamics of the system with n pairs.

$\Delta = 1 + (n-1)\beta$; therefore, from eqn 8.3, increasing n decreases the strength of competition among prey. Similarly, increasing n increases the element in the top right corner of matrix $\mathbf{C}$, $\frac{\partial x(t+1)}{\partial y(t)}$, thereby increasing the strength of predation on the prey population growth rate. Thus, increasing n has exactly the same effect as reducing competition and increasing predation in the community matrix for a single predator–prey pair (eqn 8.2). Not surprisingly, increasing predation relative to competition is destabilizing in the single predator–prey pair, so increasing predation relative to competition is similarly destabilizing when there are n predator–prey pairs. In fact, the stability of the n-pair system is well predicted using the community matrix for a single pair once this increase in predation relative to competition is incorporated (Figure 8.2f, dashed lines; in Figure 8.2e the lines coincide perfectly).

To summarize, adding species to a community only affects stability if the new species are somehow different. If species respond to the environment differently (case 1), this can confer stability, at least if the measure of stability depends on species–environment interactions (i.e. CV but not λ*). If species differ in strengths of interactions (case 2), then adding species decreases stability, at least in this model. Therefore, the effect of diversity on stability depends on how species differ. These conclusions were made by comparing communities that contain different numbers of predator–prey pairs, but nonetheless have equal overall impacts of competition and predation on prey and predator per-capita population growth rates. If instead adding species changes the average strength of interactions among species, this can change stability. In case 3, increasing predator–prey pairs decreased stability, because this increased the overall impact of predation relative to competition on prey per-capita population growth rates. Is this an effect of diversity, or is it an effect of increasing predation? From a theoretician's point of view, I would prefer to call it an effect of predation, not diversity, because the increase in predation would have identical impacts on the stability of a single predator–prey pair. Therefore, it doesn't really have anything to do with the numbers of species and the differences among them. Nonetheless, stepping outside theory into ugly reality, the importance of predation relative to competition might in fact increase in more species-rich communities. This is an empirical question. For theory, it seems to make most sense to pick those comparisons that are conceptually most informative, thereby isolating and separating effects that do not have anything directly to do with diversity.

8.4 Where to from here?

The relationship between diversity and stability is not a simple one. More accurately stated, the many relationships between diversity and stability are not simple. There are many definitions of stability, and diversity can have different effects depending on *how* species differ. From a theoretical perspective, the multifaceted nature of diversity and stability is liberating, in that theoreticians can select those definitions that give the most interesting results. The trick is to establish theoretical comparisons to isolate specific mechanisms that underlie diversity–stability relationships. While liberating for theoreticians, it is annoying for empiricists, because it might seem to make the theoretical literature on diversity and stability a moving target.

How can we address the general relationship between diversity and stability in real systems? A seemingly obvious approach is to perform experiments or analyses on a large number of disparate systems, standardizing as much as possible the definitions of stability and diversity (Cottingham *et al.*, 2001; Hooper *et al.*, 2005). If many different types of community show the same relationship between (the particular measures of) diversity and stability, then a meta-analysis across all studies will reveal a general relationship. I personally think that this approach is at best inefficient, at worst misleading. Even if it were possible to match a specific definition of stability with a suite of experimental systems for which the definition of stability makes sense, there is still the problem that diversity can have multiple effects on stability. In the simple community whose dynamics are given unambiguously by eqns 8.1, the effect of diversity on stability depends not only on how species differ, but also on what is kept the same while comparing communities that differ in diversity. Given the difficulty of assessing the relationship between diversity and stability in this simple model, doing so for large numbers of real systems seems a bit foolhardy.

An alternative is to select a specific hypothesis about diversity and stability, and then test it with detailed study of a single system (e.g. Petchey *et al.*, 2002; Morin and McGrady-Steed, 2004). This might make it possible to eliminate hypotheses. There is an increasing number of experimental studies designed to test theory on the relationship between diversity and stability. These studies, however, often encounter the following problem. A theoretical hypothesis generally has the form: if A, then B, where A is a list of assumptions, and B is the anticipated outcome. For example, in the theory built around eqns 8.1, the assumptions A might include that species–environment interactions differ among predator–prey pairs, species–species interactions are the same, and the strengths of predation and competition remain unchanged while increasing the number of predator–prey pairs. The anticipated outcome B would then be that increasing the number of pairs increases stability as measured by the CV in combined prey density. Suppose we set out to test this theory with a microcosm experiment such as that of Hairston *et al.* (1968) and we find that in fact increasing the number of predator–prey pairs decreases rather than increases stability. Does this mean that we have disproved the theory? No. The theory is a conditional statement: if A, then B. If B is false, then two things could be to blame. First, the theory could indeed be wrong; knowing A might not predict B. Second, the theory could be right, but A is not true for the specific experiments we performed. Unless A is rigorously confirmed, testing whether B is true gives no indication of the validity of the theory.

I think the best way to proceed is to tie experiments and theory together much more aggressively. There is a growing number of statistical techniques for fitting models to data (Kendall *et al.*, 1999; Wood, 2001; Ives *et al.*, 2003). If a model can be reasonably fit to data, then it will contain the mechanisms that are responsible for whatever

relationship between diversity and stability is observed, and confidence in these mechanisms can be assigned statistically. Ideally, experiments would be performed to give the greatest power in detecting, or refuting, the hypothesized mechanisms. Alternatively, models could be fit to observational or experimental data collected for some other reason. While not optimal for any single data-set, this certainly opens up far more data-sets for investigation. Given the infinite number of possible models, grounding models in specific, real systems should benefit theory. Conversely, given the difficulty of making any conclusion from a purely empirical study, theory tailored for specific systems should benefit empiricism.

While the relationship between diversity and stability is fascinating in a purely academic way, in real life this relationship does matter. It is central to such questions as: How can an ecosystem be managed to be stable? Is a given ecosystem especially sensitive or insensitive to a particular type of disturbance, such as acidification of a lake? Given the potential muddle of multiple definitions and effects, uncertainty about the diversity–stability relationship might seem overwhelming. Nonetheless, there are some clear expectations and cautions. Although the relevance of diversity to environmental management and anthropogenic degradation has often been posed in terms simply of species numbers, the number of species per se is unlikely to have a large impact on stability; if all species are the same, having more of them will not change stability. However, with more species comes greater variety among species, and that variety may play a central role in the stability of communities (Hooper *et al.*, 2005). As a simple example, if species differ in tolerances to pH, then having more species will increase the chances of having at least a few species that tolerate acidification (Frost *et al.*, 1995, 1999). This insurance hypothesis (MacArthur, 1955; Walker, 1992; Yachi and Loreau, 1999) is a heuristic description of the theoretical results underlying the stabilizing effect of increasing species diversity when species respond differently to environmental fluctuations (Figure 8.2b).

Reading literature on the relationship between diversity and stability that is now 40 years old, it is remarkable how contemporary most of the 'old' arguments and conclusions are. One could view this as depressing; we have made so little progress that most of the points discussed 40 years ago are still unresolved. I prefer to take a more positive view. That the same issues were discussed 40 years ago as today means that we are asking the right questions. True, the questions are hard. But they are becoming sufficiently well articulated that at least the questions are stable.

CHAPTER 9

Communities: patterns

Robert M. May, Michael J. Crawley, and George Sugihara

In all areas of ecology, from studies of individual organisms through populations to communities and ecosystems, there have been huge empirical and theoretical advances over the past several decades. Our guess—a testable hypothesis—is that the worldwide research community of ecologists has grown by roughly an order-of-magnitude since the 1960s, as is of course true for other areas of the life sciences. One consequence is that it is harder to put together a book like the present one. And we find it especially hard when we compare this chapter on community patterns with the corresponding chapter in the second edition of *Theoretical Ecology*. For the chapters on single populations, for example, there has been growth both in understanding the nonlinear dynamical phenomena that can arise, along with a host of well-designed field and laboratory experiments which illustrate these processes. The narrative, however, retains a unifying central thread, and much of the task of overview and compression lies in choosing good examples from an increasing panoply of choice.

For communities, on the other hand, we find so many different yet intersecting areas of growth, many of which have recently produced book-length collections of papers, that the task of choosing which topics to emphasize and which to elide is invidious. The result is necessarily quirky. Without further apology, here is an outline.

One broad area of community ecology deals with models for the dynamical behaviour of collections of many interacting species—either within a single trophic level or more generally—essentially as a scale-up of models for single and pairwise-interacting populations. This was the subject of the preceding chapter. Here, we begin by emphasizing the importance of work which views communities from, as it were, a plumber's perspective, looking at patterns of flow of energy or nutrients or other material. But we then move on quickly to other topics. These include: the network structure of food-webs (connectance, interaction strengths, etc.); what determines species' richness (niche versus null models); relative abundance of species (observed patterns and suggested causes); succession and disturbance; species–area relations; and scaling laws (with suggested connections among some such laws).

9.1 Flows of energy and material

To a first approximation, the basic equation underlying all life is very simple:

$$\Delta + nCO_2 + nH_2O \rightleftharpoons (CH_2O)_n + nO_2 \tag{9.1}$$

In the forward direction, left to right, primary producers (plants, in the most general sense) 'eat' photons, coming from the sun and carrying energy, Δ, ultimately derived from fusion reactions within it. This energy enables carbon dioxide and water to combine to produce the big molecules of biochemistry, which are essentially combinations of C, O, and H in the proportions 1:1:2. The other product of this interaction, oxygen molecules, was essentially absent from Earth's atmosphere before this process evolved. Running backwards, right to left, plants (in respiratory mode) and animals burn $(CH_2O)_n$ with atmospheric O_2 to produce energy, Δ, for metabolic purposes. To a second approximation, the elements N, P, and S are also essential for plants and animals, as components of the more efficient 'power packs and motors' that evolution has forged. In this sense, living things are made up of H, C, O, N, P, and S—the big six—in the rough proportions 3000:1500:1500:16:1.8:1. Beyond this, a sizable proportion of all the elements in the periodic table, including many so-called trace elements, are important for life, with many being crucial for some functions or in some environments, even though needed only in very small amounts.

It is consequently important that we understand both how these elements flow within ecosystems (with particular attention to how they may be lost and to the replenishment processes which maintain cycles) and how energy is produced to drive these flows of material. There are some interesting, but very rough and tentative, generalizations. One such in the early 1960s suggested that the food-chain efficiency for transfer of energy from one trophic level to the next was generally around 10% (the correct answer in the Graduate Record Examinations of the day). Subsequent studies showed that such food-chain efficiencies can vary over two or more orders of magnitude, from less than 0.1% to significantly more than 10%. Some evidence suggests such efficiencies may, other things being equal, be higher for carnivores and detritus feeders than for herbivores, possibly because biochemical conversion efficiencies are higher for animals eating other animals than for animals eating plants. Be this as it may, we here point the reader to the excellent reviews in Begon *et al.* (2006; chapters 16–18), and move on.

9.2 Food-web structure

As discussed more fully in the preceding chapter, the first generation of (excessively) simple models for exploring the dynamics of randomly assembled communities of many interacting species concluded that, as a mathematical generality, increasing either the number of species or the strength and/or number of interactions among them resulted in diminishing stability (in the sense of greater population fluctuations or less ability to handle disturbance, or both; May, 1973a, 2001). At the same time, this work strongly emphasized that real communities are most unlikely to be randomly assembled (*pace* the null models discussed later), thus redefining the agenda to a quest for how

ecological communities are in fact structured, in ways which can reconcile complexity with stability (e.g. Tregonning and Roberts 1979; Sugihara 1982).

In particular, suppose we have a community of S species where the way each population fluctuates about its equilibrium value ($x_i(t) = N_i(t) - N_i^*$ for the ith population) is determined by its interactions with other species, characterized by interaction coefficients α_{ij}:

$$dx_i/dt = \Sigma_j \alpha_{ij} x_j \tag{9.2}$$

Obviously this linearized analysis, where essentially all disturbances are assumed to be initially relatively small, is only a guide to the full dynamics. But it is nevertheless useful. In the simplest caricature, this matrix $\{\alpha_{ij}\}$ of interactions may be characterized by the average connectance of the web (the number of links connecting one species to another, expressed as a fraction of the total number of topologically possible links), C, along with the average magnitude (actually the root mean square) of the interaction between linked species, α. This interaction strength, α, is measured in relation to the intraspecific self-regulatory terms α_{ii}, assumed to be all of equal strength; that is, in effect, we put $\alpha_{ii} = -1$. It follows that, for large S, these systems will tend to be stable if

$$\alpha(SC)^{1/2} \lesssim 1 \tag{9.3}$$

and unstable otherwise. This result generalizes an earlier one, for the special case of Hermitian matrices, by the physicist Wigner (1958) (see May, 1972; also Sinha and Sinha, 2005). Note that the magnitude of the interaction strengths, α, is scaled against intraspecific interaction strengths, and that the inequality in eqn 9.3 is not exact (for the familiar Lotka–Volterra competition equation for two species, for example, the coexistence criterion $\alpha_{12}\alpha_{21} < \alpha_{11}\alpha_{22}$ is well-known, but eqn 9.3 applied too literally would give an extra factor $2^{\frac{1}{2}}$; these points are not always appreciated).

One tentative suggestion arising out of this work (May, 1973a) is that compartmentalization of food-webs may help reconcile increasing species richness (and consequently more-efficient exploitation of available resources) with stability and persistence. Early analyses within trophic levels find species richness increasing primarily by adding compartments (guilds) rather than by increasing their size (Sugihara, 1982). More detailed and recent studies suggest that model food-webs whose connectivity structure and distribution of interaction strengths conform to observed patterns do indeed tend to be more stable than their randomly generated counterparts (de Ruiter *et al.*, 1995; Neutel *et al.*, 2002; Rooney *et al.*, 2006). For a wider-ranging assessment of current 'understanding of the relationship between complexity and ecological stability', see Montoya *et al.* (2006).

9.3 Food-webs as networks

A scientific growth area in recent years has been the study of networks of interacting components within a system, as varied as human or other animals transmitting infection, proteins in cells, cells in organisms (e.g. neuronal networks), the Internet and the worldwide web, and species in food-webs. Before sketching some of this work on food-webs, some preliminaries are in order.

An old and well-studied network model is the Erdos–Renyi random graph (Bollobas, 2001; Newman, 2003), in which n nodes (or vertices) are connected by links (or edges) placed randomly between pairs of nodes. The random matrices underlying eqn 9.3 are like such graphs, but with the additional features that the directions and strengths of the interconnecting links (not just presence or absence) are specified.

A basic statistic used to characterize the structure of any large network is its degree distribution, $P(i)$. Here, $P(i)$ is the probability that a randomly chosen node will have degree i; that is, be linked to i other nodes. For the Erdos–Renyi network, the degree distribution is given exactly by the binomial distribution, or in the limit of large n by the Poisson distribution: $P(i) = m^i e^{-m}/i!$, where m is the average number of links.

Poisson and binomial distributions are strongly peaked about the average, m, with the probability of finding larger i values diminishing rapidly, as $1/i!$. That real-world networks are more complicated was driven home in 1967 by Milgram (1967), who asked 160 people in western USA to send a letter to someone (unknown to them) in Massachusetts by sending it to an acquaintance who might be able to further its journey to the target; 42 letters arrived, after an average of 5.5 hops; hence, 'six degrees of separation'. This motivated Watts and Strogatz's (1998) interesting and influential work on 'small-world' networks, which combine local clusters with occasional 'long hops'. We think this helped focus attention on degree distributions with 'fat tails', which decrease relatively slowly as i increases. One canonical such distribution is the exponential:

$$P(i) \sim e^{-\alpha i} \tag{9.4}$$

More recently, guided by data and the models discussed here, much attention has been given to so-called scale-free networks (Barabasi, 2002). These obey the power-law degree distribution $P(i) \sim i^{-\gamma}$, where γ is a constant (usually $2 < \gamma \le 3$). Such distributions have very fat tails, and also the peculiar feature that there is no characteristic number of links per node: hence scale-free.

Two other interesting statistical properties of large networks are the network diameter, d, and the clustering coefficient, K. The former is calculated by first finding the shortest path (smallest number of links) between each pair of nodes; d is then the maximum such shortest path (some biological authors take d to be the overall average shortest path). For an Erdos–Renyi random network with n nodes, d is proportional to $\log n$. Exponential networks have smaller diameters, and scale-free ones are ultra-small (with diameters scaling as $\log(\log n)$ in the usual case where $2 < \gamma \le 3$). K is the average probability that two neighbours of a given node are also neighbours of each other; that is, how densely packed with triangles a graph is. Again, this quantity is well-named, giving an intuitive sense of how 'clumpy' the network is. Many real-world networks have high K values (≥ 0.5). For the Erdos–Renyi network, $K = m/n$ is, by definition, the same for any node, regardless of its neighbours. Notice that for these random networks, the clustering coefficient, K, is identical with the connectance, C, defined above. For most other networks, however, the connectance C—which may equivalently be defined as the probability that a given pair of nodes or species are linked—differs from the clustering coefficient K.

It is important to remember that, although the degree distribution gives important information, it does not define the structure of a network uniquely (although it does define the connectance, C). Figure 9.1 shows two networks with identical $P(i)$ values, but with significantly different d and K values. More generally, Milo *et al.* (2002) have propounded a bottom-up approach which concentrates on identifying small patterns (subnets with three, four, or even five connected nodes) that are over-represented in the network. These are called network motifs and can be used to characterize distinct categories of networks (Milo *et al.*, 2002; Cattin *et al.*, 2004; Bascompte and Melián, 2005). They can also be related to the questions of compartmentalization and stability discussed above.

One major problem, long familiar to workers in this field, lies in the extent to which constituent species are lumped into functional groups, in ways that can bias analysis (Paine, 1988; Sugihara *et al.*, 1997; Solow and Beet, 1998; Bersier *et al.*, 1999). Indeed, the question of data resolution, and the inevitable arbitrary decisions made for lumping species, and for including or ignoring a linkage, are clearly problematic. For example, subsampling versus exhaustive sampling can have significant consequences for whether a network is accurately characterized as scale-free (Stumpf *et al.*, 2005). With this caveat, we ask what patterns, if any, are found in the degree distributions, diameters, and clustering coefficients of interacting species in food-webs?

Two recent analyses (Dunne *et al.*, 2002; Williams *et al.*, 2002) of 16 high-resolution food-webs from aquatic and terrestrial ecosystems (with total species numbers or nodes from 25 to 172) strongly suggest 'two degrees of separation', in the sense that 'more than 95% of species [are] typically within three links of each other' (Williams *et al.*, 2002). These degree distributions are not random Erdos–Renyi ones, although whether they are exponential, scale-free, or something else is the subject of debate (Keller, 2005; Proulx *et al.*, 2005). Interestingly, Stouffer *et al.* (2005) have shown that approximately exponential degree distributions similar to those observed can be derived from two different models developed from the cascade model of Cohen *et al.* (1990). One of these is largely based on reasonable phylogenetic constraints (Cattin *et al.*, 2004) and the implied underlying tree-structure (hierarchy) linking niches (Sugihara 1982, 1984), and the other on a simple version of a niche-overlap model (Williams and Martinez, 2000).

Another long-standing question is whether there are significant differences between terrestrial and aquatic ecological networks. The analyses of Dunne *et al.* (2002) and Williams *et al.* (2002) would suggest not, and this view is supported by recent studies of three different marine ecosystems, which 'substantiate previously reported results for

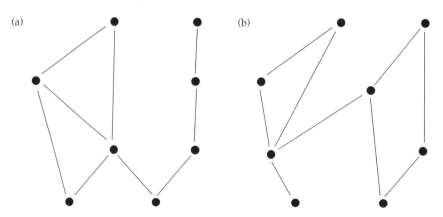

Figure 9.1 Two 8-node networks with the same degree distribution, but differing in structure (and having different diameters, d, and clustering coefficients, K). Specifically, the degree distribution $\{P(i)\}$ for both (a) and (b) is $P(4) = 0.125$, $P(3) = 0.125$, $P(2) = 0.625$, and $P(1) = 0.125$, but for (a) $d = 5$ and $K = 0.43$, whereas for (b) $d = 4$ and $K = 0.21$.

estuarine, fresh-water and terrestrial datasets, [suggesting] that food webs from different types of ecosystems with variable diversity and complexity share fundamental structural and ordering characteristics' (Dunne *et al.*, 2004).

To the contrary, a review by Shurin *et al.* (2006) contrasts aquatic and terrestrial food-webs, and concludes that there are 'systematic differences in energy flow and biomass partitioning between producers and herbivores, detritus and decomposers, and higher trophic levels'. They argue that the magnitudes of different trophic pathways differ significantly between sea and land, with the latter typically having less herbivory, more decomposers, more omnivory, and more detrital accumulation. They go on to speculate on the underlying ecological reasons for these differences: phytoplankton are smaller, have faster growth rates, and are more nutritious to heterotrophs than their terrestrial analogues; plankton food-webs tend to be more strongly size-structured than most terrestrial ones (promoting omnivory); and there are other differences in food-web architecture.

Other studies ask how the network structure of a food-web influences what happens when species are added or removed. Not surprisingly, most such studies (reviewed by Proulx *et al.* 2005 and Dunne, 2006) tend to show that removing the most highly connected species causes more knock-on extinctions than does random removal. Berlow *et al.* (2004), however, have shown that removal of low-connectivity species can have large effects, demonstrating that keystone species are not necessarily highly connected ones.

Ultimately, as seen in the previous chapter and earlier in this one, the response to disturbance must depend not only on network structure, but also on the strength of interactions. McCann's (2000) review of the existing data and experiments argues that distributions of interaction strengths are strongly biased towards weak interactions. He also observes that weak average interaction strength tends to be correlated with high variability in the strength. McCann makes it plain that, although most species' invasions have a weak impact on ecosystems, removal of or invasion by a single species—and not necessarily a strongly interacting one—can have huge effects on an

ecosystem (see also Fagan, 1997; Ives and Cardinale, 2004; Bascompte *et al.*, 2005). In particular, Rooney *et al.* (2006) suggest that real food-webs are structured in such a way that top predators are disproportionately important as regulators of stability-conferring energy flows. This has the unhappy corollary that our tendency to eradicate top predators can seriously damage 'the very structures and processes that…confer stability on food webs'. More generally, the book *Ecological Networks* (Pascual and Dunne, 2006) contains an excellent collection of recent reviews.

In short, despite much theoretical and empirical progress, along with careful refinement and redefinition of questions about food-web patterns, much remains to be done. One very broad generalization, first proposed on theoretical grounds in the early 1970s (see May, 2001), does however seem to have survived: 'increases in diversity cause community stability [measured, for example, by aggregate biomass variability following disturbance] to increase, but population stability to decrease' (Tilman, 1999).

9.4 How many species?

Related to the above discussion, but certainly not isomorphic with it, is the truly fundamental question of what determines how many species are to be found in a given place. Why is it, for example, that we find roughly 700 species of breeding birds in North America, rather than 7 or 70 000? And is 200 such species in Britain roughly what we might expect, given the relative areas of the regions, or not?

Looking at this question more locally, MacArthur (1972) asked why there are five species of warblers co-occurring in trees in Vermont. Why not more or fewer similar species? What, in other words, are the 'limits to similarity'? This pointed to a line of attack on the problem that began by asking, within a single trophic level, about competitive coexistence. The underlying thought was that, if this question could be resolved, the road to answering the larger question of limits to similarity lay in identifying, as it were, the number of niches. At much the same time MacArthur's mentor, Hutchinson (1961), posed the paradox of the plankton: phytoplankton communities generally exhibit high species numbers despite apparently limited opportunities for niche partitioning of their resource. Maybe fluctuations in habitat conditions or resource abundance kept things shuffling around, nullifying any 'one niche, one species' oversimplification. As we shall now see, and as was foreshadowed in chapter 6, several lines of past and present work suggest that, rather than being dichotomously opposed, such niche and null/neutral ideas may represent extremes of a continuum, both usually contributing to what is actually seen.

A simple metaphor for competing species is shown in Figure 9.2. Here the bell-shaped Gaussian curves represent the utilization functions for each of two species, aligned along some one-dimensional resource (amount of seeds as a function of seed size, for example). Each species has a preferred location on the resource axis, a spread of characteristic width, w, about this optimal choice of seed size, and the two species' optima are separated by a distance, d. In this caricature, the resource-utilization functions define the species' niches, and we can ask what are the limits to similarity by asking how small d can be, in relation to w, yet have the two species persist together. Robert MacArthur put this precise question to one of us (RMM) on our very first meeting in 1971. The initial thought was that, if we studied the dynamical behaviour of a set of n such species competing along a one-dimensional resource

continuum, stability considerations may put some limit on niche overlap, as defined by d/w. For a set of n species uniformly spaced at intervals d along the resource continuum and obeying the Lotka–Volterra competition equations, we can calculate the competition coefficients from the overlap in the uniformly spaced ultilization functions of Figure 9.2: the competition coefficient between species i and j, whose mean utilizations are separated by $|i-j|d$, is $\alpha^{(i-j)^2}$, where $\alpha = \exp(-d^2/4w^2)$. We can then obtain exact analytic expressions for the eigenvalues of the linearized interaction matrix which characterizes the stability properties (May and MacArthur, 1972; May 1973a, 2001). These eigenvalues, however, always have negative real parts, corresponding to the system returning to its initial state following disturbance, no matter how small d/w. On the one hand, the dominant eigenvalue (whose corresponding eigenvector represents the sum of all species' populations) has a real part which becomes increasingly negative as species numbers, n, increase. This corresponds to increasing stability of the total population. On the other hand, however, the $n-1$ eigenvalues corresponding to 'internal modes of vibration' of the system (which characterize the fluctuation levels of individual populations) have negative real parts which creep close to zero as n increases, indicating increasing fluctuations and diminishing stability as n increases. That is, as n increases, with consequent increase in niche overlap and decrease in d/w, individual populations

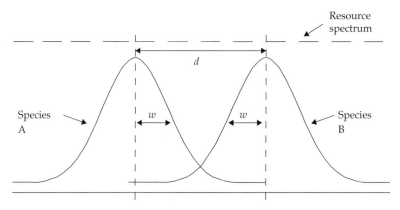

Figure 9.2 The dashed line indicates some resource spectrum, say amount of food as a function of food size, which sustains two species whose bell-shaped utilization functions are as shown, characterized by a standard deviation w and an average separation d.

are less stable, although the total population fluctuates less.

If stochastic noise is then added to the resource spectrum, these high fluctuation levels which occur once d/w is much less than one will lead to stochastic extinction of some species, tending *very* approximately to suggest a limit $d/w \geq 1$. That is, the average difference between adjacent species' utilization of resources (d) should not be too much smaller than the range of utilization *within* either species (w). Subsequent studies emphasize the approximate nature of this work and its sensitivity to the exact form in which environmental stochasticity is introduced into the utilization function, while other work has generalized the models and the conclusions (Sasaki, 1997).

A very important extension of this work is by Scheffer and van Ness (2006; see also Nee and Colgrave, 2006). They use the above model for competing species with Gaussian utilization functions, the essentials of which are indicated in Figure 9.2, but now employ extensive computer simulations to explore what happens when a large number of species are placed along the resource spectrum and allowed to 'evolve'. Obviously, the analytic techniques used by May and MacArthur (1972) could not handle the complexities of this system; interesting studies of this kind were essentially impossible in the early 1970s. Scheffer and van Ness find that such an assembly of species will organize itself—along a continuous resource spectrum with no discontinuities—into clumps of species with very similar niches (highly overlapping utilization functions) and a large difference between neighbouring clumps. At first sight, the results seem surprising. It can, however, be understood intuitively by recognizing that the simulations tell us that a species located at an intermediate point between clump A and clump B will find it easier to focus its competition on either A or B, rather than confront two different competitive tussles with A and B simultaneously. This work is fleshed out with some very nice empirical evidence, which supports the conclusions. There are interesting parallels between this result and previous work (which used algebraic topology and earlier data) showing niche assembly by adding species competing within single guilds, as distinct

from bridging different guilds (Sugihara, 1982, 1984). Scheffer and van Ness further point out similarities with Hotelling's (1929) suggestion that competition among companies or political parties will often result in convergence rather than differentiation.

Even earlier, Hutchinson (1959) had observed that the competitive exclusion principle, which stated that species making their living in identical ways cannot coexist, was unhelpful. Emphasizing the more meaningful question of limits to similarity, he catalogued many examples, drawn from both vertebrates and invertebrates, of sequences of competitors in which the average individuals in successive species have weight ratios of around 2; this implies ratios of about 1.3 between the typical linear dimensions (for example beak length) of successive species. This magic 1.3 ratio was no newcomer to the biological literature, Dyar (1890) having noted that successive larval instars of many insects have weight ratios of 2, and linear ratios of 1.3. Following Hutchinson, many other examples were tabulated, eventually culminating in a *jeu d'esprit* by Horn and May (1977) documenting Hutchinson–Dyar sequences of length ratios of 1.3 among musical instruments (recorders, stringed instruments) and even children's tricycle/bicycles and cooking implements. Incidentally, notice that even if the $d/w < 1$ rule was much more precise than it could ever claim to be, it still would not explain the Hutchinson–Dyar rule, unless you could also explain why the standard deviations of utilization functions were always around 30% of the mean separation, d.

A more fundamental questioning of this entire enterprise arose in the late 1970s and 1980s with the first advent of null models and neutral theories in ecology. This work, in my opinion, was very valuable in bringing statistical rigor to the search for patterns in the balance of nature, and putting an end to some excesses of enthusiasm (see, e.g., Strong *et al.*, 1984). The essential idea here is that you cannot simply claim to identify a pattern in data, but rather must carefully demonstrate the statistical significance of the alleged pattern, discriminating it against coincidence or selective focus or other biases. This movement, however, quickly developed some excesses of its own.

The basic and excellent idea is to construct a null hypothesis by reshuffling the data in some appropriately random way, and determining whether the putative pattern survives or not. If it does, you should probably worry. There can, however, be non-trivial difficulties in constructing an appropriate null model (in statisticians' jargon, one has to beware of type II error: failure to detect significant pattern when, in fact, it exists), and the 1980s offered occasional examples of remarkably silly null models. As an illustrative example, consider the case where two species are claimed to be so similar that coexistence is impossible, as evidenced by the checkerboard pattern found in studies of an archipelago of n islands with one, but only one, of the two species found on every island (species A on m, species B on $n - m$). Suppose you construct a null hypothesis by reshuffling the data, assigning the species at random among islands, but (naturally) subject to the constraint that there are still n islands and species A is on m of them, species B on $n - m$. Obviously the checkerboard pattern remains in this null model, the only difference being the species occur on different islands. Do you conclude there is no ecological significance in the alleged pattern? Of course not. You recognize this null model as wholly inappropriate; you should have put species A on any one island with probability m/n, species B with probability $1 - m/n$, to get a pattern with some islands empty, some with one species, and some with two, wholly different from the observed distribution. Yet, at the height of this particular controversy in the 1980s, there were some notable examples of inappropriate null models which did things essentially equivalent to the above example.

A brilliantly elegant send-up of these occasional excesses is by Feinsinger et al. (1981), who constructed null hypotheses by randomly re-assorting the musical notes in Bach's fugues. They then used standard statistical tests to determine whether the actual fugues differed significantly from these randomly assembled sequences of the same notes. Musicologists may be astonished to learn they did not! Statisticians could reassure them that appropriate null models cannot always be lifted off the shelf. For a general review of this subject, see Harvey et al. (1983).

9.5 Neutral community ecology

Over the subsequent four decades, Hutchinson's (1961) paradox of the plankton has been documented in many other contexts. There are over 1000 tree species within a 52-ha plot of tropical forest in Borneo (Wills et al., 2006), and more than 40 species of grasses and forbs in a 10 m × 10 m plot in the Park Grass Experiment at Rothamsted Experimental Station, UK (Crawley et al., 2005). The question abides, how can such levels of coexistence be explained?

Building on his seminal study which has tracked every tree (more than 200 000 woody plants with stem diameter greater than 1 cm) in a 50-ha site of Barro Colorado Island in Panama over the past two decades, Hubbell (2001, 2006) has stood convention on its head. Instead of looking for differences between species in order to understand coexistence, he asked what would be observed if species were exactly the same as one another? It seems that this extraordinary assumption may actually predict some of the most conspicuous patterns in large-scale ecology (Bell, 2001; Hubbell, 2001).

Hubbell's neutral model is a modern descendent of the null-hypothesis movement discussed above, differing in that it adds tunable parameters which relate to possible diversity-maintaining mechanisms. It is neutral in that it assumes that all the species have exactly the same fecundity, independent of their surroundings, so that recruitment is directly proportional to the relative abundance of surviving adult plants. The species compete for space, and—if this were the only thing going on—all but one of the species in the system would follow a random walk to extinction. The shorter the generation time, or the higher the mean death rate, and the smaller the size of the local community, the more rapidly species will be lost. Local extinction is not permanent, however, because Hubbell's local community is embedded in a much larger metacommunity, with a degree of coupling between the local community and the metacommunity ensuring that biodiversity is preserved. Species in the regional metacommunity have much lower extinction rates than in the local community, and species go extinct in the

metacommunity only on the timescale of speciation (which is usually thought of as outside the
timescales normally associated with modelling
plant population dynamics). By the same token,
new species must appear both locally and regionally by speciation.

A serious difficulty with this neutral model is its
extreme sensitivity to the neutrality assumption.
Once differences between species are allowed,
then competitive exclusion is likely to occur
rapidly. Bell (2001) distinguishes between the
'weak version' and 'strong version' of Hubbell's
ideas. Given the difficulties, he nevertheless argues
that the weak version of the neutral model is
useful as a null hypothesis. The strong version
says that Hubbell's suggestions really do correspond to what is happening in the real world. If
true, then we will be able to predict community
processes such as the rate of local extinction, the
flux of species through time, and the turnover of
species composition in space, all in terms of simple
parameters such as dispersal rates and local community sizes.

As is true in other areas of ecology and evolution, it is not easy to distinguish between these
neutral theories of diversity and older niche-
oriented 'adaptationist theories' (Bell 2001). As will
be seen below in this chapter, for example for
patterns of species' relative abundance (SRA),
different assumptions about causal mechanisms
can lead to community patterns which differ only
in subtle details (or in the number of adjustable
parameters). That being said, where attempts have
been made to confront the neutral model with
long-term field data from tropical forest or coral
reef sites, the neutral model has often been found
wanting (McGill, 2003; Wooten, 2005; Dornelas *et
al.*, 2006; Pandolfi, 2006; Wills *et al.*, 2006). One
problem is that the real communities have too few
extremely rare species. But this could be because
rare species are more prone to extinction, and
when they go locally extinct they take longer to re-
immigrate than do common species. Alternatively,
populations of rare species might grow differen-
tially faster into higher abundance categories as a
result of some form of rare species advantage
(although there are obvious problems reconciling
this with the basic assumption of neutrality).

Another test assesses rates of turnover caused by
ecological drift among hypothesized neutral species against observed rates of turnover among
South American and European passerine birds; the
observed rates are much faster than can be plausibly explained by neutral drift (Ricklefs, 2006).
More generally, the June 2006 issue of *Ecology*
contained a series of papers in a special feature on
neutral community ecology (Naeem, 2006).

As emphasized in Chapter 6 (see also Tilman,
1999), stochastic niche theory can resolve many of
the apparent differences between neutral theories
and classic niche-oriented approaches to understanding community structure. Indeed, despite its
many oversimplified elements, the earlier model of
May and MacArthur (1972; May, 1973a) did just
that: a purely deterministic model for competition
along a resource continuum gave no limits to niche
overlap; when environmental noise, and consequent population fluctuations, was acknowl-
edged, rough limits emerged (with details
dependent on the nature and magnitude of the
environmental stochasticity). But which species
survive and which drift to extinction in these old
models is a matter of chance.

Another recent exploration of this theme again
employs the metaphor of many species with
Gaussian utilization functions competing along a
resource continuum, and shows explicitly that
niche-oriented and neutral theories are two
extremes of a continuum. The degree of apparent
community structure—niche rather than neutral—
depends on many factors, including number of
species (diversity), degree of niche overlap, dispersal abilities, and assumptions made about
environmental variability (Gravel *et al.*, 2006). The
theme is re-echoed, with variations, by Bonsall *et
al.* (2004b), who show how 'organization dominated by niche structure and organization through
chance and neutral processes' can operate simultaneously, with varying weights, depending on the
interplay between population dynamics (ecology)
and life-history trade offs (evolution).

9.6 Relative abundance of species

Turning from how many species, what about SRA?
For any particular group of S species, we may

express the number of individuals in the ith species, N_i, as a proportion, p_i, of the total number of individuals, $N = \Sigma N_i$:

$$p_i = N_i / N \tag{9.5}$$

This information may then be displayed on a graph of relative abundance versus rank, as exemplified by Figures 9.6 and 9.7, below. These figures show the patterns in the relative magnitudes of constituent populations, from most to least abundant. Equivalently, the information may be incorporated in a distribution function, $S(N)$, as in Figure 9.3; here $S(N)dN$ is the number of species comprising between N and $N + dN$ individuals.

Some of the main distributions that have been discussed in connection with natural systems are presented below. For a more complete account, see Magurran (2004) or Begon *et al.* (2006).

9.6.1 Lognormal distributions

Once the community consists of a relatively large assembly of species, the observed distribution of

SRA, $S(N)$, is very commonly lognormal. That is, there is a bell-shaped Gaussian distribution in the logarithms of the species' abundances. This lognormal distribution has been documented for groups of organisms as varied as diatoms, moths, birds, or plants, provided always that the sample is large enough to contain a good number of species (see, e.g., May, 1975 or Begon *et al.* 2006). Figure 9.3 provides a typical example.

What may be the explanation for these pervasive patterns? It seems likely that the relative abundances within a largish group of species will be governed by the interplay of many more-or-less independent factors. It is, moreover, in the nature of the dynamics of interacting populations that these several factors should compound multiplicatively. The statistical central limit theorem applied to such a product of factors implies a lognormal distribution. Such a broad-brush statistical argument similarly suggests lognormal patterns for the distribution of wealth in large countries, or for people or gross domestic product among nation states; such is in fact the case.

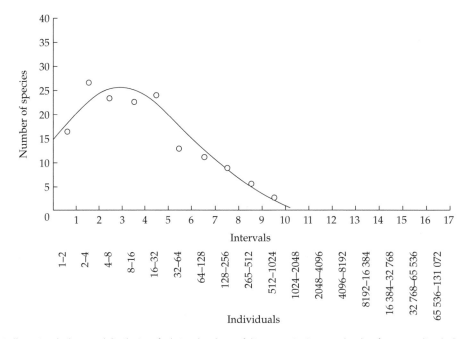

Figure 9.3 Illustrating the lognormal distribution of relative abundance of diatom species in a sample taken from an undisturbed community in Ridley Creek, Pennsylvania. The abundances are, as indicated, plotted as logarithms to the base 2: for further discussion, see the text. After Patrick (1973).

We would emphasize, not without irony, that this explanation for the observed lognormal patterns of SRA is the ultimate neutral or null model, invoking nothing more than the central limit theorem applied to multiplicative factors.

More specifically, ecologists usually write the lognormal distribution as

$$S(R) = S_0 \exp(-a^2 R^2) \qquad (9.6)$$

Here S_0 is the number of species at the mode of the distribution, having populations N_0; a is an inverse measure of the width of the distribution ($a^2 = 0.5/\sigma^2$, where σ is the standard deviation); and, following the idiosyncratic convention established by Preston's (1948) early work, R expresses the abundance as a logarithm to base 2, so that successive intervals or octaves (music rearing its head again in this chapter) along the x axis correspond to population doublings:

$$R = \log_2(N/N_0) \qquad (9.7)$$

The two parameters S_0 and a specify the distribution uniquely. In particular, the total number of species S is, to a good approximation,

$$S \approx \pi^{1/2} S_0 / a \qquad (9.8)$$

One further useful parameter may be introduced. Define R_{max} as the expected R value, or octave, for the most abundant species, and define R_N to be the location of the peak of the distribution in total numbers in individuals (i.e. the distribution of $NS(N)$). The parameter γ is then defined to be

$$\gamma = R_N / R_{max} \qquad (9.9)$$

Any pair of the three parameters S_0, a, and γ may now be used uniquely to characterize the distribution. The three are approximately related by $a^2 \gamma^2 (\ln S_0) \approx 0.12$ (May, 1975).

The purpose of this brief frenzy of notation is to discuss two empirical 'laws', which provoked much speculation in earlier years. The first is the observation, first made by Hutchinson (1953), that the parameter a usually has a value around 0.2:

$$a \approx 0.2 \qquad (9.10)$$

The second is Preston's (1962) canonical hypothesies, which says that the usual value of γ is around unity:

$$\gamma \approx 1 \qquad (9.11)$$

Preston's rule fits a large body of data, some of which is presented in Figure 9.4. In addition, it will be seen later that species-area relations derived from it seem to fit a lot of data.

Returning to the approximate relation between S and S_0 given by eqn 9.8, along with that among a, γ, and S_0 given immediately below eqn 9.9, we can obtain the rough relationship $a^2 \gamma^2 \ln(aS/\pi^{\frac{1}{2}}) = 0.12$, connecting a, γ, and S. The corresponding rough relation among a, γ, and the total number of individuals, N_T, has a similar form, $G(a, \gamma) a^2 \gamma^2 \ln(N_T) =$ (constant), with the function $G(a, \gamma)$ varying more slowly with a, γ than $a^2 \gamma^2$ (May, 1975). This amounts to the observation that the two parameters a and γ, which characterize the lognormal distribution, can be derived from the total numbers of species and individuals, S and N_T: to a good approximation, a and γ both depend on S and N_T only as $(\ln S)^{\frac{1}{2}}$ and $(\ln N_T)^{\frac{1}{2}}$. This means that big changes in S and N_T give only relatively small changes in a and γ. So it is not surprising that an enormous range of communities, with S ranging from 20 to 10 000 and N_T ranging from $10S$ to $10^7 S$, are characterized by values of a in the range 0.1–0.4 and of γ in the range 0.6–1.8 (May, 1975).

These observations led May (1975) to suggest that the enigmatic rules (eqns 9.10 and 9.11) are merely mathematical properties of the lognormal distribution for SRA (with the lognormal itself reflecting nothing more than statistical generalities).

Although this may be the correct explanation for the rule in eqn 9.10, Sugihara (1980) has shown that the canonical hypothesis, eqn 9.11, seems to be obeyed too accurately to be explained by May's arm waving. Figure 9.4 is Sugihara's plot of the number of species, S, against the standard deviation of the logarithmic abundance, σ (Sugihara, 1980). For a lognormal distribution, any specified value of the parameter γ leads to a unique relation between S and σ; these relations are shown in Figure 9.4 for $\gamma = 1.8$, 1.0, and 0.2. It does appear that natural communities conform more closely to

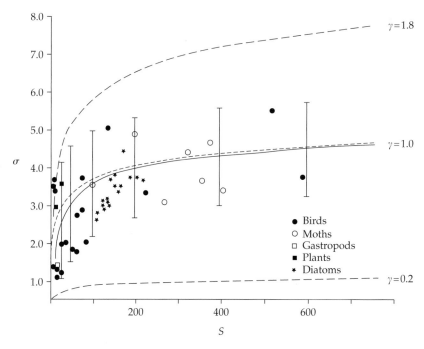

Figure 9.4 The data points show the number of species, S, and the standard deviation of the logarithms of the relative abundances, σ, for various communities of birds, moths, gastropods, plants, and diatoms (as labelled). The dashed lines show the relations between S and σ for communities in which relative abundance is distributed lognormally, with $\gamma = 1.8$, 1.0, or 0.2; $\gamma = 1$ corresponds to Preston's 'canonical lognormal'. The solid line is the mean relation predicted by Sugihara's model of sequential niche breakage, and the error bars represent two standard deviations about this mean. After Sugihara (1980).

the canonical relationship, eqn 9.11, than can be explained by mathematical generalities alone.

9.6.2 Other SRA distributions

Sugihara (1980) has also suggested a biological mechanism that will, among other things, produce the observed SRA patterns. The multidimensional 'niche space' of the community is imagined as being like a hypervolume, which is sequentially broken up by the component species, such that each of the S fragments denotes the relative abundance of a species. This sequential broken stick model is biologically and mathematically very different from MacArthur's broken stick model in which a stick is broken simultaneously into S pieces. The sequential breakage pattern (with any fragment being equally likely to be chosen for the next breakage, regardless of size) seems more in accord with evolutionary processes, and the patterns of relative abundance

thus generated are unlike those of the MacArthur broken stick model. The solid line in Figure 9.4 shows the mean relation between S and σ predicted by Sugihara's model and the error bars show the range of two standard deviations about the mean. Of course, the fact that this model provides a remarkable fit to observed distribution patterns does not prove the model to be correct; other biological assumptions could, and arguably do, also fit observed distributions. We would strongly emphasize, however, that Sugihara's sequential broken stick model contains no adjustable parameters, whereas the corresponding SRA relations derived from Hubbell's neutral models have several adjustable parameters. Although the SRA distributions given by the sequential broken stick are similar to those for the canonical lognormal, the former are significantly more left-skewed than the latter, which tends to accord with subsequently observed patterns (Nee et al., 1991).

More concretely, Sugihara's niche hierarchy model can be thought of as a metaphor for a niche-overlap dendrogram, in which niche space is pictured literally as a branching tree of niche similarities and differences (Sugihara, 1980). As foreshadowed above, this has other interesting consequences (Sugihara *et al.*, 2003). If the relative abundance of species within some taxonomic group is indeed determined by sequential subdivision of niche space, then there clearly may be associations between the relative abundance of a species and its location on a niche dendrogram.

Consider first a caricature of the various roles played by species within their niche space, depicted as a branching tree. Suppose the group is beetle species, making their living on a literal tree: one major stem of the niche dendrograms might be

predation on other insects, another herbivory. Within herbivory, one branch might correspond to eating leaves, another mining into the bark. And on any one branch of the dendrograms—at each niche interface—further subdivisions arise, and so on. Although this sequential niche splitting can occur in many ways, the end result will be that some species are represented by small twigs, others by big branches, of the niche dendrograms. This can imply correlations between the relative abundances of species and their location on the niche dendrogram. Sugihara *et al.* (2003) compare the resulting predictions with data from 11 communities (encompassing fishes, amphibians, lizards, and birds) where both abundances and dendrograms have been reported. They found significant correlation between theory and data. Figure 9.5 illustrates one of these.

Figure 9.5 Illustrating the discussion in the text, this figure shows the relationship between the (lower) histogram of species abundances and the (upper) dendrogram of ecological similarity for a community of co-occurring waterfowl. Note that the species located on relatively highly subdivided branches tend to be less abundant. Courtesy of Benoît Renevey, www.naturecommunication.ch, using data from Sugihara *et al.* (2003).

9.7 Succession

Another SRA distribution is the *geometric series*. This gives distributions similar to those commonly observed in the early stages of succession, when communities often tend to consist of a handful of weedy, pioneer species that happen to get there first. The species arriving first may tend to grow rapidly, pre-empting a fraction, *k*, say, of the available space or other governing resource before the arrival of the next species, which in turn will tend to preempt a similar fraction of the remaining space before the arrival of the next, and so on. The consequent pattern will show up, very roughly, as a geometric series when the relative abundances are arranged hierarchically. Typical examples are seen towards the left-hand side of Figures 9.6 and 9.7.

As succession proceeds, things become more complicated, and a multitude of ecological dimensions are likely to be relevant to the ultimate composition of the community. As discussed above, this tends to lead to a lognormal distribution of SRA: on a plot of abundance against rank this produces the sort of *S*-shaped curve seen for

later successional stages in Figure 9.6, with a preponderance of 'middle-class' species. The later, lognormally distributed, community tends to be an egalitarian socialist society compared with the feudal hierarchy characteristic of early succession.

Something akin to a reversal of these successional patterns takes place when mature communities become polluted. This has been demonstrated particularly for diatoms in streams and lakes subject either to enrichment by waste, heat, sewage, or other organic materials, or to toxic pollution by heavy metals or other poisons (Patrick, 1973, 1975). The pristine equilibrium diatom community essentially always shows the classic lognormal distribution seen in Figure 9.3. When polluted, the community typically shows a pattern in which a few species become exceptionally common, with their relative abundances tending to exhibit a geometric series on a rank/abundance plot.

Another beautiful example is provided by the experimental Park Grass Plot at Rothamsted. These plots were set aside in 1856, and each was subjected to some specified treatment, such as the

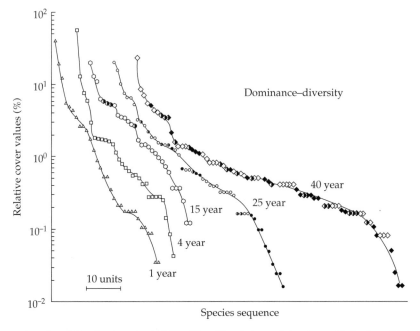

Figure 9.6 Patterns of species relative abundance in old fields of five different stages of abandonment in Southern Illinois. The patterns are expressed as the percentage that a given species contributes to the total area covered by all species in the community, plotted against the species' rank, ordered from most to least abundant. Symbols are open for herbs, half-open for shrubs, and closed for trees. After Bazzaz (1975).

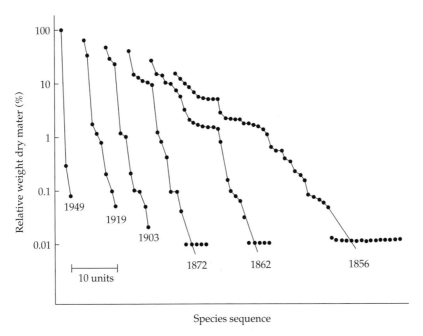

Figure 9.7 Changes in the patterns of relative abundance of species in an experimental plot of permanent pasture at the Park Grass Plot, Rothamsted Experimental Station, UK, following continuous application of nitrogen fertilizer since 1856. Species with abundance of $< 0.01\%$ were recorded as 0.01%. Notice that time runs from right to left; the patterns look like the successional patterns of Figure 9.6, run backwards in time. After Kempton (1979).

withholding or overapplication of certain fertilizers. The resulting changes in the relative abundances of the grass species present have been monitored over the past century and a half. One set of results is shown in Figure 9.7. Tilman (1980b) has made a thorough and perceptive analysis of the Park Grass data, showing the patterns exhibited in Figure 9.7 to be typical. This figure illustrates the arguments developed here, with the overfertilized plot showing changing patterns of relative abundance of species which are strikingly like succession in reverse.

9.8 Species–area relations

One of the earliest accomplishments of theoretical ecology was the discovery of a relationship between the number of species—plants, birds, beetles, or whatever—on a given island and the area of that island. First clearly enunciated by MacArthur and Wilson in their influential *Theory of Island Biogeography* (1967), such species–area

relations were largely phenomenological, based on observations.

The islands described by such species–area relations may be real islands in the ocean, or virtual islands such as hilltops (where the surrounding lowland presents a barrier to many species), lakes, or wooded tracts surrounded by open land. In such island groups, plotting the number of species, S, in a particular taxonomic category against the area, A, results in a straight line on a log-log plot, which corresponds to a power-law relation of the form

$$S = cA^z \tag{9.12}$$

The constant c is characteristic of the taxonomic group, but the dimensionless exponent z typically has a value roughly around 0.2–0.3. Table 9.1 gives a summary of earlier studies (May, 1975); more recent compilations are similar (Begon *et al.*, 2006). This relation is often summarized as saying that a 10-fold increase in island area approximately doubles the number of species.

Table 9.1 Values of z in eqn 9.12, as deduced from observations on various groups of plants and animals in various archipelagoes. For original references, see May (1975).

Organism	Location	z	Source
Beetles	West Indies	0.34	Darlington
Reptiles and amphibians	West Indies	0.30	Darlington
Birds	West Indies	0.24	Hamilton, Barth, Rubinoff
Birds	East Indies	0.28	Hamilton, Barth, Rubinoff
Birds	East–Central Pacific	0.30	Hamilton, Barth, Rubinoff
Ants	Melanesia	0.30	MacArthur and Wilson
Land vertebrates	Lake Michigan Islands	0.24	Preston
Birds	New Guinea Islands	0.22	Diamond
Birds	New Britain Islands	0.18	Diamond
Birds	Solomon Islands	0.09	Diamond
Birds	New Hebrides	0.05	Diamond and Mayr
Land plants	Galapagos	0.32	Preston
Land plants	Galapagos	0.33	Hamilton, Barth, Rubinoff
Land plants	Galapagos	0.31	Johnson and Raven
Land plants	World-wide	0.22	Preston
Land plants	British Isles	0.21	Johnson and Raven
Land plants	Yorkshire nature reserves	0.21	Usher
Land plants	California Island	0.37	Johnson, Mason, Raven

MacArthur and Wilson's *Theory of Island Biogeography* went on to explain, in qualitative terms, why we should expect small and more isolated islands to have fewer species than large ones close to mainland sources. They did this by looking at the balance between new species arriving on the island and existing inhabitants being lost. As the curves in Figure 9.8 indicate, we would expect rates of addition to the existing species pool to be relatively higher for islands closer to major sources of immigration. And—other things being equal—we would expect rates of loss, caused by demographic or environmental fluctuations or other mechanisms, to be higher for the smaller populations found on smaller islands. Figure 9.8 shows how the balance between such processes of gain and loss will tend to result in more species on large, close islands than on small, distant ones; $S_2 > S_1$.

Although the terms null and neutral lay decades in the future in 1967, we think that the theory of island biogeography represents a rough draft for Hubbell's (2001) important neutral community ecology, as described above. Note that MacArthur and Wilson made no assumptions whatsoever about which particular species would be found on a given island, nor about the precise mechanisms determining which species would be lost. In this sense, their ideas parallel Hubbell: species are lost (for Hubbell, the main mechanism is stochastic fluctuations, essentially the same way as surnames are lost in human populations; MacArthur and Wilson are less, not more, specific); species arrive (for Hubbell, from adjacent metacommunities or by speciation; for MacArthur and Wilson, only from metacommunities). The difference, of course, is that, as we shall now see, Hubbell's ideas give a much richer account of the details of gain and loss, based on extensive analysis of large data-sets.

First, however, note that the empirical eqn 9.12 can be derived by assuming the relative abundance of the species of interest obey a lognormal distribution, and that the total number of individuals in this species group, N, is proportional to the island's area, A. A lognormal distribution of SRA gives a relation between S and N, which depends only on one other parameter characterizing the distribution. This relation is complicated, but for $S > 20$ or so it can be well approximated as $S \sim (\text{constant})N^z$, where $z = 1/(4\gamma)$ for $\gamma > 1$ and

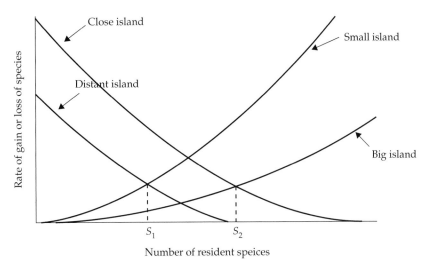

Figure 9.8 As discussed in the text, this figure (based on MacArthur and Wilson, 1967) illustrates in general and qualitative terms why we would expect small, distant islands to contain fewer species than large ones closer to mainland areas.

$z = (1 + \gamma)^{-2}$ for $\gamma < 1$, with γ defined by eqn 9.9 above (May, 1975). If indeed N is proportional to A, we get the species–area relations of eqn 9.12, with z given in terms of γ. If we particularize to the canonical lognormal, $\gamma = 1$ and so $z = \frac{1}{4}$, a value often used to summarize the range of observed data, and the basis of the rule that a 10-fold reduction in area leads to a halving of the number of species. More generally, remember that, in discussing possible explanations for the canonical lognormal, we saw that an enormous range of S and N values are encompassed by γ lying in the range from 0.6 to 1.8, which corresponds to z in the range 0.12–0.41.

Preston (1962) earlier observed that his canonical lognormal, combined with $N \sim A$, gave eqn 9.12, although his numerical simulations gave an exponent $z = 0.262$. He also emphasized that the power law of eqn 9.12 was an approximation for large A (more strictly, large N), with S showing a downturn against A on log-log plots for small islands. Such a downturn is indeed a feature of most data sets (see Figure 9.9), which also tend to show other departures from eqn 9.12 if very large areas are involved (the curves can flatten, or sometimes rise, if continued to continental area scales).

Any theoretically derived species–area relations will have an adjustable parameter, corresponding

to c in eqn 9.12, that essentially depends on overall species richness, setting the scale for the vertical axis in Figure 9.9. Beyond this, the canonical lognormal gives a theoretical prediction with *no* adjustable parameters. More generally, other lognormal distributions give one-parameter fits to species–area relations data. Sugihara's sequential broken stick model, highlighted in Figure 9.4, gives fits very similar to those from the canonical lognormal, and it also has *zero* adjustable parameters.

Harte *et al.* (1999) have offered an explanation for the exact power-law species–area relations of eqn 9.12 by assuming self-similarity: the fraction of species found in an area A, which is also found in one-half A, is independent of A. This attractive assumption certainly supports the power law, but it has the disadvantage that it gives no reason for why the z exponent is so consistently in the range around 0.2–0.3; nor does it agree with the observed departure from a pure power law at low values of A.

Hubbell and co-workers begin with an especially valuable collection of data on the diversity of tropical tree species within each of five 50-ha study sites, in India, Panama, Thailand, and two in Malaysia. Motivated by the work of Harte, Plotkin *et al.* (2000) calculated, for each plot, a spatial persistence function, $a(A)$, which describes the

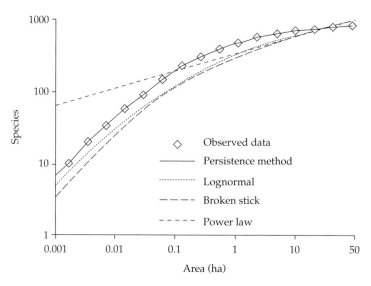

Figure 9.9 The species–area relation observed by Plotkin *et al.* (2000) in the Pasoh tropical forest site in Malaysia compared with the predictions of four theoretical models. Specifically, all four models have a scaling parameter *c* that essentially depends on overall species richness. For the Plotkin persistence method, the curve (which involves at least two adjustable parameters) is taken from their paper. Harte *et al.*'s (1999) self-similarity assumption gives a pure power law, with an arbitrary exponent *z*; here, $z = 0.25$. The canonical lognormal and the sequential broken stick graphs have uniquely determined shapes (which asymptotically give power laws with $z = 0.25$), and thus have no adjustable parameters beyond *c*. After May and Stumpf (2000).

average fraction of species present in area *A* that are found (or 'persist') in one-half *A*. For Harte's self-similarity assumption, *a(A)* is constant. To the contrary, Plotkin *et al.* found that the persistence function depends on *A*, in a way that seems fairly similar for each of their 50-ha plots. Using this, and other information about the movement of species among communities within the metacommunity defined by Hubbell, they estimate the empirical *a(A)* curves for each plot. On this basis, they then estimate the shape of the species–area relations for that region. Interestingly, they find that the species–area relations estimated in this way for any one of their plots gives a good description (to within 5–10%) of the shape of the species–area relations for any of the other four. That is, the five species–area relations *S* have similar shapes, although the absolute number of species for a given value of *A*, which depends on the parameter *c*, varies significantly among them. Provided *A* is not too large, these species–area relations curves are a modification of eqn 9.12, $S = cA^z \exp(-kA)$, with *c* and *z* as before and *k* an additional parameter.

As Figure 9.9 shows, the resulting species–area relations curve (which is constructed using at least two adjustable parameters) is very similar to those given by the zero-parameter canonical lognormal and sequential broken stick models. All three dip at small *A* values, and are not fully asymptotic to the linear log-log relation of eqn 9.12 even at the largest *A*-value. The data presented by Plotkin *et al.*, which are the outcome of a long-term research programme coordinated by the Smithsonian Institution's Centre for Tropical Forest Science, are immensely useful. As emphasized by these authors, the similarities in their species–area relations shapes, despite the differences in overall species richness among their five plots, suggest that we can estimate the diversity of tree species in other unstudied tropical places on the basis of sampling in just one relatively small area. Harte's assumption produces the pure power-law species–area relations of eqn 9.12 with the single free parameter *z*. Note that Plotkin *et al.*'s (2000) generalization of Harte's assumption—their 'persistence function', *a(A)*—is derived phenomenologically, from observed clustering patterns of trees; future work could possibly provide more basic understanding, ultimately leading to fewer adjustable parameters in the model.

Clearly all such studies of species–area relations have potential application to conservation planning. It is well to remember, however, that the theoretical basis for much work on species–area relations is the dynamics of turnover of plants and animals on real or virtual islands. This has clear similarities to asking about subplots within a larger tropical forest plot, but it is not exactly the same question. Some other studies are not so much about species-area relations, but rather about sampling effects (May, 1975). Before taking phenomenological rules and theoretical ideas about species–area relations and applying them to problems in conservation biology, such as the fragmentation of tropical forests, we would like to see more careful discussion of the similarities and differences between these two ecological situations (Watling and Donnelly, 2006). A further problem, not always recognized, is that quantitative estimates of the rate of loss and gain of species from real or virtual islands, by local extinction and re-immigration, can depend on the census interval. If you are estimating the rates at which lights wink on and off, you need to be careful that they do not wink off and on again while you are not looking (Diamond and May, 1977, 1981).

9.9 Scaling laws

One of the appealing things about the physical sciences is the existence of invariance principles and conservation laws, which provide the basis for powerful simplicities and generalizations. A particularly notable example of the use of dimensional arguments was given in the 1950s by G.I. Taylor, the leading fluid dynamicist involved in the Manhattan Project at Los Alamos, NM, USA. In an atomic explosion, there is an essentially instantaneous release of a large amount of energy, E, from what is effectively a point source. The subsequent spherical shock wave propagates into the surrounding air, of density ρ, with the pressure behind the early-stage wave front being vastly larger than the air pressure. It follows that the only physical factors determining the radius of the spherical shock wave front, R, are E, ρ, and the elapsed time, t. In terms of the basic scaling dimensions of mass, length, and time (M, L, T), these three independent variables have dimensions $[E] = ML^2T^{-2}$, $[\rho] = ML^{-3}$, and $[t] = T$; R has dimensions $[R] = L$. To get the scaling relation between R (dimension L) and t (dimension T), we eliminate M among $[E]$, $[\rho]$, and $[t]$ to get $L^5 \sim T^2$. This implies $R \sim t^{2/5}$ or a straight line with slope 1 when $\ln R$ is plotted against $(2/5)\ln t$. Taylor used the data from a series of high-speed photographs of the fireball expanding over the test site in Nevada to verify this result, and then further used the y-axis intercept of this line to estimate $E \sim 10^{21}$ erg. He published this simple and elegant analysis in 1950, causing a furore among the military bureaucracy; although the film was not classified, the energy-release figure was top secret (for a more detailed account, see Barenblatt, 1996).

These ideas about scaling have made their way into several areas of biology, mainly at the level of the physiology and behaviour of individual organisms. But there are some interesting examples in ecological contexts, and we end this chapter by touching on some of them. One of the earliest such examples is in the only paper Hutchison and MacArthur (1959) published together. They were interested in why there are so many more species of small vertebrates than big ones. The paper is not easy to read, but it did anticipate more recent interest in fractal scaling and home-range size (Hastings and Sugihara, 1993; Haskell et al., 2002). Hutchinson and MacArthur hypothesized that, at the upper end of the size spectrum, species' home-range area, H, would scale with individuals' characteristic length, L, as $L^2 \sim H$, and that the total number of species in the size class L would scale inversely with H, $S(L) \sim 1/H(L)$. The result is the suggestion that $S(L) \sim L^{-2}$.

Subsequent analyses of species-size distributions for various taxa were made by May (1978b), who—fitting by eye, given the nature of the data—found humped distributions of $S(L)$ against L for individual taxa, and also for all terrestrial animals combined, with $S(L)$ very roughly decreasing as L^{-2} to the right (large L) side of the peak (which itself occurs around 1 cm for all animals combined, arguably indicating lack of taxonomic knowledge

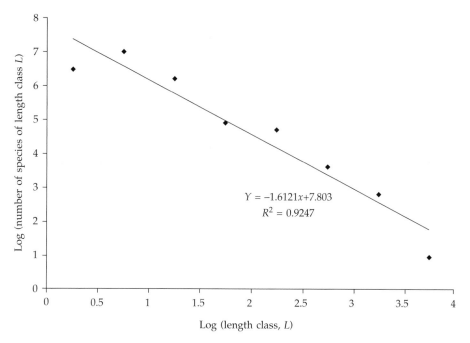

Figure 9.10 A log-log plot of number of species in length class L, $S(L)$, versus L for terrestrial vertebrates. Data are from May (1978), and the regression line has a slope of roughly $-3/2$ (more exactly, $y = -1.61x + 7.80$), consistent with the scaling implied by the seemingly unrelated eqns 9.12 and 9.14. After Southwood *et al.* (2006).

below this, rather than anything else). Figure 9.10 shows a more careful analysis of these data for terrestrial vertebrates. The regression line has a slope of -1.6 (which is also true if the low-lying first and last points are removed). In round numbers, we might express this relation for terrestrial vertebrates as

$$S(L) = (\text{constant})L^{-3/2} \qquad (9.13)$$

In an altogether different study, Burness *et al.* (2001) looked at data from continents and oceanic islands over the past 65 000 years, to see how the mass of the largest animal, M_{max}, varied with the average area where it was found, A. They found that, to a good approximation,

$$M_{max} = (\text{constant})A^{1/2} \qquad (9.14)$$

Returning to eqn 9.12, which relates the total number of species, $S = \Sigma S(L)$, to area, A, it can be observed that any one of eqns 9.12, 9.13, or 9.14 can, despite their seemingly different derivations, be derived from the other two (Southwood *et al.*,

2006). For instance, suppose we are told the biggest species has characteristic length L_{max}; this corresponds to $S(L_{max}) = 1$, whence the constant in eqn 9.13 scales as $L_{max}^{3/2}$. The upper length scale, L_{max}, is essentially irrelevant to the integral over all size classes of $S(L)$, $\int S(L)dL$, whence

$$S \sim L_{max}^{3/2} \qquad (9.15)$$

But $M_{max} \sim L_{max}^{3}$, which via eqn 9.14 gives $L_{max} \sim A^{1/6}$. Substituting this into eqn 9.15 thus leads us from eqns 9.13 and 9.14 to the SRA eqn 9.12 with the canonical lognormal value of $z = 1/4$ (Southwood *et al.*, 2006). In similar fashion, we can deduce eqn 9.13 from eqns 9.12 and 9.14, or eqn 9.14 from eqns 9.12 and 9.13.

On a different but related tack, Finlay *et al.* (2006) have documented self-similarities in distributions of relative abundance and in species-area and species-size relations among insect species on spatial scales ranging from several hectares to the entire planet. They used a database of more than 600 000 insect species from a wide variety of places.

In particular, the rank-abundance plots (characterizing SRA) for the world total (601 958 species), North America (60 592), and the UK (11 260) completely coincide when they are rescaled to allow for the gross discrepancy in species numbers; these patterns are different on scales of order 100 ha or less. Finlay *et al.*'s analysis also indicates that a possible reason for the 'falling away' of species numbers of terrestrial invertebrates at smaller sizes (characteristic length less than 1–10 mm) may be caused as much by significantly wider geographic distribution of smaller species as by taxonomic neglect (see also Fenchel and Finlay, 2004).

A variety of other interesting approaches to identifying and codifying possible patterns in the amazing diversity of life on local, regional, and global scales is to be found in the book *Scaling Biodiversity*, edited by Storch *et al.* (2006). As in all such quests, it is good to keep in mind Einstein's dictum, which translates roughly as 'seek simplicity, and distrust it'.

CHAPTER 10

Dynamics of infectious disease

Bryan Grenfell and Matthew Keeling

10.1 Introduction

Host–pathogen associations continue to generate some of the most important applied problems in population biology. In addition, as foreshadowed in Chapter 5 of this volume, these systems give important insights into the dynamics of host–natural enemy interactions in general. The special place of pathogens in the study of host–natural enemy dynamics arises partly from excellent long-term disease-incidence data, reflecting the public health importance of many infections. However, we argue that host–pathogen dynamics are also distinctive because the intimate association between individual hosts and their pathogens is often reflected with particular clarity in the associated population dynamics. Throughout this chapter we focus in parallel on the population dynamics of host–pathogen interactions and the insights that host–pathogen dynamics can provide for population biology in general.

Population-dynamic studies of infectious disease have a long history, which predates the modern foundations of ecology (Bernoulli, 1760). During the twentieth century, the preoccupation of population ecologists with the balance between extrinsic and intrinsic influences on population fluctuations and the role of nonlinearity and heterogeneity (Bjørnstad and Grenfell, 2001) find strong parallels in epidemiological studies of human diseases (Bartlett, 1956; Anderson and May, 1991). In terms of the ecological effects of parasitism, the traditional view held that 'well-adapted' parasites would not have a consistent impact on the ecology of their hosts (Grenfell and Dobson, 1995). The 1970s saw a new departure, when Anderson and May pointed out the potential

of infectious agents to exert nonlinear—regulatory or destabilizing—influences on the population dynamics of their hosts (Anderson and May, 1978, 1979; May and Anderson, 1978, 1979). There has since been an explosion of work on the population biology of human, animal, and plant pathogens. This work spans a huge range: from highly applied to basic theoretical work; from within-host to the metapopulation scale; from short-term population dynamics to long-term evolutionary processes. In this chapter we first outline the simple theory of epidemiological models; we then refine this picture to illustrate the potential impact of pathogens on the population dynamics of their hosts, as well as aspects of host–pathogen interactions which provide important insights into more general ecological dynamics. A particular focus of this review is the calibration of simple and complex epidemiological models against the rich databases available for many host–parasite systems.

Before the 1970s, most theoretical studies in epidemiology focused on the dynamics of epidemics in constant host populations. Although this approximation is reasonable for acute infections in human populations, it will not do if we wish to consider the impact of infectious disease on the dynamics of natural populations. This issue was addressed by Anderson and May (Anderson and May, 1979; May and Anderson, 1979), who proposed a unified framework for capturing the dynamics of parasites and their impact on the ecology of their hosts. They divided parasites into two broad groups: microparasites (mainly viruses, bacteria, and other micro-organisms) and macroparasites (mainly helminths and arthropods). Despite their names, the main differentiation between the two groups is a functional one derived from their life history. For microparasites, which reproduce rapidly within the host, the resulting theoretical framework is based upon keeping track of the number of *hosts* of different types (e.g. susceptible to infection, infected, or recovered from infection). In contrast for macroparasites, which generally reproduce more slowly via multiple transmission stages, it is more important to model the number of *parasites*.

The study of the population dynamics of infectious diseases has seen several major advances over the past decades (Anderson and May, 1991; Hudson *et al.*, 2001). In brief, we can partition these developments into six basic categories, each of which brings epidemiological models closer to reality.

1 Host heterogeneities. The most basic disease models assume that all individuals are identical, with equal risks of infection and subsequent transmission. However, motivated by observations of sexually transmitted infections, it was realized that the heterogeneity in the number (and type) of contacts could play a crucial role in the transmission dynamics—with a highly connected core group responsible for maintaining the infection (Hethcote *et al.*, 1982; Anderson and May, 1991; Morris, 1993; Lloyd-Smith *et al.*, 2005).

2 Within-host dynamics. The basic compartmental models for infectious diseases treat all infected hosts equally—leading to a constant transmission rate throughout an exponentially distributed infectious period. In reality, the transmission rate varies during the infectious period, which itself is better approximated as a normal or gamma distribution (Keeling and Grenfell, 1997). Such details of pathogen biology can have a profound effect on epidemic dynamics (Grossman, 1980) and the impact of control mechanisms (Fraser *et al.*, 2004; Wearing *et al.*, 2005). This work has strong parallels with metapopulation theory in ecology, with simple compartmental models being equivalent to Levins metapopulations, whereas more complex models describe the 'within-patch' dynamics. Capturing the within-host behaviour can itself be seen as a question of population dynamics (Nowak and May, 2000); there is currently substantial interest in modelling the interaction of the pathogen and the host's immune system, how this generates the observed host-level behaviour, and how within-host dynamics translate to the population level (Grenfell *et al.*, 2004).

3 Parasite genetic heterogeneities and evolutionary dynamics. Microparasite evolutionary dynamics are some of the most spectacular manifestations of host–pathogen heterogeneity, as well as particularly clear examples of rapid evolution in action. The evolution of parasite drug resistance and immune escape also represent some of the major problems in the fight against disease (Frank, 2002). We return to these issues below. Another major challenge is to understand the complex interplay between host and parasite genetic heterogeneity, and interpret how this will be reflected in the epidemiological and evolutionary dynamics.

4 Temporal forcing. The transmission of many pathogens is affected by external forcing at seasonal and other temporal scales. Seasonal effects have been particularly well characterized in the transmission of childhood infections, where school terms and holidays impact upon the level of transmission; this forcing in turn interacts with the host–pathogen epidemic dynamics producing the potential for rich and complex dynamical behaviour (Schaffer and Kot, 1985a, 1985b; Olsen *et al.*, 1988; Rand and Wilson, 1991). School terms are not the only source of forcing; more generally, climatic fluctuations influence transmission, leading to long-period oscillations (for example, the influence of the El Niño southern oscillation on cholera dynamics; Koelle *et al.*, 2005).

5 Spatial structure. A fundamental feature of most infectious disease transmission is that it only occurs between individuals in 'close' contact. Therefore diseases tend to spread locally in space, with proximity acting as a major risk factor (Bartlett, 1960; May and Anderson, 1984; Cliff and Haggett, 1988; Murray, 1989; Cliff *et al.*, 1993; Grenfell *et al.*, 2001; Smith *et al.*, 2002). Again this work has clear parallels with ecological dynamics where the locality of dispersal can have profound effects (see Chapters 4 and 5 in this volume). More recently attention has focused on the network structure of contacts, explicitly modelling the local interactions that can lead to disease transmission (Keeling and Eames, 2005).

6 Stochasticity. The random transmission of infection and the discrete nature of the host population has a fundamental, and often very relevant, influence on disease dynamics. Stochastic models provide us with a method of capturing these two essential elements. As such, demographic stochasticity is most important when dealing with low numbers of cases—when it can give rise to local extinctions of the disease (Bartlett, 1960; Keeling and Grenfell, 1997). As described below, stochasticity can have a major impact on the dynamics of both micro- and macroparasites.

All these refinements allow us to capture more of the known dynamical behaviour of an infectious disease. However, a recurring theme in the chapter is that simple models are often remarkably successful at generating robust dynamical insights, which can inform more detailed models.

We structure our tour of host–parasite population dynamics by considering progressively more realistic assumptions about host ecology and the biological details of the host–parasite interaction. We begin by outlining a basic model for the dynamics of a microparasite invading a closed and constant host population. Though simple, this scenario allows us to derive the crucial thresholds for parasite invasion and spread, which apply universally across more ecologically realistic situations. We then explore how childhood infections epitomize the nonlinear spatio-temporal dynamics of ecological population cycles. Turning to the ecological impact of parasitism, we revisit classical models for the impact of parasites on host-population regulation and the special dynamical role of heterogeneities in the host–parasite relationship. A concluding case study on foot and mouth disease illustrates a range of other features of host–pathogen dynamics, particularly the synthesis of simple and complex models of population dynamics.

10.2 Microparasite models: the simple epidemic

The essential functional feature of microparasites is that they reproduce directly within their hosts (Anderson and May, 1991). Thus, for many

diseases, we can capture the essential dynamics via a *compartmental model*, where hosts are divided between different infection categories. The most studied form is the family of susceptible $\rightarrow$ infected $\rightarrow$ recovered (SIR) models for strongly immunizing infections. Individuals are recruited into a previously uninfected susceptible class, S (often via births). They then become infected (I) by contact with infected individuals, before finally moving into an immune-recovered class, R (Bartlett, 1956; May, 1980; Mollison, 1995). The simplest realization of the SIR model describes the dynamics of the *simple epidemic*—an outbreak of a non-fatal infection directly transmitted between hosts in a closed population, without host demographic changes (May, 1980):

$$\frac{dS}{dt} = -\beta IS$$
$$\frac{dI}{dt} = \beta IS - \gamma I \qquad (10.1)$$
$$\frac{dR}{dt} = \gamma I$$

The model assumes a well-mixed population, with homogeneous random mixing between individuals. As with the simplest predator–prey models (see Chapter 3 in this volume), this leads to a bilinear ($I \times S$) interaction term for the infection process, controlled by the infection parameter, β. This form assumes that transmission depends upon the density of susceptible and infectious individuals (as discussed further below; Begon et al., 2002). For simplicity (eg Bjornstad and Grenfell (2002)) we allow β to subsume the rate at which susceptible and infectious individuals make sufficiently close contact to allow a chance of transmission, and the probability of transmission of the infectious agent given such a contact. After infection, individuals move into the recovered class, R, at the recovery rate, γ.

A key epidemiological parameter leaps out of this model: consider an epidemic sparked by a single infected individual introduced into a total susceptible population of density N. At the start of the epidemic (when $S \cong N$) the level of infection, I, can only increase ($dI/dt > 0$) if $\beta N > \gamma$; this leads naturally to a definition of the *basic reproduction ratio* or *basic reproductive number of infection*:

$$R_0 = \beta N / \gamma \qquad (10.2)$$

Since $1/\gamma$ is the average duration of infection, R_0 can be interpreted as the total number of secondary cases produced by one infected individual when introduced into a population of N susceptibles. As the epidemic proceeds, and more individuals recover and are therefore immune, the effective reproduction ratio, R, declines with the proportion of susceptibles, $s = S/N$; that is, $R = R_0 s = R_0 S/N$. The basic and effective reproduction ratios provide a powerful framework for exploring the dynamics and control of epidemics. In particular, an infectious agent can only invade a susceptible population if $R_0 > 1$; if this criterion is satisfied the subsequent epidemic will only continue to increase while $R > 1$; when the proportion of susceptibles is reduced by the epidemic below $1/R_0$, the number of cases must start to decline. An endemic infection, steadily maintained at a constant level, will have an effective reproduction ratio $R = 1$; but $R = R_0 s^*$, where s^* is the equilibrium fraction who are susceptible, whence R_0 can be estimated as $R_0 = 1/s^*$.

10.2.1 Control by vaccination

In essence, the simple epidemic extinguishes itself by reducing its supply of susceptibles below $1/R_0$. The same threshold also applies to control of immunizing pathogens by vaccinating with an inactivated or attenuated pathogen (Anderson and May, 1991); reducing the proportion of susceptibles below $1/R_0$ will keep R below unity and prevent invasion of the infection. Note that the resulting critical proportion of susceptibles that require vaccination to prevent an epidemic ($p_c = 1 - 1/R_0$) is less than unity; this reflects the concept of *herd immunity* in which not all susceptible individuals in a population need to be protected to prevent a large-scale epidemic—logistical constraints then determine whether this level of vaccination can be achieved.

10.3 Host vital dynamics and long-term behaviour of the SIR model

The simple SIR epidemic described above seals its own fate by depleting its susceptible fuel. However, if we allow for the added realism (in most populations) of a steady replenishment of susceptible individuals via births (Anderson and May, 1991), further, *recurrent* epidemics become possible. This brings us to the SIR model with vital dynamics:

$$\frac{dS}{dt} = \mu(N - S) - \beta IS$$
$$\frac{dI}{dt} = \beta IS - (\mu + \gamma)I \qquad (10.3)$$
$$\frac{dR}{dt} = \gamma I - \mu R$$

Again, the total host population size is constant and infection is not assumed to affect host survival or reproductive rates. However, this model replaces the previous assumption of a 'closed' host population, without vital dynamics, by assuming 'background' birth and death rates of hosts, controlled by a per-capita death rate, μ. Although eqn 10.3 is a general formulation, we focus in particular on 'acute' infections, such as measles, where the duration of infection is relatively brief (i.e. $\gamma \gg \mu$). A numerical simulation with measles parameters (Figure 10.1a) immediately illustrates the impact of susceptible replenishment by births—a series of recurrent epidemic oscillations (with an approximately 2-year period for measles parameters), slowly damping to a stable equilibrium abundance of susceptible and infected individuals (May, 1980). Essentially, the nonlinear feedback between infected and susceptible densities, along with the upper limit on susceptible abundance, N, stabilizes their interaction. As shown by the S and I nullclines for the model (Figure 10.1b), these dynamics are closely analogous to the dynamics of Lotka–Volterra predation models with density-dependent limitations on prey abundance (Rosenzweig and MacArthur, 1963).

The period of these damped model oscillations close to equilibrium captures the tendency for measles epidemics to be strongly biennial in the era before vaccination in developed countries (Bartlett, 1957; Anderson et al., 1984; see Figure 10.3a, below). However this comparison of model and data also illustrates a significant question: what additional biological feature do we need to translate the weakly damped epidemics of the

(a)

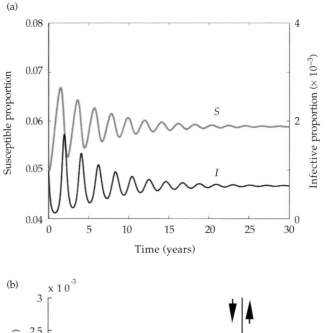

(b)

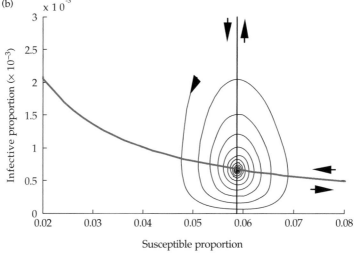

Figure 10.1 Deterministic behaviour of the SIR model including vital dynamics; we use epidemiological parameters for measles (Earn *et al.*, 2000). (a) A numerical simulation showing the damping of susceptible (*S*, black line) and infected (*I*, grey line) proportions towards their deterministic equilibrium values. (b) Corresponding nullclines (zero growth contours) for *I* and *S* at $S_0 = (\mu + \gamma)/\beta$ and $I_0 = (N - S_0)/\beta S_0$; the arrows show direction vectors for *S* (black line) and *I* (grey line) and the spiral corresponds to the simulation in (a).

model into the dramatic sustained oscillations of the observed series?

10.3.1 Seasonal forcing and sustained epidemic oscillations

The above question parallels a long-standing population-dynamic debate about the cause of sustained host–natural enemy cycles of small

mammals, game birds, and forest insects (Bjørnstad and Grenfell, 2001). The key issue in ecology is generally to explain the rarity of such cycles in practice compared with basic theory; here, the issue is reversed: how do we explain the violent cycles of measles and other childhood diseases, given the long-term stability of very plausible simple models? As with ecology, much of the debate has revolved around the balance between intrinsic nonlinear dynamics, time delays, and external forcing (Bartlett, 1956, 1957, 1960; Dietz, 1976; May, 1980; Fine and Clarkson, 1982; Schenzle, 1984; Bolker and Grenfell, 1993).

For measles, Bartlett's key insight is that repeated external perturbations, either from stochastic fluctuations or seasonal variations in transmission, can throw the damped oscillations of the SIR model into sustained epidemic cycles (Bartlett, 1956, 1957, 1960). In practice, seasonal variations in infection rates appear to be the major force sustaining epidemic cycles of measles which are relatively invariant to population size and therefore stochastic effects (Fine and Clarkson, 1982; Schenzle, 1984; Bolker and Grenfell, 1993).

We illustrate this role of seasonality in Figure 10.2. Figure 10.2a shows the dynamics of our simple SIR model with vital dynamics, with the addition of an annual sinusoidal swing of 20% in the per-infective infection rate, β (inset). This repeated perturbation to the system pushes the dynamics into sustained cycles; the deterministic trajectory loops around the unforced equilibrium in SI phase space (Figure 10.2b). Furthermore, the annual period of the forcing resonates with the unforced biennial (but damped) tendency to generate focused, violent epidemics reminiscent of the observed pre-vaccination case-report data (compare Figures 10.2a and 10.3a). More broadly, the synthesis of models with data on childhood epidemics illustrates key principles of ecological dynamics in two parallel, and ultimately interlocking, directions.

10.3.1.1 Realistic epidemiological models
A major body of work has developed increasingly accurate models for assessing the performance of vaccination strategies and exploring spatio-temporal dynamics (Bartlett, 1956; May and Anderson, 1984; Schenzle, 1984; Anderson and May, 1991; Cliff et al., 1993; Babad et al., 1995; Grenfell et al., 2001). These range from complex age-structured models (Schenzle, 1984; Anderson and May, 1991; Bolker and Grenfell, 1993; Babad et al., 1995)—key for designing age-targeted vaccination strategies—to much simpler aggregate formulations (Earn et al., 2000; Bjørnstad et al., 2002). In fact, simple models (combining epidemic

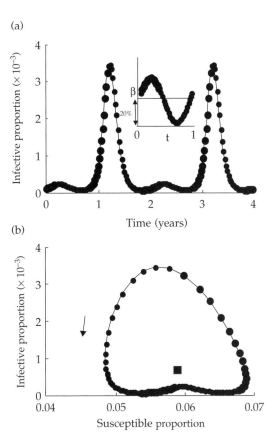

Figure 10.2 Impact of seasonal forcing of the infection rate, β on the dynamics of the simple SIR model, with vital dynamics (Figure 10.1). We use a simple sinusoidal annual forcing: $\beta(t) = \beta_0(1 + \beta_1 \cos(2\pi t))$ (t is time in years), with seasonal forcing amplitude, $\beta_1 = 0.2$ (a 20% seasonal swing). (a) Time series, with the seasonal forcing pattern shown in the inset; symbol area is proportional to forcing amplitude. (b) SI phase plane plot corresponding to (a); the arrow shows the direction of trajectories and the square is the unforced equilibrium, (S^*, I^*).

dynamics with realistic seasonal forcing functions) can capture long-term measles dynamics in large cities surprisingly well (Figure 10.3a), as long as we allow for secular changes in birth rates (which effectively tune the propensity for oscillations of different periods; Bjørnstad et al., 2002; Grenfell et al., 2002). As shown in Figure 10.3a, dynamics in these large centres are relatively insensitive to demographic stochastic perturbations. In smaller populations, below a critical community size of around 300 000 individuals, infection generally disappears in the troughs between epidemics due

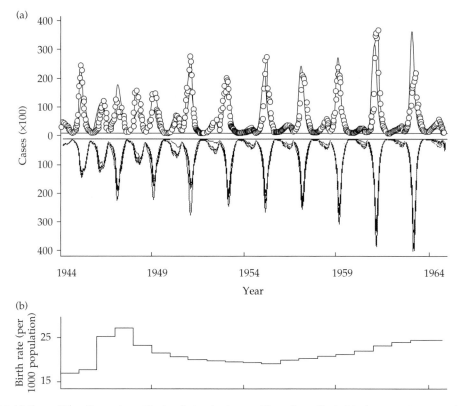

Figure 10.3 (a) Observed biweekly measles notifications for London (corrected for under-notification) in the pre-vaccination era (circles). The line shows a corresponding deterministic simulation of a forced SIR model set in a time-series framework (the TSIR model; see Bjørnstad *et al.* (2002) and Grenfell *et al.* (2002) for details and fitting procedures). The lower reflected graph in (a) shows 10 replicate stochastic simulations of the same model. This underlines that demographic stochasticity has remarkably little effect on the endemic limit-cycle behaviour of measles in large communities. (b) Corresponding annual birth rates (per thousand population) for London; from the 1950s onwards we see sustained biennial cycles; by contrast, the baby boom during the 1940s increased the recruitment of susceptible individuals, driving annual measles epidemics.

to demographic stochasticity (Bartlett, 1960). Another body of literature has used stochastic models to explore these dynamics and the spatio-temporal patterns which emerge from them (Lloyd and May, 1996; Grenfell *et al.*, 2001).

10.3.1.2 Nonlinear dynamics and chaos in epidemics
The search during the 1980s and 1990s for the imprint of chaotic dynamics in real ecological systems (May, 1976a; see also Chapter 3 in this volume) quickly revealed that few ecological systems had the necessary length and detail of time series to discern the sensitivity to initial conditions that is their characteristic fingerprint (Sugihara and May, 1990). In the mid 1980s W.M. Schaffer realized that the relatively long and densely sampled

time series of measles, and other strongly fluctuating childhood infections, might provide the key to this enigma (Schaffer and Kot, 1985a, 1985b; Olsen *et al.*, 1988). Schaffer and Kot realized that acute, self-immunizing, seasonally driven infections like measles are also especially suitable candidates for exotic dynamics—their natural oscillatory behaviour has the propensity to become chaotic at relatively high amplitudes of seasonal forcing. The resulting syntheses of models and data provided a great stimulus to our understanding of nonlinear dynamics in ecology (see Chapters 3 and 5 in this volume).

The effects of high-amplitude seasonal forcing on the deterministic dynamics of measles models are illustrated in Figure 10.4. As forcing amplitude

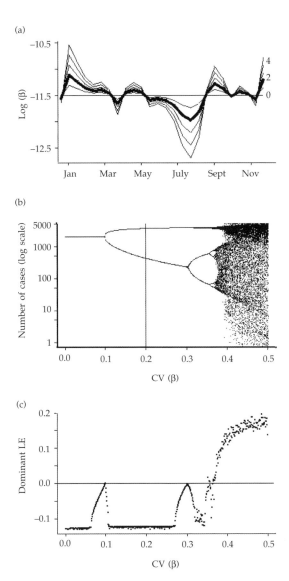

Figure 10.4 Nonlinear behaviour of the SIR model as a function of the amplitude of seasonal forcing. We illustrate the effects of increasing seasonality using the TSIR model (full details of model and figure are given by Grenfell *et al.*, 2002). (a) Observed seasonal variation in infection rate, *I*, for pre-vaccination London (bold line); other lines show increasing amplitudes of forcing used in (b) and (c). (b) Bifurcation diagram for the TSIR model as a function of increasing forcing amplitude (measured as the coefficient of variation (CV) of the patterns in a). For each forcing amplitude, we sampled annually at the start of each year to generate the points. The diagram clearly shows forcing thresholds at approximate coefficients of variation of 0.1 and 0.4, above which the dynamics move to biennial and chaotic behaviours respectively. (c) Corresponding Lyapunov exponents (LE), showing a transition to positive exponents in the chaotic region (Olsen *et al.*, 1988; Tong, 1990; Grenfell *et al.*, 2002).

increases, measles first enters a biennial regime, which dominates over a wide region of forcing/ levels (Figure 10.4b). At high levels of forcing, this behaviour gives way to a region of

high-amplitude chaos, characterized by irregular dynamics with deep inter-epidemic troughs. However, in terms of the comparison between models and data, there is still no compelling empirical evidence for chaos in epidemics, for two reasons. First, from the data perspective, non-stationarity in driving parameters such as birth rate (Figure 10.3) and the underlying stochasticity of transmission effectively reduce the ability to detect the signature of sensitivity to initial conditions (Bjørnstad and Grenfell, 2001; Bjørnstad et al., 2002). Second, the strength of seasonal forcing required for chaotic dynamics in models is stronger than that estimated from the case-reporting data. In addition, the model dynamics within the chaotic parameter region tend to generate deeper troughs with more-localized extinctions of infection—and therefore a much higher critical community size—than are seen in practice (Bjørnstad et al., 2002). Nevertheless, the combination of over-compensatory dynamics, forcing, and noise seen in SIR-type infections make measles, and other acute, strongly forced infections, strong candidates for complex dynamics (Rand and Wilson, 1991; Earn et al., 2000; Keeling et al., 2001a).

10.4 Impact of parasites on host dynamics and population regulation

The last two decades have seen an explosion of empirical evidence on the impact of emerging and established pathogens on the dynamics and evolution of their host populations (Grenfell and Dobson, 1995; Hudson et al., 2001). A significant impetus for this work was the basic theory which elucidated the potential nonlinear impact of micro- and macroparasites on population regulation of their hosts (Anderson and May, 1978; May and Anderson, 1978). To illustrate the general point, we turn to macroparasitic worms, where theory and empirical ecology come together in a particularly elegant way. The key functional properties of macroparasites are that parasites do not reproduce directly in their hosts and generally produce deleterious effects in proportion to their abundance; thus we have to track the intensity of infection in individual hosts. To test the potential for population regulation, Anderson and May (see also Crofton, 1971; May, 1977b) formulated a basic

model for a directly (i.e. non-vectored) macroparasite in an exponentially growing host population:

$$\frac{dH}{dt} = (a - b)H - (\delta + \alpha)\overline{m}H$$
$$= (a - b)H - (\delta + \alpha)P \qquad (10.4)$$

$$\frac{dP}{dt} = \frac{\lambda PH}{H_0 + H} - b\overline{m}H - vP - \alpha HE(m^2)$$
$$= \frac{\lambda PH}{H_0 + H} - (b + v)P - \alpha HE(m^2) \qquad (10.5)$$

Here, a host population (H), with per-capita birth and death rates a and b, respectively, will grow exponentially ($a > b$) in the absence of parasitism. Adult parasites (total population P) produce transmission stages into the environmental at a per-capita rate, λ; the nonlinear success of transmission is captured by the parameter H_0 in eqn 10.5. Additionally, parasites have a background death rate of v. The impact of parasites on the host is assumed to be proportional to the average parasite burden ($\overline{m} = P/H$): those hosts with more parasites suffer a greater reduction in survival and reproduction governed by per-capita rates α and δ respectively.

10.4.1 Heterogeneity and regulation

The final term in both the host and parasite equation (eqn 10.5) allows for two biological processes (Anderson and May, 1978; May and Anderson, 1979): (1) the net host mortality due to parasitism is proportional to average parasite burden ($\overline{m}$), and (2) if parasites kill their host then they themselves die. This combination of effects means that host death can have a disproportionately big effect on parasite mortality, as captured by the expectation $E(m^2)$: the second moment (mean^2 + variance) of the distribution of parasites per host. Therefore, those hosts with a greater-than-average burden are more likely to die, which in turn leads to the death of a greater-than-average number of parasites. In the overwhelming majority of situations (Anderson and May, 1978; Roberts et al., 1995) the parasite distribution is highly aggregated (variance > mean), which greatly magnifies this effect. To an excellent empirical approximation, parasite numbers per host can be assumed to be negatively binomially distributed with dispersion parameter k and variance $= \bar{m} + \bar{m}^2/k$, where $\bar{m}$ is the mean parasite burden per host ($\bar{m} = P/H$). The negative binomial has second moment $E(m^2) = \bar{m} + \bar{m}^2(k+1)/k$, thus eqn 10.5 becomes (May and Anderson, 1978):

$$\frac{dP}{dt} = \frac{\lambda PH}{H_0 + H} - (b + v + \alpha)P \\ - \alpha \frac{(k+1)P^2}{kH} \tag{10.6}$$

Ignoring for the moment the reduction in host reproduction due to parasites ($\delta = 0$), Anderson and May show that parasites can regulate the host population to a stable equilibrium as long as the parasite frequency distribution is aggregated (i.e. $k < \infty$; where $k = \infty$ yields a Poisson distribution). Again, this result has general ecological resonance—heterogeneity in encounter rates can contribute significantly to the stability of host–natural enemy interactions (see Chapter 5 in this volume). We return to the special impact of heterogeneities on host–pathogen dynamics in the case study on foot and mouth disease (section 10.5).

10.4.2 Macroparasites and host population cycles

A variety of refinements have been made to the basic host–macroparasite model (Roberts et al., 1995). The most interesting synthesis of theory and field epidemiology arises from the potentially strongly destabilizing effects of reductions in host reproduction rates due to parasitism (i.e. $\delta > 0$). We can immediately discern the qualitative effects of reproductive limitations by parasitism if we re-express the parasite equation (eqn 10.6) in terms of parasites per host; $\bar{m} = P/H$:

$$\frac{d\bar{m}}{dt} = \frac{\lambda \bar{m} H}{H_0 + H} - (a + v + \alpha)\bar{m} \\ - (\alpha/k - \delta)\bar{m}^2 \tag{10.7}$$

Note that, as we are now dealing with the average number of parasites per host, the natural loss of hosts (at rate b) is irrelevant, whereas the birth of uninfected hosts (at rate a) dilutes the average burden.

Focusing on the quadratic term in $\bar{m}$, parasite-induced host mortality (α) exerts a regulatory negative feedback in an aggregated parasite population ($k < \infty$). However, host reproductive limitations by the parasite (δ) reduce this regulatory effect; indeed, for $\delta > \alpha/k$, the quadratic term becomes positive and the system is thrown into limit cycles (May and Anderson, 1978). Heuristically, reproductive limitations by the parasite reduce host-population growth rate, but have much less strong short-term effects on parasite abundance per host—hence the reproductive effects continue until the host population reaches a lower threshold where the parasite population crashes and the hosts recover, etc. These very distinctive feedbacks again illustrate the special ecological dynamics that can emerge from the intimate relationship between host and parasite.

Probably the best empirically characterized impact of parasitism on long-term host dynamics arises in the ecology of a macroparasite: the nematode *Trichostrongylus tenuis* infecting red grouse populations in northern England and Scotland. Game bag records for the grouse show regular 5–7-year cycles since the nineteenth century. In an echo of other debates about the role of natural enemies in ecological cycles, *T. tenuis*, which has strong effects on host recruitment at high host densities, has long vied with intrinsic behavioural explanations as a candidate for causing the cycles (Hudson et al., 1998, 2001). Because birds can be treated with antihelmintic drugs, parasites are more easily 'excluded' from the system than predators usually are in other cycling host populations. This led Hudson et al. (1998) to carry out a decade-long field experiment demonstrating that cycles were prevented on treated grouse moors compared with untreated control moors where the cycles continued. The debate about grouse cycles continues (Mougeot et al., 2005). However, the work of Hudson et al. remains one of the best experimental tests of simple theory in population dynamics.

10.5 Foot and mouth disease in the UK

The 2001 foot-and-mouth epidemic had a huge and lasting impact on the UK livestock industry. However, the epidemic also precipitated a substantial change in the way that models of infectious disease are formulated and utilized. We now briefly review the epidemic situation in 2001 and the biological characteristics of foot and mouth disease before showing how these influenced the choice of modelling approach. Apart from its applied importance, the UK foot-and-mouth epidemic illustrates a number of general points about the spatio-temporal dynamics of infectious diseases, as well as the level of model detail required to address different questions.

Foot and mouth disease is one of the most rapidly transmitted of all livestock infections, and can quickly spread both within and between farms. Foot and mouth disease can infect most livestock species, although during the 2001 epidemic in the UK it was primarily cattle and sheep

farms that were most frequently affected. In total, 2026 farms reported infection with foot and mouth disease (leading to the culling of their animals), while approximately four times that number of farms had their animals culled in an attempt to stop the spread of infection. Despite this seemingly intense culling, the epidemic lasted from February until September, with a long protracted tail once the main bulk of the epidemic had died away. Figure 10.5 shows the pattern of cases and culls in space and time.

We now outline one of the models used during the 2001 epidemic (Keeling *et al.*, 2001b), focusing on how the epidemiological characteristics dictated the model structure and how models are necessary to interpret the complex nonlinear trade-offs that occur with control. The data from the UK foot-and-mouth epidemic have presented modellers with an astoundingly detailed data-set, with a large amount of information on both susceptible and infected farms, as well as a range of complex

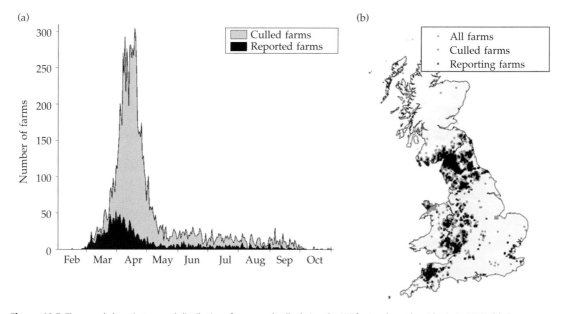

Figure 10.5 The recorded spatio-temporal distribution of cases and culls during the UK foot-and-mouth epidemic in 2001. (a) Time series of daily reported cases (black) and farms whose animals were culled as a preventative measure (grey). It is clear that the majority of farms lost was due to preventative culling, although an uncontrolled epidemic would have been far worse. (b) The spatial location of all farms (pale grey), culled farms (dark-grey circles), and farms reporting infection (black squares). The concentration of infection in Cumbria, Dumfries and Galloway, Devon, and the Welsh borders is clear (Data from DEFRA; www.defra.gov.uk/footandmouth/cases/fmdcases/map.htm).

questions. Three models were utilized during the 2001 foot-and-mouth epidemic (Keeling, 2005); they ranged from elaborate simulations (Morris *et al.*, 2001) to a system of (complex) differential equations (Ferguson *et al.*, 2001a, 2001b). These models generally agreed in the type of additions to the simple SIR models (eqn 10.1) that were required to capture the observed dynamics of foot and mouth disease.

Host heterogeneities. Variability at the host level comes from two different sources, the first is the intrinsic differences between different livestock species. In particular, cattle are far more susceptible than sheep, whereas pigs featured rarely in this epidemic; in addition, cattle are also responsible for a greater amount of transmission than sheep. From this perspective cattle can be seen as a core group, being both more at risk of infection and subsequently a greater potential for generating further cases (Anderson and May, 1991). This heterogeneity is amplified by the clustering of animals (often of the same species) within farms— essentially corresponding to assortativity of the core group. All three of the 2001 models of foot and mouth disease treated the farm as the host unit—effectively taking a metapopulation approach with each farm acting as a susceptible patch for the infection to colonize. This leads to the second form of host heterogeneity, the differences between farms. Primarily this variability has been attributed to the number and species of livestock within the farm, although farm geography, farm fragmentation, management practices, and livestock breed could all play a significant role. As an example, the model of Keeling *et al.* (2001b) assumes that for farm i the susceptibility (σ_i) and transmissibility (τ_i) are proportional to the number of cattle (N_i^C) and sheep (N_i^S) present:

$$\sigma_i = s_C N_i^C + s_S N_i^S \quad \text{and}$$
$$\tau_i = t_C N_i^C + t_S N_i^S \tag{10.8}$$

where the species-level parameters, s and t, reflect both species-level heterogeneities and species-level farming practices.

Within-farm dynamics. The standard SIR-type models (eqns 10.1 and 10.3) implicitly assume that all infected hosts (in this case farms) recover at the same rate irrespective of the time since infection— leading to exponentially distributed infectious periods, with hosts in the tail of the distribution contributing to the majority of transmission. Observations of foot and mouth disease suggest that this is not the case. Infected farms usually pass through an exposed or latent period of around 4 days before they start shedding virus; it is then around a further 5 days before signs of infection appear and control measures can be applied. Therefore, the simplest assumption (used by Keeling *et al.*, 2001b) is to model the exposed and infectious periods as discrete intervals of a constant length. A more realistic assumption is to pick the periods from gamma or normal distributions that most closely match the observed variance, including the fact that early signs of infection in sheep can often go unnoticed (Ferguson *et al.*, 2001a).

Spatial structure. The 2001 epidemic was primarily confined to four regions of the country: Cumbria, Devon, the Welsh borders, and Dumfries and Galloway; with Cumbria suffering the most cases. This spatial distribution, seeded by initial movements of infected animals, is a clear indication of the important role of localized transmission once a movement ban had been imposed. Further analysis refines this concept, showing that, after the initial spread through markets, 75% of all new cases were within 3 km of a previously reported infection, and 85% were within 5 km. Again all three models incorporate the consequences of spatial structure to some degree. Keeling *et al.* (2001b) chose to capture this effect through a localized transmission kernel, using the known location of each farm in the UK. The rate of transmission between infected farm i and susceptible farm j is modelled as:

$$\text{Rate}_{ij} = \tau_i \sigma_j K(d_{ij}) \tag{10.9}$$

where the kernel, $K(d_{ij})$, is a decreasing function of the distance between farms.

Stochasticity. Although not all models incorporated the stochastic transmission of infection, it is clear that during the initial and latter stages of the epidemic (when the number of infectious cases were low) that random effects could play a major role. Stochasticity is therefore vitally important in predicting the end point of the epidemic; this has strategic importance, as epidemic duration is a major factor in the financial costs of an outbreak. Keeling *et al.* (2001b) therefore utilized a probabilistic approach, using the rates given above to calculate the daily probability that a given susceptible farm would be infected:

$$Prob(\text{farm } j \text{ infected})$$
$$= 1 - \exp\left(-\sum_{\text{Infectious farms, } i} \text{Rate}_{ij}\right) \quad (10.10)$$

This term has many echoes of the simple βIS term in the SIR model (eqn 10.1).

The above description of the disease dynamics and modelling approach has clear resonance with metapopulation concepts in ecology; here farms play the role of patches or islands of various sizes and quality, with colonization between islands being a function of their separation. The only discrepancy with standard metapopulation models is the rapid culling of farms once the infection is detected.

It is important to consider why models were so successfully used during the 2001 epidemic. This stems from three basic elements. The first is the large amount of data available on the epidemic and culls, but also on the location and heterogeneity of susceptible farms. This provided a detailed set of initial conditions and the ability to accurately parameterize the necessary epidemiological processes. Secondly, the nature of transmission is relatively simple, even though the precise mechanism of some infection events is still uncertain. This was especially so after the movement ban was imposed. For human (or wildlife) diseases the movement and mixing of humans (or animals) greatly complicates the potential transmission routes—consider trying to model all human social contacts that could lead to disease spread (Ferguson *et al.*, 2005). In contrast, the sessile nature of farms simplifies the spatial dynamics, and has more in common with models of the spread of plant pathogens. Finally, the models were driven by well-posed and non-intuitive

questions. Most importantly, although it is clear that the pre-emptive culling of livestock near infected farms can reduce the number of cases, is this reduction sufficient to lead to a total saving of livestock in the long term? In more precise terms, is the sum of culled and infected farms (or livestock) less than the number of infected farms (or livestock) in an uncontrolled epidemic? This is a complex nonlinear question that requires the use of accurate, well-parameterized models.

Despite the inherent differences between the three models that were used during the 2001 epidemic, remarkably similar conclusions were drawn, as follows.

1 The culling of animals on infected premises and on dangerous contacts (those farms thought to be at high risk of infection) should continue.
2 The culling on infected premises and dangerous contacts should take place as rapidly as possible and hence reduce the duration of the infectious period.
3 Further localized culling would reduce the total number of farms lost by simultaneously removing infected farms and the susceptible farms that are at the greatest risk due to their proximity to infectious sources.

The uniformity of these conclusions lies in the use of models that all agree with the observed epidemic pattern, and which agree on the initial state of the population and the effectiveness of the controls. Figure 10.6 shows the predicted epidemic profiles under a range of culling strategies; while prompt implementation is always advantageous, the complex trade-off between control of the epidemic and loss of livestock due to culling is clear.

Future epidemics of foot and mouth disease in the UK would probably also feature vaccination as a major control measure (Keeling, 2005; Keeling *et al.*, 2006). The same messages about promptness apply as for culling, although the main dynamic complexity of 'vaccinate to live' policies involves how to optimize spatial vaccination strategies rather than trade-offs between culling and slaughter (Keeling *et al.*, 2003). Analyses of the potential of vaccination for foot and mouth disease reveal particularly clearly the need for both spatially detailed stochastic epidemic models and

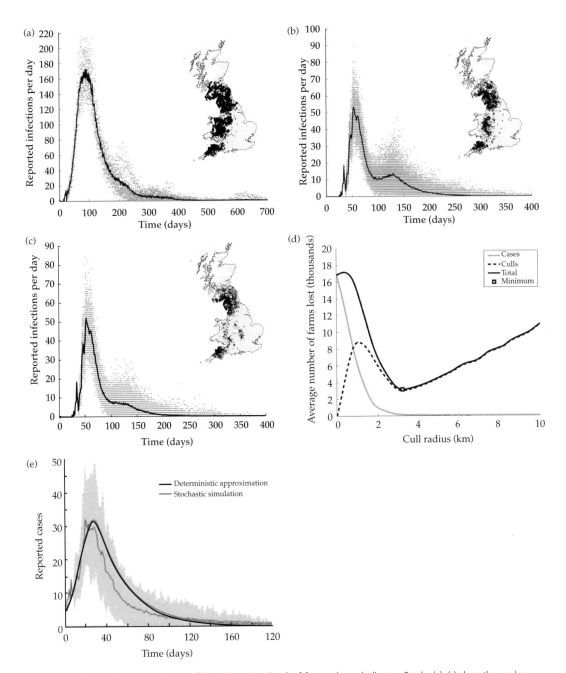

Figure 10.6 Model results from simulations of the 2001 UK outbreak of foot and mouth disease. Graphs (a)–(c) show the number of daily reported cases from 100 simulated epidemics (grey dots) together with the mean for different control options; the inset in each case shows the spatial distribution from one replicate using the same scheme as Figure 10.5b. All control options follow the 2001 temporal pattern, ramping up from low initial values and slow response times. Graph (a) is when only farms reporting infection have their livestock culled (IP culling), graph (b) includes the additional culling of dangerous contacts (akin to contact tracing), and graph (c) also includes culling of contiguous premises (akin to local spatial culling). Finally, graph (d) exemplifies the difficult trade-off between sufficient and overwhelming control: whereas culls within a large radius dramatically reduce the epidemic size, beyond some radius (about 3.2 km) the cull starts to be detrimental (data kindly provided by Dr Mike Tildesley). Graph (e) compares simulations of the effects of mass reactive vaccination (300 000 cattle per day), generated by repeated iterations the spatial stochastic simulation model described in the text (grey area and mean trajectory of persisting epidemics), compared with a simple core-group epidemiological model (black line); see Keeling et al. (2003; box 1) for full details.

more simple formulations to help interpret them. Figure 10.6e (Keeling *et al.*, 2003) compares two models from opposite ends of the spectrum between detail and simplicity. The grey bars show the range of results from 100 stochastic simulations of the full spatial model defined by eqns 10.8–10.10 with culling of infected premises and mass vaccination. The black line is the output of a simple deterministic model that has a core group of farms that are both more susceptible and more infectious. The simple and the detailed models agree on many features of the epidemic—in particular its initial rise and overall shape. However, only the stochastic simulation model is able to capture the long, low tail of the epidemic (which, in reality, dragged on to September 2001). Although small, this tail was of enormous practical importance to the UK because it determined the date from which calculations about the resumption of trade were made. Thus simple models can be invaluable for understanding how heterogeneity impacts the time course and the control of this epidemic. But for some features only a detailed model is adequate to capture observed patterns. The key to good modelling is knowing when simple models will suffice; this relies in turn on an understanding of host–pathogen natural history and a sound statistical comparison of model results with epidemiological data.

10.6 Conclusions

The study of infectious-disease dynamics has developed enormously since the 1976 and 1981 editions of *Theoretical Ecology*. A comprehensive coverage of all these developments is impossible in this short chapter. Instead, we have used a review of simple models, then more complex derivatives of them, to illustrate the place of parasitism in ecological dynamics. We argue that, in addition to their applied importance and impact on host dynamics, host–parasite relationships throw a special light on host–natural enemy interactions in general. We have focused on the synthesis of models and data for foot and mouth disease, childhood epidemics, and macroparasites in red grouse; however, a number of other examples could have illuminated general questions in

ecological dynamics. Here we summarize the chapter's conclusions and point out key areas for future work.

10.6.1 Model complexity and data

Epidemiological models of specific systems are most successful when calibrated by high-quality data and a good quantitative understanding of the natural history of infection. This was the case for both foot and mouth disease and childhood infections such as measles in England and Wales. Specifically, these systems provided both an estimate of patterns of host susceptibility at epidemiologically meaningful spatio-temporal scales (farms for foot and mouth disease; cities for measles) and disease-incidence data. Together, these elements allow a key calibration: how the (imperfectly known) host-contact network determines the pattern of transmission. In contrast, recent models of smallpox and pandemic influenza (Halloran *et al.*, 2002; Ferguson *et al.*, 2005)—while often well constructed using cutting-edge modelling techniques—are necessarily hampered by a lack of epidemiological data and therefore an incomplete quantitative understanding of the precise transmission network. The public health and veterinary importance of infectious disease is likely to lead to an explosion in epidemiological and 'denominator' data collection—especially of pathogen sequence data (Kuiken, *et al.*, 2005). This should help modelling efforts; indeed models, and associated statistical developments (Keeling *et al.*, 2004), will be key for designing such sampling schemes and interpreting their results. Using models to extrapolate the fate of true emerging infections is, however, always going to be a difficult task, hedged with caveats. The use of a family of models—simple as well as complex—is key, for predicting the potential for epidemics as well as for helping to manage outbreaks once they have begun.

10.6.2 Dangerous liaisons

The intimate association of parasites and their host 'habitat' gives these interactions a special place in the ecology of host–natural enemy systems. As discussed above, host–parasite interactions often involve and invoke heterogeneities with special dynamic consequences. Furthermore, parasites are sometimes easy to remove at the individual or local scale by treating or vaccinating their hosts. The resulting ability to manipulate the system (as in the red grouse experiments) or observe manipulations (such as the start of vaccination in childhood diseases) can enhance population-dynamic understanding. In particular, such experiments provide insights into a wider range of phase and parameter space than would normally be observed, giving clues to the dynamics away from the narrowly confined standard attractor. Such experiments (especially the implementation of control strategies) are arguably more readily performed and controlled in infectious diseases than in other ecological populations. The simple obligate interaction between some parasites and their hosts has also allowed researchers to tease apart the complex relationships between seasonality, stochasticity, and this natural-enemy interaction. Finally, the often strong effects of microparasites on host demographic rates also give them a special place in community dynamics (see also Chapter 5

in this volume). For instance, during wildlife, plant, or even human epidemics an RNA virus can cause community effects completely out of proportion to its miniscule biomass.

The importance of infectious diseases is likely to mean an ever-increasing focus on their dynamics over the next decade. As well as issues touched on above, there are a number of important areas for future development in disease dynamics. For instance, how do we characterize and model transmission network structure (Keeling and Eames, 2005; May, 2006)? How do we move beyond interactions of single hosts and parasites to model, for example, the interaction of multiple pathogens in the same host underlined by recent studies (Graham *et al.*, 2005)? However, arguably the biggest questions and opportunities arise in synthesizing population kinetics and evolutionary dynamics of host–parasite associations across scales (Nowak and May, 2000; Frank, 2002; Grenfell *et al.*, 2004). As well as their applied importance, the evolutionary dynamics of fast-evolving pathogens, such as influenza, provide a wonderful model for evolutionary ecology.

CHAPTER 11

Fisheries

John R. Beddington and Geoffrey P. Kirkwood

11.1 Introduction

The depletion of fish stocks on a global scale is well documented. The United Nations Food and Agriculture Organisation collects statistics on fisheries from all states and, despite obvious shortcomings in the data, a clear picture has been available for some time. Garcia and Grainger (2005) have succinctly documented the position from the latest available date: in 2003, only 3% of stocks were underexploited and 26% moderately exploited, while 52% were fully exploited, 16% were overfished, 7% were depleted, and 1% were recovering from earlier depletion. These global statistics mask two important phenomena. The first, highlighted by Pauly *et al.* (1998), is that fisheries are increasingly focusing on species lower down in the food-web and the second, highlighted by Myers and Worm (2003, 2005), is that large predatory fish have been particularly reduced in abundance.

Both of these analyses are somewhat flawed. In the case of Pauly *et al.* there are two problems: the first is that the metrics used for the mean trophic level are presented as simple numbers with no estimates of error or indeed sensitivity. In such a situation, the changes in mean trophic levels are hard to interpret, particularly where the mean trophic level changes by at most around 10% over four decades. The second problem has been highlighted by a recent paper by Essington *et al.* (2006). They point out that in the periods when according to the analysis of Pauly *et al.* the mean trophic level was declining, in most cases catches of apex predators and indeed all upper trophic levels increased (an exception is the North Atlantic).

In the case of the Myers and Worm analysis, they used the catch per unit of effort (CPUE) as an

index of abundance. As discussed later in this chapter, there are problems with this, but more importantly for some key apex predators, in particular large tunas, the CPUE declines in the early stages of the fishery, where catches are small, but remains relatively stable under a regime of much higher catches. In such a situation, the interpretation that the CPUE reflects changes in abundance is clearly problematic.

Despite these concerns on details, the general conclusions of both sets of authors are probably correct; indeed, neither phenomenon is particularly surprising, as the simplest economic and ecosystem modelling would predict the initial targeting of large, valuable, slow-growing species by fishers being succeeded by switching to more abundant, smaller, faster-reproducing species. Nevertheless, it is reasonable to ask whether fisheries science provides a sufficient level of understanding of the processes involved to guide management in a more efficient stewardship of marine resources in the future.

11.2 Basic assessment of single-species fisheries

In its current paradigm, fisheries science is still dominated by the single-species formalism developed by Beverton and Holt (1957). These models of fish population dynamics are either age- or size-structured, with multiple cohorts. A typical age-structured model is as shown below. It represents a special case of the general formulation sketched in Chapter 3 in this volume.

Suppose fish first recruit to the potentially fishable population at age r. If $N_{a,y}$ is the number of fish of age a in the population at the start of year y,

M_a is the instantaneous rate of natural mortality of fish of age a, and $F_{a,y}$ is the instantaneous rate of fishing mortality experienced by fish of age a during year y, then

$$N_{r,y} = f(B_{y-r-1}) \tag{11.1}$$

$$N_{a+1,y+1} = N_{a,y}\, e^{-(M_a+F_{a,y})}$$
$$\text{for } a = r, 1, \ldots, p \tag{11.2}$$

$$N_{p,y} = N_{p-1,y-1}e^{-(M_{p-1}+F_{p-1,\,y-1})} + N_{p,y}\, e^{-(M_p+F_{p,y})} \tag{11.3}$$

Here $f(.)$ is the *stock-recruit relationship* relating the numbers of young fish entering the fishery for the first time each year to B_y, the *spawning stock biomass* $r+1$ years ago, and p is the age of the so-called plus group of fish of age p or greater.

Growth in length is typically described by a von Bertalanffy curve (von Bertalanffy, 1938) and converted to mass via a length/weight relationship. If L_a is the length of a fish of age a, and W_a is its mass, then

$$L_a = L_\infty(1 - e^{-K(a-t_0)}), \quad W_a = \gamma L_a^\delta \quad \text{and} \tag{11.4}$$

$$B_y = \sum_{a=r}^{p} N_{a,y}\, W_a O_a \tag{11.5}$$

here O_a is the proportion of fish mature at age a.

Finally, annual catches at age in numbers, $C_{a,y}$, are given by the Baranov (1918) catch equation:

$$C_{a,y} = \frac{F_{a,y}}{M_a + F_{a,y}}(1 - e^{-(M_a+F_{a,y})})N_{a,y} \tag{11.6}$$

and the annual biomass yield by

$$Y = \sum_{a=r}^{p} C_{a,y} W_a \tag{11.7}$$

From the above set of equations, it is not clear that there are any sources of density dependence in the model dynamics. In most applications in marine fisheries, density dependence is assumed to occur only in the stock-recruit relationship. Two forms of stock-recruit relationship are commonly used: those due to Beverton and Holt (1957) and those due to Ricker (1954). The former, relating recruitment R to spawning stock biomass S, takes the form

$$R = \frac{\alpha S}{1 + \beta S} \tag{11.8}$$

This was derived under the assumption that the mortality experienced between eggs being spawned and subsequent recruitment is made up of both density-independent and density-dependent effects, with the pre-recruits inhibiting themselves by competition for food or space (Quinn and Deriso, 1999). The alternative Ricker formulation is

$$R = \alpha S e^{-\beta S} \tag{11.9}$$

where now the pre-recruit fish are inhibited by the spawning stock, such as through cannibalism (Quinn and Deriso, 1999). Shepherd (1982) developed a form which incorporates both these models as special cases.

The models above make no allowance for density dependence in other key biological parameters of the fishable population. In reality, given the intimate connection between food availability and growth and survival, there are compelling grounds to expect density dependence in growth, onset of maturity and survival. For some of the better-studied marine fish stocks there is also good empirical evidence of substantial trends over time in growth rates, and related changes in age or length at first maturity (e.g. Beverton and Holt, 1957; Cook and Armstrong, 1984). Further, since the primary cause of natural mortality in the younger age classes of a fish stock is likely to be predation, the rates of natural mortality would be expected to vary with the abundance of predators and also with age/size, since predation is usually strongly size-specific.

In addition to density dependence, it is likely that the fishing process itself may lead to changes in vital rates. Many fishing gears are strongly size-selective. Substantial periods of heavy size-selective fishing pressure would therefore be expected to exert selective pressure on growth rates and also on the timing of the onset of maturity (Roff, 2002). Evidence of such effects has been demonstrated for a number of stocks (e.g. Bering Sea pollock

(Laevastu, 1992), Newfoundland cod (Hutchings, 1999), North Sea plaice (Rijnsdorp, 1993)) and an explanatory model has been proposed by Law and Rowell (1993).

In practice, however, it is normally difficult enough to get single reliable estimates of the principal biological parameters for marine fish stocks, let alone characterize density-dependent or ecosystem-related changes in these over time. In particular, the natural mortality rate has proved highly resistant to estimation from data and it is commonly assumed to be both age- and density-independent. In consequence, most models of the dynamics of marine fish stocks take the form as described above. Fortunately, a possible reality check on imperfectly estimated biological parameters is available from life-history theory. Using this, Charnov (1993) and later Jensen (1996) have proposed the natural mortality rate (M), the age (T_m) or length (L_m) at first maturity, and the von Bertalanffy growth parameters L_∞ and K:

$$MT_m = 1.65, \quad \frac{M}{K} = 1.5, \quad \frac{L_m}{L_\infty} = \frac{2}{3} \qquad (11.10)$$

Recently it has been shown that these 'invariant' properties derive in part from a direct lack of independence among plotted variables—a familiar statistical sin (Nee et al., 2005). However, in this case, the estimation of natural mortality and the parameters of growth curves are derived independently using rather different data. Furthermore, Jensen's derivation of the relationship between M and K relies simply on the existence of an inflexion point in the growth curve and is supported by the available data. Whatever the ultimate development of life-history theory, currently use of these relationships within fisheries appears to remain both valid and useful.

For freshwater fish stocks, especially those subject to culture-based fisheries or aquaculture, the opportunity exists to observe growth and mortality over a much wider range of densities than is possible for most marine fish. Models incorporating density-dependent growth and size-dependent mortality have been developed by Lorenzen (1995, 2005) and Lorenzen and Enberg (2002).

Returning briefly to stock and recruitment, in the early part of the twentieth century it was believed that marine fish stocks were effectively inexhaustible. Even after Beverton and Holt (1957) and after the development of the theory of stock and recruitment, the enormous fecundity of many fish stocks lead to the view that fish stocks were highly resistant to overfishing. This was further encouraged by the appearance of plots of estimates of spawning stock biomass and resultant recruitment that suggested maintenance of average recruitment levels despite large reductions in spawning stock sizes. Subsequent undeniable empirical evidence and analyses of comprehensive databases of stock and recruitment data, such as that collated by Myers et al. (1995), have now led fisheries scientists to realize that fish stocks are all too vulnerable to overfishing. Indeed, a meta-analysis of these data by Myers et al. (1999) demonstrated that the maximum reproductive rate of a fish stock, measured by the number of subsequent spawners produced per spawner each year, generally ranged between 1 and 7 across fish populations, with an average of 3. This suggests that the capacity for population growth and vulnerability to over-exploitation of fish stocks is similar to those of many mammals (Beddington and Basson, 1994). In contrast to mammals, however, the highly stochastic nature of the recruitment process (Rothschild, 1986) imposes additional risks for populations that have been reduced to low levels.

The age-structured models above have typically been applied to temperate fish species that can be aged, usually by counting annuli on hard parts such as otoliths ('earplugs'). For many tropical species, however, this is not possible. In the absence of reliable estimates of age structure, it is often necessary to revert to simpler biomass dynamic models (Hilborn and Walters, 1992) that take a form similar to the familiar logistic model. A discrete-time version proposed by Pella and Tomlinson (1969) takes the form

$$B_{t+1} = rB_t\left(1 - \left[\frac{B_t}{K}\right]^z\right) - C_t \qquad (11.11)$$

where B_t is the vulnerable population biomass at the start of year t, C_t is the catch biomass during year t, r is the intrinsic growth rate, K is the population carrying capacity, and z is a phenomenological shape parameter or fudge factor allowing for maximum sustainable yield (MSY) for the population to occur elsewhere than at half carrying capacity. A simpler earlier model first proposed by Schaefer (1954) fixes z at 1. The earlier Ricker equation (eqn 11.9), originally proposed in studies of salmon recruitment, is another example. At first glance, it may seem that such a simple caricature of the true dynamics would not be suitable for use in quantitative assessments of fish stocks. However, simulation studies (e.g. Ludwig and Walters, 1985) have demonstrated that such models can be surprisingly robust, particularly in circumstances where reliable estimates of the many parameters of the age-structured models are difficult to obtain. The simpler biomass dynamic models tend to break down in cases where there is large inter-annual variability in recruitment, and, less obviously, when the vulnerable population varies over time through changes in targeting by different fishing fleets. As discussed later, these models have proved particularly useful in wider bio-economic studies of fisheries-management strategies.

11.3 Managing single-species fisheries

The principal mechanisms available to fishery managers to control fishing are limitations on catch and/or fishing effort, closed areas and seasons, and various controls on the catch of juveniles. For the purpose of this explanation, we concentrate here on the dynamics and management of a fishery targeting a single species. In this context, the critical issue is control of the amount of catch taken from the stock each year. From the point of view of the stock, it is not really relevant whether this is achieved by setting a total allowable catch or achieving the same end by restricting fishing effort. Thus for the time being we will ignore this distinction, but return to it below when considering the economics of fishing and issues of compliance. Traditionally, closed areas and seasons, as well as controls on catches of juveniles, have been measures designed to ensure that fishing is restricted to appropriate segments of the population, or to protect spawning grounds and nursery areas. These are also important topics, but they will not be covered here; for a full account, see

Quinn and Deriso (1999). More recently, however, closed areas (now renamed marine protected areas) have emerged as a potentially valuable tool for fishery management, particularly in an ecosystem context, and we will return to them below.

In order to set appropriate limitations on catches from exploited fish stocks, fishery managers rely on advice from fishery scientists. In its simplest form, this scientific advice is based on a stock assessment. Given a model of the dynamics of the fish stock and estimates of its key biological parameters, estimates of historical and current abundances are calculated by fitting the selected model to available data on historical catches and time series of estimates of relative or absolute abundance. Forward projection from estimated current abundance under different levels of catch or fishing mortality can then allow determination of appropriate catch limits that will meet management objectives. Put this way, the process seems straightforward, but in practice it is anything but that.

Historical catch data are obviously fundamental. Yet the collection of even such basic data can be fraught with difficulties. Especially with the recent increased pressure to reduce catches, for fisheries managed by total allowable catches there are strong incentives for misreporting of catches and the extent of discarding of undersized fish is rarely well documented. There are two principal sources of relative abundance data: fishery-dependent data such as CPUE and fishery-independent data from research surveys. Nominal CPUE data are simple to collect, but identification of measures of effort such that CPUE is really an index of abundance is extremely difficult (Clark, 1985, Cooke and Beddington, 1985). Data from research surveys are more likely to provide unbiased indices, but their level of precision is often rather low. Naturally, if age-structured dynamics models are to be fitted, then these data need to be broken down by age class.

Estimation methods have become increasingly sophisticated over time, mirroring the increase in available computing power. The simplest assessments rely on fitting biomass dynamic models to total catch and CPUE data using nonlinear statistical methods. The most widely used approach to fitting age-structured models is via a number of variants of virtual population analysis (Quinn and Deriso, 1999), which were developments of cohort analysis (Pope, 1972). These are currently being slowly replaced by various integrated statistical catch-at-age analysis methods that also incorporate auxiliary information, such as catch-length frequencies, recruitment indices, and results from tag-recapture experiments (e.g. stock synthesis (Methot, 1990), COLERAINE (Hilborn et al., 2003), and CASAL (Bull et al. 2005)). Some also take the spatial structure of stocks into account, incorporating data on fish movements and oceanographic processes (Bertignac et al., 1998).

A notable feature of recent fishery stock assessment methods is the use of Bayesian estimation methods (McAllister and Ianelli, 1997; McAllister and Kirkwood, 1998) These methods are also incorporated into integrated assessment methods such as COLERAINE and CASAL. Bayesian methods provide a much more natural setting for the incorporation of prior but uncertain information on key population parameters, such as that arising from meta-analyses of data from the same or related species elsewhere (Myers et al., 2001). They also allow a more rigorous treatment of risk and decision analyses of alternative management strategies.

The more complex assessment methods such as those described above not only have formidable data requirements, but also require a level of analytical sophistication not available to many fishery-management agencies in developing countries. To address this problem, Beddington and Kirkwood (2005) have developed techniques based on the life-history relationships described above (eqn 11.10). These methods use estimates of the growth parameters, the length at first capture, and the 'steepness' (Mace and Doonan, 1988) of the stock-recruitment relationship to estimate the MSY and the fishing mortality rate producing the maximum yield. This allows sustainable yields and fishing capacity to be estimated from relatively sparse data, such as those available for fisheries in developing countries. It also can provide a quick reality check on the results of more complex analyses.

Identification of an appropriate catch limit for a fishery requires not only estimation of the current stock status in relation to its unexploited state, but also specification of an 'optimum' state. Armed with both of these, stock projections under different catch levels will allow estimation of a catch or sequence of catches that will move the stock towards that optimum level. For many years there was no doubt that this optimum catch level should be the so-called MSY. Indeed, maintaining fish stocks at MSY is still enshrined in the United Nations Law of the Sea Convention and the United Nations Straddling Stocks Agreement and the recent summit on Sustainable Development in 2002 called for action to 'maintain or restore stocks to levels that can preserve the maximum sustainable yield'.

Many critical articles have been written about MSY. Its origins lie in analyses of the deterministic logistic or Schaefer (1954) population-dynamics model. That model apparently suggests that a maximum yield of $rK/4$ can be taken ad infinitum from a population at half its carrying capacity K. The first obvious criticism is that this is not even a stable equilibrium of the deterministic model, and in a truly stochastic context a constant catch equal to the deterministic MSY is definitely not sustainable (Beddington and May, 1977). These criticisms seem to us to be a little pedantic; for any given stochastic model of population dynamics there is no conceptual difficulty in identifying some maximum catch level that is sustainable over a specified period, even allowing for the uncertainties in the assessment. Rather, the problem with MSY is that it cannot be the sole management objective. Almost all sets of fishery management objectives at least impose a requirement to prevent the stock from falling to dangerously low levels, in addition to seeking to maximize the catch. Simultaneous consideration of these two objectives has given rise to the concept of biological reference points (Mace, 1994) and discussion of them in the wider context of sustainability (Quinn and Collie, 2005).

The first change was to move from catch targets to resource targets and to introduce several types of reference point: limit, threshold, and target. The second is that all of these are considered in an appropriately stochastic setting. MSY, or rather the fishing mortality rate corresponding to MSY (F_{MSY}), is now considered as a limit reference

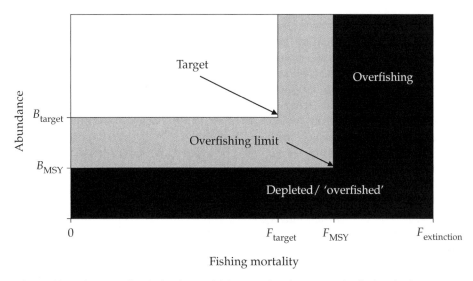

Figure 11.1 The possible combinations of stock abundance and fishing mortality after Quinn and Collie (2005). White areas are sustainable, grey areas problematic. and black areas are unsustainable.

point, which the fishing mortality rate should not be allowed to exceed. In order to minimize the probability of stock collapse, a limiting lower value of spawning stock biomass (B_{lim}) and a corresponding upper value of fishing mortality rate (F_{lim}), have been identified. An absolute limit for fishing mortality rate is $F_{extinction}$, the fishing mortality rate that will drive a population to extinction (Mace, 1994). In stock assessments carried out under the auspices of the International Council for the Exploration of the Sea for European fish stocks, further, more conservative 'precautionary approach' levels of biomass have also been identified, in recognition of the fact that both current and limit levels of fishing mortality rates and biomasses are imperfectly estimated. Target levels are typically defined in terms of seeking to preserve appropriate levels of spawning stock per recruit (typically 35–45%; Quinn and Deriso, 1999; see Figure 11.1). This is still a developing area of research, as ideally one would wish to incorporate estimation of reference points corresponding to specified levels of risk into Bayesian stock assessments and risk analyses.

If the most modern assessment methods are used and proper attention is given to more precautionary targets and limits for levels of fishing, will this be sufficient to ensure sustainable management of fish stocks? If we were starting from scratch, and the degrees of precaution actually applied were sufficiently great, the answer is probably yes. After all, it does not require a welter of sophisticated population biology, mathematics, and statistics to realize that if catch levels were kept at sufficiently low levels then they can be sustainable and the risk of over-exploitation can be kept to very low levels. But, as indicated above, we are not starting from scratch. Many of the world's most important commercial fish stocks are currently over-exploited and a number severely so. What is needed for these stocks is a series of recovery plans, inevitably requiring further substantial reductions in catches and consequent additional pain for an already over-stretched fishing industry. The inevitable political consequence of this is pressure to set catch limits as high as can be considered compatible with management objectives. In such circumstances, even the best assessments may still be not be enough. The reason for this is that stock assessment and the provision of scientific advice is only a part of a much wider fishery-management system involving the collection of imprecise data, analysis of those data for a stock assessment, the translation of the resulting scientific advice into management regulations, and the often imperfect compliance with those regulations. Furthermore, this system sometimes involves substantial time delays. Last, but not least, the models for the underlying population dynamics may be based on incorrect assumptions.

To address these wider issues, a simulation technique known as management-strategy evaluation has been developed. Originally devised by scientists advising the International Whaling Commission to develop a revised management procedure for baleen whales (Kirkwood, 1997), the process can be illustrated by Figure 11.2. Below the water (represented by the thick line) the fish go through their annual cycle (including being fished) according to their 'true' population dynamics. Naturally we do not know this, but the opportunity exists to incorporate 'operating models' for the fish stock that are far more complex than usually assumed in stock-assessment models (e.g. incorporating spatial structure, inter-species interactions, and alternative hypotheses about the true dynamics). From this population, each year catches are taken and sampled, as are biological data, and these and similar historical data are input to the stock assessment, which usually will assume a much simpler dynamical model. A recommended catch is determined from the assessment according to a specified catch-control rule and then translated into regulations and implemented by the fishermen. This sets the catch for the next year and the annual cycle starts again. Typically, this process is simulated for a number of years into the future, and each of these simulations repeated a large number of times. Output consists of the annual outcomes for the true fish stock and these outcomes are compared with those desired by management.

Using this approach, the performance of a variety of alternative management strategies (control rules) under different operating models can be compared. A number of such studies have now been carried out (e.g. Butterworth *et al.*, 1997; Punt *et al.*, 2002; and in an ecosystem management context by Sainsbury *et al.*, 2000). One recent study evaluated the likely success of recovery plans for a variety of North Sea roundfish stocks (Kell *et al.*, 2005). North Sea cod simulations provided an extreme example. Under the then-proposed recovery plan these simulations revealed that in many cases the stock eventually collapsed after an initial brief recovery. This prediction was not substantially changed even under the assumption of a perfect assessment. In this case, the principal cause was the (unavoidable) long time delay between initial collection of data, its incorporation into assessments, translation into regulations, and subsequent detection of the actual impact on the population. The total time lag can be 5 years.

The understanding generated by the various developments of the Beverton and Holt formulism is substantial. In particular, the explicit consideration of uncertainty and the unavoidable time delays in the assessment process allow for the

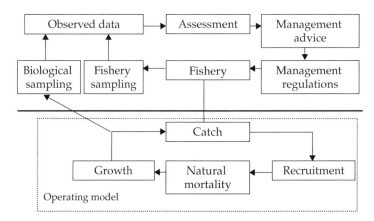

Figure 11.2 Annual flowchart for management-strategy evaluation. The thick black line represents the surface of the sea. Processes occurring underwater represent the true dynamics of the fish stock and are unknown to the stock assessors and managers above water. See text for details.

possibility of robust management strategies which incorporate these constraints. Nevertheless, the successful use of such procedures depends on the ecosystem and the economic system within which the fishery operates. These are addressed in the remaining sections of this chapter.

11.4 Multispecies fisheries and ecosystem considerations

Clearly the single-species paradigm of stock assessment addressed above is vulnerable to assumptions about the ecosystem and the environment. In principle, the effects of ecosystem changes and climatic change can be incorporated into the single-species formalism by allowing the demographic parameters of the basic model to vary with changes in the climate and the ecosystem. Such an approach can be implemented directly for environmental change if the necessary data on demographic and environmental variables are available. Manifestly this is rarely the case, but there have been successful attempts to use this approach (e.g. Agnew et al., 2004).

However, the problem of incorporating ecosystem changes is both more difficult and more subtle. Clearly individual targeted species inhabit an ecosystem with other targeted and untargeted species. The interaction of fishing on this complex will directly affect the abundance of the targeted species and indirectly affect both targeted and untargeted species by the mechanisms of competition and predation.

There have been a number of approaches to addressing this issue. The earliest and simplest have used models of a Lotka–Voltera type, similar to the single-species biomass models discussed above. These models have given useful insights into the constraints of joint harvesting of predator and prey (Beddington and May, 1980) or the effect on competition of targeted harvesting (Clark, 1985). Such models have been used in developing a framework for multispecies management in the Antarctic under the Convention for the Conservation of Antarctic Marine Living Resources (May et al., 1979). However, they do not allow the level of prediction to be made that provides recommendations on catch limits of individual species.

Another approach is that of Kerr and Dickie (2001), whose size spectrum models ignore species and treat individuals on the basis of size and trophic levels. These models provide interesting insights into the properties of the system, but again do not permit quantitative prediction of a type that is relevant to fisheries management. Similar constraints occur in the studies of very large food-web models explored by Yodzis (1998), who looked at the implications of different ecosystem structures in predicting the effects of fishing.

A much less general approach has been the development, primarily within the framework of the International Council for the Exploration of the Sea (ICES), of multispecies virtual population analysis (MSVPA; Pope, 1991; Hollowed et al., 2000; Vinther, 2001). These models are aimed at teasing out the magnitude of predation mortality by using data on diet linked in with more conventional techniques of virtual population analysis. These approaches have been relatively successful in estimating predator-induced mortality at different ages, which have then been incorporated into single-species assessments carried out by ICES.

In a similar way, Collie and Gislason (2001) have developed biological reference points for species conditional on predator abundance. These approaches have the merit of leading directly to a prediction on the effects of different catch limits, yet they do use some of the more subtle ecosystem effects such as those reviewed by Mangel and Levin (2005). The main limitation of the MSVPA approaches and similar techniques, such as those of Stefansson and Palsson (1998), are that they cannot predict ecosystem shifts; for example, the replacement of an apex predator or the inhibition of recovery of a fish stock from depletion by the expansion of competitive species.

An attempt to address these limitations is the multispecies modelling approach of ECOPATH and ECOSIM (Walters et al., 1997; Christensen and Walters, 2000; Pauly et al., 2000). This approach essentially models the interactions of prey and predators within an ecosystem by a set of differential equations similar to the Lotka–Volterra forms, but with a modification that only a proportion of the prey are vulnerable to predation. These take the form (Walters and Kitchell, 2001):

$$\frac{d(N_i - V_{ij})}{dt} = -v_{ij}(N_i - V_{ij}) + v'_{ij}V_{ij} \qquad (11.12)$$

$$\frac{dV_{ij}}{dt} = +v_{ij}(N_i - V_{ij}) + v'_{ij}V_{ij} - a_{ij}V_{ij}N_j \qquad (11.13)$$

Here, for prey i and predator j, V_{ij} is the component of the prey abundance vulnerable to predator j, and $(N_i - V_{ij})$ is the corresponding non-vulnerable prey abundance, a_{ij} is the per-capita consumption rate of prey i by predator j, and N_j represents the number of predator group j. v and v' are prey vulnerability parameters.

There are many versions and ongoing developments of these models; Plaganyi and Butterworth (2004) provide a detailed critical review. The advantages of these modelling approaches are that they provide a framework for exploring particular ecosystems at a level of complexity that permits the discovery of established phenomena such as depensation as well as detailed assessment of unsuspected changes produced by fishing. The main disadvantage lies in the high level of parameter uncertainty which renders the approach problematic in providing detailed predictions. The potential for a significant improvement over single-species approaches almost certainly lies in a complementary approach, a conclusion reached by Plaganyi and Butterworth (2004).

One of the important implications of considering an ecosystem approach to fishery management is that the role of marine reserves has been revisited in some detail. Protected areas have been a commonplace of fisheries management for many years, but recently there has been a development to provide large marine protected areas as a precautionary tool for providing protection to marine ecosystems. Hilborn *et al.* (2004) consider the circumstances in which marine reserves can help fishery management and conclude that their use should be evaluated on a case-by-case basis. A somewhat more optimistic view of their applicability is given by Roberts *et al.* (2005). There are calls from a number of bodies for a protection of some 20–30% of ocean area (Roberts and Hawkins, 2003). Whether this is sensible, as Hilborn *et al.*

(2004) question, or feasible from a practical viewpoint is problematic given the difficulties of policing the more limited management measures discussed below.

Perhaps the most comprehensive large-scale study of fishing effects on a marine ecosystem has been carried out by Sainsbury (1988), who conducted a long-term experiment on Australia's North West Shelf ecosystem. Using an adaptive management approach and Bayesian techniques, the most appropriate of four candidate ecosystem models for explaining substantial changes in species composition and community structure was found to be one that incorporated impacts of fishing on the seabed habitats. Following this study, restrictive zoning of seabed trawling in the North West Shelf was implemented and a valuable trap fishery established in its place.

11.5 Economic considerations

The application of stock assessment, whether in a single- or multispecies context, is geared to providing advice on the catch levels that are sustainable. As discussed above, it is possible to explore the whole management structure and so arrive at assessments which allow for the uncertainties and time delays in the process. In this section we explore some of the practical details of this process and how economic considerations can undermine the management activities.

Fisheries management operates by the process of regulation and enforcement. Regulation of fisheries can include specifying catch levels, minimum size limits, closed areas to protect breeding stocks or other ecosystem considerations, restrictions on access to the fishery, and levels of allowable effort.

Enforcement, whether in port or by at-sea inspection, or by means of sophisticated satellite-based vessel-monitoring systems, is aimed at ensuring that regulations are obeyed. The problem lies in the economic motivation of the fishermen. Whereas individuals will have differing attitudes, regulations will tend to be broken when the profit to be made from breaking them is significantly greater than the product of the probability of being caught times the penalty.

For many fisheries, the probability of capture is low, sometimes extremely low, and courts often tend to limit fines to some small multiple of the benefit. This results in widespread abuse of catch limits and a substantial trade in illegal landings. Similar considerations explain the widespread operation of illegal, unregulated, and unreported fishing.

A parallel problem to the issue of regulation is overcapitalization, which is also driven by the economic motivation of fishermen. The analysis of open-access resources is well known. Clark (1985) describes how the pursuit of profit attracts entrants to the fishery and this leads inexorably to the dissipation of profit, so that fishermen operate at a level where the resource is overexploited and vessels cover only their marginal costs. This is not an abstract idea; between 1970 and 1990 world fishing capacity grew eight times faster than fish landings and various estimates of overcapacity indicate a loss of around US$50 billion per annum (Garcia and Newton, 1997; Garcia and Grainger, 2005). What is less well known is that even in fisheries where entry is regulated, vessel owners will, if in the short-term additional profit is realizable, increase their capital spend on improved vessel efficiency so that overcapacity is still generated.

The existence of overcapacity means that the fishing community will be very resistant to the management process and will have few concerns for the long-term sustainability of the fishery compared to their immediate economic priorities. This pressure from the industry has all too often been reflected in a lack of political will to take difficult management decisions.

Attempts have been made to reduce overcapacity by buyback programmes. These have largely been both expensive and unsuccessful as vessel owners have adjusted their strategies of investment to benefit from the programmes while conserving or increasing their fishing power (Clark *et al.*, 2005).

The key problem is that the motivation of fishermen is for short-term competitive activity to increase their share of the catch or, in a regulated fishery, their share of the total allowable catch.

The accepted way of dealing with this problem is to transfer the property rights from the management authority (the state) to individual vessel owners. In a sense these individual fishing quotas (IFQs) give fishers a more direct 'ownership of the future' and have value only as long as the resource is exploited sustainably, which means that the owners of IFQs will have no incentive to overexploit nor significantly increase their investment beyond a level where their quota can be caught efficiently. There are real issues of compliance here as motivations to cheat remain, but evidence exists that a degree of self-policing by the industry tends to follow such allocation. A number of successful implementations of such schemes have now been achieved in Iceland, New Zealand, Australia, and Chile.

This is not the case in the rest of the world. At present, the vast majority of fisheries are either open access or have partially restricted access. Few are regulated and enforced efficiently. To achieve an improvement in the poor state of world fisheries requires not only better science, but better economic management.

11.6 Concluding remarks

In this chapter, we have attempted to review the current thinking in fisheries ecology and management. The single-species paradigm is well developed and forms the basis of the vast majority of practical assessments of the status of commercially exploited stocks. The need for a more ecosystem-based approach is well recognized, but its practical application is currently difficult given the serious problems of parameter proliferation and estimation. The future is likely to be increasingly dominated by considerations from community ecology, but these are most likely to be implemented in adjustments to the single-species paradigm. By contrast, a full appreciation of the economic motivation of the stakeholders in fisheries is going to be needed if the understanding of fisheries and their management is to be improved.

CHAPTER 12

A Doubly Green Revolution: ecology and food production

Gordon Conway

12.1 Introduction

Ecology has informed and underpinned agricultural production since the first faltering steps in domestication and cultivation. When someone (probably a woman) living in the Fertile Crescent carried seeds of wild wheats and barleys from the great natural cereal stands of the region and sowed them near her house she initiated the process of domestication. She also began the process of crop cultivation, creating what were to become ecologically complex, home gardens. Similarly swidden agriculture was based on imitations of ecological processes that would create a sustainable form of agriculture. The first articulation of this concept was not for many thousands of years later. The great Roman writer and agriculturalist of the first century BC, Marcus Terentius Varro, wrote as follows (Hooper and Ash, 1935):

Agri cultura est 'Non modo est ars, sed etiam necessaria ac magna; eaque est scientia, quae sint in quoque agro serenda ac facienda, quo terra maximos perpetuo reddat fructus'

Agriculture is 'not only an art but an important and noble art. It is, as well, a science, which teaches us what crops are to be planted in each kind of soil, and what operations are to be carried on, in order that the land may regularly produce the largest crops.' (Varro, *Rerum Rusticarum* I, III)
Not only does Varro place crops in their environment but the phrase *quo terra maximos perpetuo reddat fructus* (which can be translated as 'that the land yields the highest in perpetuity') struck me, when I first came upon it in one of the little red Loeb Classical Library translations, as an extraordinarily clear, elegant, and concise definition of

sustainability. In this chapter I want to illustrate how ecological concepts illuminate the building blocks of agriculture—gardens, swiddens, pastures, orchards, and fields—and provide a basis for the continuing challenge of feeding everyone in an increasing population.

12.2 Agroecosystems

The transformation of an ecosystem into an agroecosystem involves a number of significant changes. The system itself becomes more clearly defined, at least in terms of its biological and physico-chemical boundaries. These become sharper and less permeable, the linkages with other systems being limited and channeled. The system is also simplified by the elimination of much of the natural fauna and flora and by the loss of many natural physico-chemical processes. However, at the same time, the system is made more complex through the introduction of human management and activity.

An example of an agroecosystem that illustrates these points is the ricefield (Figure 12.1). The water-retaining dyke, or bund, forms a strong, easily recognizable boundary, while the irrigation inlets and outlets represent some of the limited outside linkages. The great diversity of wildlife in the original natural ecosystem is reduced to a restricted assemblage of crops, pests, and weeds. The basic ecological processes, such as competition between the rice and the weeds, herbivory of the rice by the pests and predation of the pests by their natural enemies remain, but are now overlain by the agricultural processes of cultivation, subsidy, control, and harvesting.

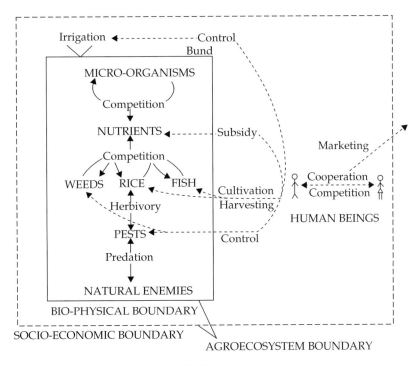

Figure 12.1 The ricefield as an agroecosystem. From Conway (1985).

It is this new complex agro-socio-economic-ecological system that I call an agroecosystem. Essentially the same systems diagram as shown in Figure 12.1 can be drawn for farms, villages, or watersheds, but the increasing complexity of the interactions makes a simple representation difficult, if not impossible. However, this complexity, at least in terms of its dynamic consequences, can be captured by four system properties which together describe the essential behaviour of agroecosystems (Conway, 1985, 1987). These are productivity, stability, resilience, and equitability. They are relatively easy to define (Figure 12.2), although not equally easy to measure. They are each described below.

• *Productivity* is the net increment in valued product per unit of resource (land, labour, energy, or capital). It is commonly measured as annual yield or net income per hectare or man-hour or unit of energy or investment.
• *Stability* is the degree to which productivity remains constant in spite of normal, small-scale fluctuations in environment variables, such as

climate, or in the economic conditions of the market; it is most conveniently measured by the reciprocal of the coefficient of variation in productivity.
• *Resilience*[1] can be defined as the ability of a system to maintain its productivity when subject to stress or shock. A stress is here defined as a regular, sometimes continuous, relatively small and predictable disturbance, for example the effect of growing soil salinity or indebtedness. A shock, by contrast, is an irregular, infrequent, relatively large, and unpredictable disturbance, such as is caused by a rare drought or flood or a new pest. Lack of resilience may be indicated by declining productivity but equally, as experience suggests, collapse may come suddenly and without warning.
• *Equitability* is a measure of how evenly the productivity of the agroecosystem is distributed among its human beneficiaries. The more equitable

---

[1] In the original paper Conway (1985) and subsequently I referred to this property as sustainability but in practice sustainability has come to mean a much more embracing concept. Resilience, similar in definition to the concept of C.S. Holling, is more appropriate (Holling, 1973).

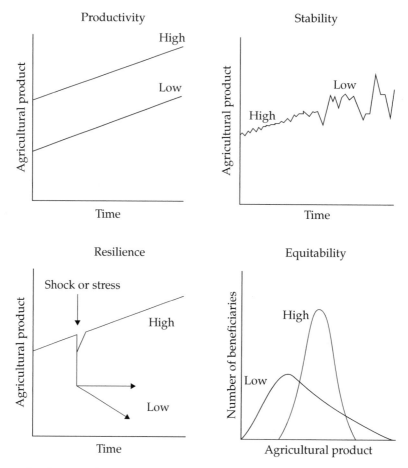

Figure 12.2 The properties of agroecosystems.

the system the more evenly are the agricultural products—the food, income, or resources—shared among the population of the farm, village, region, or nation. It can be represented by a statistical distribution or by a measure such as the Gini coefficient.

These four properties are essentially descriptive in nature, summarizing the status of the agroecosystem. But they can also be used in a normative fashion, as indicators of performance, and in this way can be employed both to trace the historical evolution of an agroecosystem and to evaluate its potential, given different forms of land use or the introduction of new technologies.

Experience shows that in agricultural development there is almost inevitably some degree of trade-off between the different system properties. New forms of land use or new technologies may have the immediate effect of increasing productivity, but this is often at the expense of lowered values of one or more of the other properties. Agricultural development typically involves a progression of changes in the relative values of these properties, successive phases of development producing different priorities. Thus the Roman agriculture described by Varro, which was based on the Roman villa, was not very productive (cereal yields were about 1 t/ha), but it was relatively stable and resilient. After all, it lasted several hundred years and spread throughout much of Europe and the Mediterranean. However, being largely based on slave labour, it was not what we would call equitable. Subsequent agricultural revolutions have greatly increased the productivity—by 5-fold in terms of average cereal yields—but often with a loss of stability and resilience.

12.3 Home gardens

As an illustration it is useful to contrast two of the building blocks of agriculture—the home garden and the cereal field. Home gardens are one of the oldest forms of farming system and may have been the first agricultural system to emerge in hunting-and-gathering societies (Hoogerbrugge and Fresco, 1993). Today, home or kitchen gardens are particularly well developed on the island of Java in Indonesia, where they are called *pekarangan* (Soemarwoto and Conway, 1991). Their immediately noticeable characteristic is their great diversity relative to their size: they usually take up little more than half a hectare around the farmer's house. Yet, in one Javanese home garden 56 different species of useful plants were found, some for food, others as condiments and spices, some for medicine, and others as feed for the livestock: a cow and a goat, some chickens or ducks, and fish in the garden pond. Much is for household consumption, but some is bartered with neighbours and some is sold. The plants are grown in intricate relationships with one another: close to the ground are vegetables, sweet potatoes, taro, and spices; in the next layer are bananas, papayas, and other fruits; a couple of metres above are soursop, guava, and cloves, while emerging through the canopy are coconuts and timber trees, such as *Albizzia*. So dense is the planting that to a casual observer the garden seems like a miniature forest.

The diversity is in contrast to the adjacent, much simplified ricefield systems where the only crop is rice, perhaps with some edible weeds and fish. Closer analysis shows the high diversity in the home garden is matched by high levels of productivity, stability, resilience, and equitability. A comparable ricefield has higher gross income and higher production of staple foods, but its other indicators are considerably lower (Table 12.1).

12.4 Swidden

A common method of cultivation in the developing countries, particularly in the tropical uplands, is swidden cultivation (otherwise known as shifting cultivation or slash and burn). Forest is cleared and burned and a crop grown for several successive years until yields fall too far. The land is then allowed to revert to forest and the farmer opens up (shifts to) another piece of land. Once the natural fertility has recovered, the regrowth is cut and the cycle repeats itself. Resilience here is crucially dependent on the respective lengths of the cropping and the fallow periods. In the example in Figure 12.3, when up to eight crops are grown in the cropping phase, the system goes back to mature forest but beyond eight crops it degrades to an unproductive grassland. This is an example of the threshold and breakpoint, or tipping point, phenomenon discussed in Chapter 3 in this volume.

Grazing systems can behave in a similar fashion (Noy-Meir, 1975). Increasing the number of livestock on a range raises productivity but also stresses the vegetation. At a certain intensity of stress, which is often very close to the maximum livestock carrying capacity, the vegetation collapses and the grazing system moves to a new level of productivity, much lower than before.

Table 12.1 System properties of the home garden when compared with a rice field (Soemarwoto and Conway, 1991).

	Home garden	Rice field
Productivity	Higher standing biomass; higher net income (lower inputs); greater variety of production	Higher staple production; higher gross income
Stability	Year round production; higher year-to-year stability	Seasonal production; vulnerable to climatic and disease variation
Resilience	Maintenance of soil fertility; protection from soil erosion	Heavy pest and disease attack
Equitability	Home gardens ubiquitous; barter of products	Product to landowners

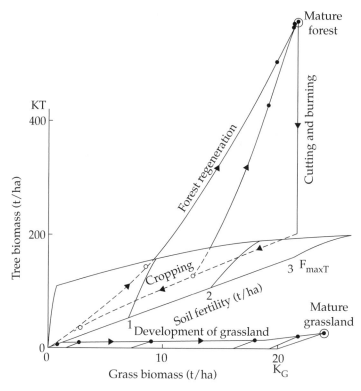

Figure 12.3 Resilience of a swidden cultivation system. K_G and K_T are maximum biomass of grass and trees under fallow, respectively; F_{maxT} is maximum soil fertility achievable under 100% trees. From Trenbath *et al.* (1990).

12.5 The Green Revolution

The trade-offs I have described in the previous sections are even more stark in the case of the Green Revolution that occurred in the 1960s and 1970s.

The Green Revolution was one of the great technological success stories of the second half of twentieth century. The introduction of new, short-statured varieties of wheat and rice along with irrigation and packages of fertilizer and pesticides dramatically increased yields. Overall food production in the developing countries kept pace with population growth, both more than doubling. However, these vast expanses of monocropped cereals required tight control to maintain their stability and they were prone to pest and disease epidemics. The new rices, in particular, were attacked by devastating outbreaks of bacterial blight and brown planthopper.

There is now also evidence of increasing production problems in those places where yield growth has been most marked. For example, in the Punjab, although wheat yields are still growing, this achievement is now being seriously threatened (Randhawa, no date). Of greatest concern is the growing scarcity of water. In some of the most intensively cultivated districts the ground water table has dropped to a depth of 9–15 m and is falling at about 0.5 m a year. This and other, albeit largely anecdotal, evidence from Luzon, Java, and Sonora suggests there are serious and growing threats to the sustainability of the yields and production of the Green Revolution lands (Pingali and Rosegrant, 1998).

There is also widespread evidence of declines in the rates of yield growth (Figure 12.4; Mann, 1999). A combination of causes is responsible (Cassman, 1999; Pingali and Heisey, 1999). In parts of Asia declining prices for cereals are causing farmers to

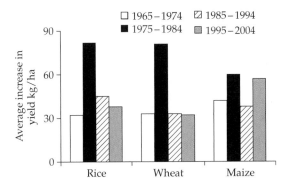

Figure 12.4 Average annual increase in developing-country cereal yields by periods (United Nations Food and Agriculture Organization, 2006).

invest more in higher-value cash crops. But more importantly there has been little or no increase in yield ceilings of rice and maize in recent years. A third factor is the cumulative effect of environmental degradation, partly caused by agriculture itself (Conway and Pretty, 1991). Virtually all long-term cereal experiments in the developing countries exhibit marked downward trends in yields.

The initial higher production helped reduce food prices in real terms by over 70% and this benefited the poor, who spend the highest proportion of their income on food. Yet today there are still some 800 million people who live a life of permanent or intermittent hunger and chronic undernourishment (United Nations Food and Agriculture Organization, 2001). A high percentage of the hungry are women and children; more than 150 million children under 5 years of age are severely underweight. Hunger and health intersect here—children who are malnourished are more vulnerable to infections and disease. In the developing countries, 11 million children under 5 years die each year, and malnourishment contributes to at least half of these deaths (UNICEF, 2001).

The Green Revolution helped Asia and many of the Asian poor but it by-passed sub-Saharan Africa. There the situation is especially dire. Food production per capita in most African countries has declined over the past decade, reflecting rapid population growth (averaging 3% per year) and low yields resulting from depletion rates for soil nutrients that far exceed replenishment (average losses in many countries exceed 60 kg NPK/ha per

year) and crop losses caused by pests, diseases, and abiotic stresses, such as drought (Henno and Baanante, 1999). Unlike in Asia, where average crop yields have increased substantially, average cereal yields in Africa have remained stagnant at only about 1 t/ha for the last three decades (Figure 12.5).

Some argue that hunger is simply a matter of poverty. If the poor had higher incomes, they could purchase the food they need, and it would be produced to satisfy their demand. There is truth in this. But there are no signs of large-scale manufacturing investments in Africa that would dramatically increase incomes. The reality is that most African families are farm families. It is only through greater agricultural production (and the development of renewable natural resources generally) that poor Africans can produce enough food and other farm products to stimulate rural economies and so achieve higher incomes. In theory the industrialized countries could feed the world. However, this would require several hundred million tonnes of food aid, many times what is supplied now. It would place heavy burdens on both the donors and the recipients. The environmental costs for the developed countries would be high, and for the developing countries the availability of free or subsidized aid in such large quantities would depress local prices and add to existing disincentives for local food production. More importantly this scenario implies that a large proportion of the population in the developing world would fail to participate in global economic growth.

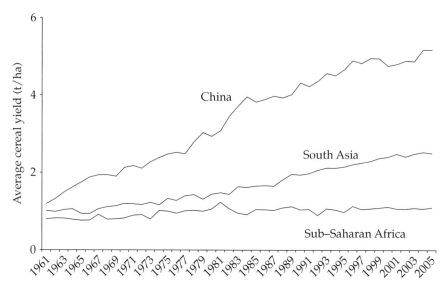

Figure 12.5 Cereal yield trends; developing countries of Asia compared with sub-Saharan Africa, 1960–2005 (United Nations Food and Agriculture Organization, 2006).

As numerous studies have shown, agricultural development is a necessary precursor to larger economic and social development (Delgado *et al.*, 1998; Department for International Development, 2005). The question is, what kind of agricultural development? I argue that, like the Green Revolution, it has to be based on science and technology. Yet it has to be different in the technologies that are used, because science has advanced and because the circumstances in Africa demand a different set of technologies.

12.6 A Doubly Green Revolution

I believe these arguments, when taken together, point to the need for a second Green Revolution, yet a revolution that does not simply reflect the successes of the first. The technologies of the first Green Revolution were developed on experimental stations that were favoured with fertile soils, well-controlled water sources, and other factors suitable for high production. There was little perception of the complexity and diversity of farmers' physical environments, let alone the diversity of the economic and social environment. The new Green Revolution must not only benefit the poor more directly, but also must be applicable under highly diverse conditions and be environmentally sustainable. In effect, we require a Doubly Green Revolution, a revolution that is even more productive than the first Green Revolution and even more green in terms of conserving natural resources and the environment (Conway, 1997). Over the next three decades it must aim to repeat the successes of the Green Revolution, on a global scale, in many diverse localities and be equitable, sustainable and environmentally friendly. In effect this can be translated as seeking to minimize the trade-offs between the agroecosystem properties (Figure 12.6).

The complexity of these challenges is daunting, in many respects of a greater order of sophistication than has gone before. Yet, I am an optimist: in part, because of the potential of two key, recent developments in the biological sciences. The first is the development of modern ecology with its sophisticated understanding of population and ecosystem processes. The second is the emergence of molecular and cellular biology, which, with their associated technologies, are having far reaching consequences on our ability to manipulate living organisms.

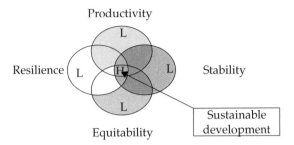

Figure 12.6 Trade-offs between livelihood properties. The small area overlapping high values (H) represents what we commonly refer to as sustainable development. L, low values.

12.6.1 Ecology and agriculture

Ecologists have long been fascinated by what makes ecological systems stable and resilient (May, 2001; Ives, 2005; Begon *et al.*, 2006). In general there is a supposition that the key factor is diversity—measured in its simplest form as species richness. Many theoretical and empirical studies have attempted to illuminate the relationships between diversity, on the one hand, and stability and resilience, on the other. The results have often been apparently contradictory, sometimes even reversing the cause-and-effect relationships. For example, diverse systems may be relatively fragile and only in stable environments can they be maintained. In unpredictable environments the communities may be more resilient yet simpler. What is clear from these studies is that it is not the sheer variety of the species present that affects stability and resilience, but their nature, their function in the systems and the relationships they have with one another.

It is also clear that 'natural communities represent not random assemblages of species but rather collections of species that can coexist' (Ives, 2005). In many respects this is even more true of agroecosystems. The diversity of crop and livestock species and of their varieties and breeds are present because human beings have recognized that they can coexist. Humans have also recognized that there are strong benefits in their coexistence. People do not live only on the calories provided by staple crops; they need sources of proteins, vitamins, minerals, and other nutrients. A diverse agroecosystem provides for a diverse and healthier diet. Farmers also recognize that different species can benefit each other. Trees and shrubs can provide shade for herbs, legumes can provide nitrogen, and livestock furnish manure. Mixtures of crops can also deter pests, and when a disaster strikes the farm—a drought or cyclonic storm—the more diverse the farm the more likely that something will survive.

The home garden is a good example. Part of the reason for the minimal trade-off in the home garden is the deliberately inbuilt diversity that helps stabilize production, buffers against stress and shock and contributes to a more valued level of production. But equally important is the intimate nature of the home garden. The close attention that is possible from family labour ensures a high degree of stability and resilience and the link between the garden and the traditional culture leads to an equitable distribution of the diverse products.

A Doubly Green Revolution seeks to exploit these relationships, through a variety of ecologically based approaches to the central processes of agriculture, for example in pest control and nutrient management, as I will now discuss.

12.6.2 Integrated pest management

In essence, integrated pest management seeks to bring about a sustainable and economically viable form of pest control that relies as far as possible on natural regulating factors and minimizes the use of damaging pesticides. Integrated pest management looks at each crop and pest situation as a whole and then devises a programme that integrates the various control methods in the light of all the factors present. As practised today, it combines modern technology, including the application of synthetic, yet selective, pesticides and the breeding into crops of pest resistance, with natural methods of control, including agronomic practices and the use of natural predators and parasites. The outcome is sustainable, efficient pest control that is often cheaper than the conventional use of pesticides.

One of the earliest examples was described in the first edition of this book (Conway, 1976). Another highly successful example is integrated pest management developed for the brown planthopper and other rice pests in Indonesia (Kenmore, 1991; Stone, 1992; Gallagher *et al.*, 1994). Under the programme, farmers are trained to recognize and regularly monitor the pests and their natural enemies, which include predators such as wolf spiders. They then use simple, effective rules to determine the minimum necessary use of pesticides. The outcome has been a reduction in the average number of sprayings from over four to less than one per season, while yields have grown from 6 to nearly 7.5 t/ha. Rice production in Indonesia increased 15% while pesticide use declined 60%, saving US$120 million a year in subsidies. The total economic benefit to 1990 was estimated to be over $1 billion. The farmers' health improved and one—not insignificant—benefit has been the return of fish to the ricefields.

Sometimes integrated pest management can be achieved solely by the use of natural enemies. A recent example is the biological control of the mealybugs (*Phenacoccus manihoti*) that infest cassava in Africa (Herren, 1996). The mealybugs came from South America and an international research centre known as the Centro Internacional de Agricultura Tropical (CIAT) found a parasitic wasp, *Epidinocarsis lopezi*, in the Paraguay River basin that maintains the mealybugs there at very low levels. The wasp was introduced into Africa and is now spreading fast, effecting satisfactory control.

Often, however, pesticide use is necessary and the challenge is to minimize any adverse effects. Another major pest of African crops is the parasitic weed *Striga hermonthica*, which sucks nutrients from the roots of maize, sorghum, and other crops. The weed is readily controlled by an herbicide, imazapyr, but this kills the crops. Recently, a mutant gene in maize has been discovered that confers resistance to the herbicide and this is being bred into local maize varieties (African Agricultural Technology Foundation, 2006). Their seed is then dipped into the herbicide before being planted. This kills the parasitic spores in the ground allowing the maize to grow while minimizing the environmental impact of the herbicide. Early trials are showing increases in yield from 0.5 to over 3 t/ha.

12.6.3 Integrated nutrient management

The next challenge is to extend the principles of integration established in integrated pest management to other subsystems of agriculture: to nutrient conservation, and to the management of soil, water, and other natural resources, such as rangeland.

African soils are eroding and losing nutrients fast. The losses of nitrogen per hectare often exceed the amounts a prosperous western farmer would put on his land each year. The losses far exceed the replenishment African farmers can afford. They pay some of the highest fertilizer prices in the world—whether in US dollars or grain equivalents (Mwangi, 1997). Prices in western Kenya are $400/ton of urea, in contrast to $90/ton in Europe (Sanchez, 2002). On average—and many use none at all—African farmers use fertilizer at only 10 kg/ha, whereas European farmers use over 200 kg/ha. This means that Africans must make as much use as possible of organic sources of nutrients, and apply them in an integrated fashion with inorganic fertilizers.

One route is through highly integrated crop/ livestock systems, where soil structure and nutrients benefit both from livestock manure and the nitrogen-fixing capacity of forage crops. Careful ecological management of crop/livestock systems can create virtuous circles: 'Cowpea thus feeds people and animals directly while also yielding more milk and meat, better soils through nitrogen fixation, high quality manure, which, used as fertilizer, further improves soil fertility and increase yields' (International Livestock Research Institute, 1999a). Forages identified by the International Livestock Research Institute for intercropping have led to wheat-yield increases of 30–100% and up to 300% increases in fodder protein while fixing 55–155 kg N/ha (International Livestock Research Institute, 1999b).

Often such forages are legumes, whose nitrogen-fixing capacity is the key to improving soil fertility (the Romans used lupins for this purpose). There are numerous examples of mixed cropping systems, sometimes based on tree legumes, sometimes on legumes grown as cash crops. A recent, highly productive system involves growing groundnuts and maize, alternating two rows of each. Yields of the maize can be over 5 t/ha whereas ground nuts achieve 1 t/ha or more (Langat et al., 2000). Sometimes, as in the case of the legume *Desmodium* intercropped with maize, there can be a double benefit since the *Desmodium* helps to destroy *Striga* (Hassanali et al., 2006).

12.7 Ecology in the seed

The foregoing examples are often effective and meet the requirements of a resilient agriculture, but they often have the drawback of being highly labour-intensive. It used to be thought that developing countries had plenty of labour and hence this was not an insurmountable problem. But even when labour is apparently available, demand is often very seasonal and it is not available when needed. Moreover, in Africa in particular, the growing HIV/AIDS crisis is making rural labour once again very scarce. The answer would appear to be building characteristics that promote not only productivity but also the other three characteristics—stability, resilience, and equitability—into the crop seed itself. Seeds can be made available to farmers cheaply or at no cost, providing they are produced by government or public–private partnerships. And in the case of self-pollinating crops such as rice, farmers can keep the seeds after harvest for the next season.

Our capacity to build ecology into the seed is largely a consequence of modern biotechnology. The Green Revolution depended on working to blueprints of desirable new plant and animal types through painstaking conventional plant breeding. Biotechnology offers a faster route. It is probably the only way to ensure that yield ceilings are raised, excessive pesticide use is reduced, the nutrient value of basic foods is increased, and farmers on less-favored lands provided with varieties better able to tolerate drought, salinity, and lack of soil nutrients (Conway, 2005).

Modern agricultural biotechnology consists of three practical processes:

- Tissue culture, which permits the growth of whole plants from a single cell or clump of cells in an artificial medium.
- Marker-aided selection, based on our ability to detect the presence of particular DNA sequences at specific locations in an organism and link these to the presence of genes responsible for particular traits.
- Genetic engineering, based on recombinant DNA technology, which enables the direct transfer of genes from one organism to another.

12.7.1 Tissue culture

Tissue culture has so far provided the greatest benefits to poor farmers. Tens of thousands of farmers in Africa are now growing food crops produced using this form of biotechnology. Tissue culture speeds up the dissemination of vegetatively propagated crops like cassava, sweet potato, and banana that traditionally have low multiplication ratios. With tissue-culture-based micropropagation, hundreds or even thousands of seedling plants can be produced from a single, superior mother plant. Done properly under sterile conditions, tissue culture has the added advantage of eliminating nearly all diseases from the regenerated plantlets. In East Africa, where banana is a staple crop, micropropagation of improved and disease-free banana seedlings through tissue culture is improving food production and generating income for small-scale farmers (Wambugu and Kiome, 2001).

To date, the most dramatic achievements have been with rice. Using anther culture and another tissue-culture technique, embryo rescue, scientists have crossed the high-yielding Asian rice *Oryza sativa* with the African rice *Oryza glaberrima*. The progeny of such crosses usually have low fertility, but in the Ivory Coast, using anther-culture techniques developed in China, African scientists at the African Rice Centre (WARDA) have been able to produce crosses that combine the high yields of the Asian rice with the weed-competitiveness and drought-tolerance of African rice (Jones, 1999). The new varieties are producing yields of up to 3 t/ha when only 1 t/ha was possible before. Unsurprisingly they are spreading rapidly; from village to village, not only through West Africa but also in parts of East Africa.

12.7.2 Marker-aided selection

With marker-aided selection it is possible to identify segments of the plant genome that are closely linked to the desired genes, so the presence of the trait can be determined at the seedling or even the seed stage. This makes it possible to achieve a new variety in four to six generations instead of 10.

Maize streak virus, the most serious disease of maize in Africa, affects 60% of the planted area and causes an estimated 37% yield loss, roughly production losses of 5.5 million t/year (Jeffers, 2001). Excellent genetic resistance to maize streak virus has been known for over 20 years, but it has not been widely deployed in local maize varieties because few national breeding programmes can afford to maintain the insect colonies and other infrastructure necessary to measure for resistance against insect-vectored viral diseases. Now, using genetic markers on the molecular map of maize, it is possible to identify the precise location of the resistance gene and, using the DNA markers flanking the gene, to backcross it into numerous well-adapted local varieties without expensive disease screening.

Marker-aided selection is particularly useful for breeding drought-tolerance, which typically occurs as the result of a number of different traits— deeper roots, early flowering, osmotic changes— working together. Breeding for it is a particularly difficult and slow process using conventional techniques, but markers are now permitting combinations of these traits to be accumulated in new varieties.

12.7.3 Genetic engineering

Genetic engineering moves genes between organisms, including those that do not cross in nature. The resulting plants, called transgenic or genetically modified organisms (GMOs), have been the focus of most of the controversy over biotechnology so far. For some good reasons, including those connected with early episodes of corporate haste and arrogance, people in many countries are suspicious of genetically modified organisms and are hostile to their use, especially in food.

Despite the opposition, since 1996 there has been a steady increase in the worldwide area planted with transgenic crops, with nearly 90 million ha harvested in 21 countries in 2005 (Figure 12.7). Over 8.5 million farmers grew transgenic crops in 2005, 90% of whom were small-scale farmers in developing countries, with the vast majority in China.

For poorer farmers the main benefits so far have come from growing cotton that is resistant to insects (Conway, 2005). The resistance is conferred by introducing gene constructs derived from the bacterium *Bacillus thuringiensis*. Last year in China some 6.4 million farmers grew over 3.3 million ha of *B. thuringiensis*-modified cotton. They have been able to reduce the number of pesticide applications substantially, obtaining higher yields and benefits estimated at $330–400 more per hectare (Huang *et al.*, 2002). There have been comparable results in South Africa and Mexico (Ismael *et al.*, 2001).

Genetic engineering is also being used to produce transgenic food plants with nutritional traits important to developing countries. The best example to date is the successful inclusion of β-carotene, the precursor of vitamin A, into the grain of rice to produce so-called golden rice. β-carotene is present in the leaves of the rice plant, but conventional plant breeding was unable to move it into the grain. Scientists at the Swiss Institute of Plant Sciences in Zurich successfully transferred one bacterial gene and two daffodil genes into the grain, and the added transgenes resulted in the synthesis of nutritionally significant levels of β-carotene in the grain (Ye *et al.*, 2000). Currently breeders in several Asian countries are transferring the genes into locally adapted rice varieties.

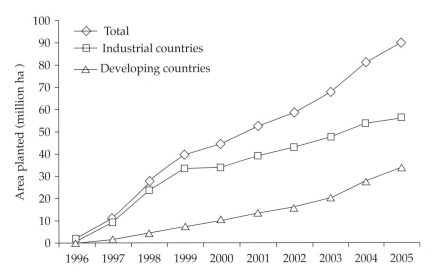

Figure 12.7 The global area of genetically modified crops, 1996–2005 (James, 2005).

The potentials for genetic engineering are almost endless. But alongside the benefits are risks; some real, some imagined. An impassioned debate in Europe is raising genuine concerns about ethics, environment, and the potential impact on human health (Royal Society, 1998; Nuffield Council on Bioethics, 2003). The developed countries are clearly better equipped to assess such hazards. They can call on a wide range of expertise and most have now set up regulatory bodies and are insisting on closely monitored trials to try and identify the likely risks before genetically engineered crops and livestock are released to the environment. So far, few developing countries have put such regulation in place. My personal belief is that the hazards are often overstated, but if the evident benefits are to be realized for the developing countries it is the responsibility of all involved to ensure that the hazard assessments are as rigorous as they are in developed countries.

More important than the potential hazards, at least to my mind, is the question of who benefits from biotechnology. So far the focus of biotechnology companies has been on developed country markets where potential sales are large, patents are well protected, and the risks are lower. But agribusiness is now turning their attention to the developing countries, and are embarking on an aggressive policy of identifying and patenting potentially useful genes. Part of the answer to this challenge lies in public–private partnerships whereby genomic information and technologies are donated to public plant breeders and agreements are struck that ensure new varieties of benefit to poor farmers in the developing countries are freely available.

12.8 Participation

The first Green Revolution essentially targeted some of the best-favoured lands in the developing countries. The land holdings were often reasonably large, flat, and well watered. In these circumstances it was relatively easy for agricultural extension workers to promote a simple, uniform package of seeds, fertilizers, and pesticides. The targets of the Doubly Green Revolution are very different—millions of small farmers inhabiting an extraordinary diversity of land, soil, and climatic types. There are no simple messages. Indeed, every farm requires its own special set of recommendations. In this context, the traditional top-down approach will not work and the only way forward is to involve farmers closely in the analysis, design, and experimentation processes.

The first Green Revolution started with the biological challenge inherent in producing new, high-yielding food crops and then looked to determine

Figure 12.8 Farmers as Ecologists. Reproduced with permission from Professor Robert Chambers.

how the benefits could reach the poor. But this new revolution has to reverse the chain of logic, starting with the socio-economic demands of poor households and then seeking to identify the appropriate research priorities.

In Rwanda a 5-year experiment involved farmers very early in the breeding process (Sperling and Scheidegger, 1995). Beans (*Phaseolus vulgaris*) are a key component of the Rwandan diet and there is an extraordinary range of local varieties: over 550 have been identified. Farmers (mostly women) are adept at developing local mixtures which breeders have difficulty in bettering. In the experiment, farmers assessed 80 breeding lines over 3 years, using their own criteria to reduce the number of lines. The farmers tagged favoured varieties on the station with coloured ribbons. A set of 20–25 lines was then taken to field trails on the farmers' plots. They then chose the best performers and were responsible for multiplying and diffusing them to their neighbours.

Yet this will not be enough. Experiments in many parts of the developing world are showing very effective ways of involving farmers right at the beginning, in the design of new varieties and in the breeding process itself. Participation has long been a slogan of development. For the first time we now have effective techniques to make it a reality. Under the heading of Participatory Learning and Action (PLA) there is a formidable array of methods which permit farmers to analyse their own situations and, most importantly, to engage in productive dialogue with research scientists and extension workers (Scoones and Thompson, 1994; Chambers, 1997, 2005). PLA arose in the late 1980s out of earlier participatory approaches by combining semi-structured interviewing and diagram-making drawn from the classical tools of ecology; for example, maps, transects, and seasonal calendars. It enables rural people to take the lead, producing their own diagrams, undertaking their own analyses and developing solutions to problems and recommendations for change and innovation. Maps are readily created by simply providing villagers with chalk and coloured powder and no further instruction other then the request to produce a map, of the village, watershed, or farm. People who are illiterate and barely numerate can construct seasonal calendars using pebbles or seeds (Figure 12.8). Pie diagrams—pieces of straw and coloured powder lain out on an earthen floor—are used to indicate relative sources of income. Such diagrams not only reveal existing patterns but point to problems and opportunities and are seized on by rural people to make their needs felt.

PLA has now spread to most countries of the developing world, and been adopted by government agencies, by research centres and university workers as well as by non-governmental organizations. In some ways it has been a revolution, a set of methodologies, an attitude and a way of working which has finally challenged the traditional top-down process that has characterized so much development work. Participants from outside find themselves, usually unexpectedly, listening as much as talking, experiencing close to first hand the conditions of life in poor households and changing their perceptions about the kinds of interventions and the research needs that are required. In every exercise the traditional position of rural people being passive recipients of knowledge and instruction has been replaced by the creation of productive dialogues.

Recently I visited a village in an upland watershed in Orissa, India. As part of a 'drought-proofing' project the villagers proudly produced a portfolio of maps and analyses of their local area, including maps of the landforms, the holdings, and the crops and analysis of the households, including their income status. The last of the presentations was a list of project priorities that they had formulated. The first six had been accomplished and they (expectantly) pointed out that they would like help for the next on the list. In some ways this is the ultimate example of demand-led development.

12.9 Conclusion

I firmly believe we can provide food for all in the twenty-first century. But there is no simple or single answer. It is not just a matter of producing more or enough food. If hunger is to be banished the rural poor have either to feed themselves or to earn the income to purchase the extra food they require. This requires a new revolution in agricultural and natural-resource production aimed at their needs. And this cannot be achieved by ecology or by biotechnology alone, or by a combination of these. It requires participatory approaches as well, involving farmers as analysts, designers, and experimenters. If we can bring all three approaches together, then we can feed the world in a way that is not only equitable but also stable and resilient.

CHAPTER 13

Conservation biology: unsolved problems and their policy implications

Andy Dobson, Will R. Turner, and David S. Wilcove

13.1 Introduction

A plot of the number of parks and other terrestrial protected areas established around the world over the past 100 years exhibits near-exponential growth (Figure 13.1), with marine parks following a similar trend. This is a testament to the growing recognition of the importance of sustaining natural systems worldwide. Yet, at the same time an expanding human population and the desire of all people for a more prosperous life have resulted in unprecedented rates of deforestation and habitat conversion. Accompanying these changes has been the spread of invasive, non-native species (including new disease organisms) to virtually all parts of the globe. With recent assessments placing 12% of the world's birds, 23% of mammals, and 32% of amphibians in danger of extinction (Baillie et al., 2004), conservationists feel a justifiable sense of panic.

Any attempt to measure the full extent of the current biodiversity crisis is made immensely more difficult by our astounding lack of knowledge about the species that share this planet with us. For example, we do not know within an order of magnitude the number of species currently present on Earth (May, 1988, 1992; Novotny et al., 2002); estimates range from 3 to more than 30 million species, of which only 1.5–1.8 million have been described to date. Not surprisingly, our inventory of the more charismatic groups of organisms, such as birds, mammals, and butterflies, is vastly more complete than our inventory of insects, arachnids, fungi, and other less conspicuous but no less important groups.

If we ask the logical follow-up question—what proportion of known (described) species is in danger of extinction?—we run into a similar barrier. While organizations like the World Conservation Union (IUCN) have prepared reasonably complete assessments for a few groups, notably the charismatic vertebrates, most species are too poorly known to assess. Even within the USA only about 15% of the species catalogued to date are sufficiently known to be given any sort of conservation rank, such as endangered or not endangered (Wilcove and Master, 2005); among invertebrates that value drops to less than 5%. Compounding this shortfall of data is an equally serious shortfall of money. Most conservation programs, especially those in developing countries, are woefully underfunded (Balmford et al., 2003).

Under these circumstances, conservation must be efficient and effective. In this chapter, we explore several ways in which theoretical ecology is contributing to the efficiency and effectiveness of contemporary conservation efforts. We begin with a discussion of the challenges associated with determining what conservation measures (generally framed in terms of the amount and distribution of protected habitat) are necessary to ensure the long-term survival of an endangered species. We use as our example the case of the grizzly bear (Ursus arctos) in Yellowstone National Park, USA. Few endangered animals have attracted greater attention than this widely admired and widely feared species. We then consider the challenges associated with creating a network of reserves to protect multiple species of concern,

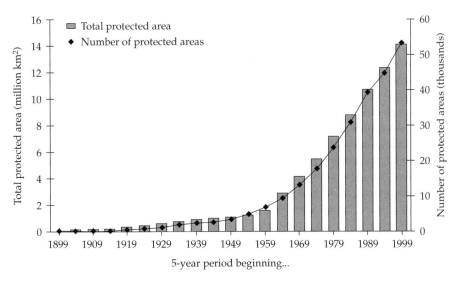

Figure 13.1 The total area of land set aside as protected areas (bars) and the total number of protected areas created in the time intervals shown (black diamonds). Data from WDPA (2004).

examining in detail some of the recent advances in the field of systematic reserve design. Having considered various theoretical challenges associated with protecting individual species or groups of species, we turn to the task of measuring and protecting the essential services provided by species, a subject of growing importance to conservation worldwide.

13.2 Protecting individual populations and species: the case of the grizzly bear

With varying degrees of enthusiasm and success, people have been trying to save populations of declining species for centuries. As far back as 1616, a rapid decline in the number of cahows (*Pterodroma hasitata* a type of seabird) prompted authorities in Bermuda to enact a proclamation prohibiting 'the spoyle and havock of the Cahowes, and other birds, which already wer almost all of them killed and scared away very improvidently by fire, diggeing, stoning, and all kind of murtherings' (Matthiessen, 1987). Nearly four centuries later, the cahow is still with us, although it remains one of the world's rarest birds, with fewer than 100 pairs nesting on a total of 1 ha of rocky islets off the coast of Bermuda (www.birdlife.org). The viability of this population is a function of many interacting factors, including the

amount, distribution, and quality of habitat, and community interactions between cahows and their prey, predators, competitors, and disease agents. These factors, in turn, are tied to a host of sociological issues that determine what is, and isn't, possible in terms of conservation.

A combination of direct persecution and habitat loss due to agriculture, logging, road building and oil and mineral exploration eliminated grizzly bears from most of their range in the western USA south of Alaska. In 1975, the US Fish and Wildlife Service declared the grizzly bear to be a threatened species in the coterminous USA. One of its last strongholds was in Yellowstone National Park and the surrounding national forests, but even here the population was small (~200 adults) and declining (Craighead *et al.*, 1995). Thirty years later, the Yellowstone population now hovers around 600, and the US Fish and Wildlife Service has announced its intention to remove the Yellowstone grizzlies from the list of endangered and threatened species. Are 600 grizzlies enough to ensure the long-term persistence of the species in Yellowstone? More generally, what constitutes a safe or viable population size for a species like the grizzly bear, and what factors other than the size of the population must be evaluated in deciding whether or not it has recovered? These are the sorts of questions that can be tackled via population viability models.

13.2.1 Population viability analysis

Our ability to model the dynamics of single populations and interacting sets of populations has grown enormously over the past two decades, providing conservation biologists with a set of powerful, new tools for developing conservation plans and designing nature reserves for endangered species. Indeed, population viability analysis has become an essential component of endangered species conservation efforts in many countries (Shaffer, 1990; Boyce, 1992; Burgman *et al.*, 1993). The most basic population-viability-analysis model simply considers the birth and death rates of the species deemed to be in danger of extinction. Thus, we could write a simple expression for the population B_t of Yellowstone's grizzlies over time as

$$B_{t+1} = sB_t(1+b) = \lambda B_t \qquad (13.1)$$

Where s is the annual survival of the bears, b is their annual fecundity, and λ is the annual rate of population increase. Determining the viability of the population essentially comes down to determining whether λ is greater than unity. Of course, the situation is more subtle than this because survival may vary from year to year and will also be different for older bears compared with cubs or yearlings. Because any variance in the birth and death rates will reduce the potential for long-term persistence, many early population viability analyses focused on obtaining accurate estimates of demographic rates and their underlying variability. However, these estimates of variability are confounded by statistical sampling procedures that, by definition, tend to give broad statistical confidence limits when data are scarce. This is often the case for rare and endangered species. With grizzly bears, several years may elapse before an individual enters sexual maturity, so a more detailed model would need to include a lag of several years in the birth term and perhaps also include some stochastic variation in birth and death rates.

More significantly, assuming that survival and fecundity are independent of climate and food resources ignores the bear's dependence upon other species in its ecosystem and its vulnerability to variation in other environmental factors that determine the bear's survival and fecundity. Alternatively, fecundity and survival may decline as the bear population becomes inbred due to genetic isolation, a real possibility in the case of the Yellowstone grizzlies, which have been isolated from any other grizzly populations for nearly 50 years. (Indeed, the federal government is contemplating occasionally adding grizzlies from outside Yellowstone to reduce the effects of genetic isolation.) Population viability analyses include these additional details in a variety of ways, with the emphasis placed on each depending on the interests of the person running the analyses and the strident responses of people who review them.

13.2.2 Simple models and confusing data

One might assume that the best-quality data available for a population viability analysis for grizzly bears are the data collected from radio-telemetry studies. This technique is widely used in wildlife research; indeed, it was first used by Frank and Lance Craighead to study grizzlies in Yellowstone (Craighead *et al.*, 1969, 1995). Radio-telemetry allows bears to be monitored with little human interference. It also makes it far easier for scientists to relocate females in order to monitor the number and survival of their cubs.

When these data were initially used to estimate the intrinsic growth rate of the Yellowstone grizzlies they presented a gloomy prognosis: the annual growth rate of the bear population was consistently less than unity, which suggested that the population was in a state of terminal decline following the closure of the garbage dumps where they had grown accustomed to feeding (Cowan *et al.*, 1974; Knight and Eberhardt, 1985; Eberhardt *et al.*, 1986). However, aerial surveys of grizzlies were also used to monitor the population, and they showed a consistent upward trend in grizzly numbers, particularly with respect to the key index of number of independent females with cubs (Figure 13.2a and b)–as well as an increase in litter size. Why, then, was the state-of-the-art radio-telemetry data giving a different answer than the

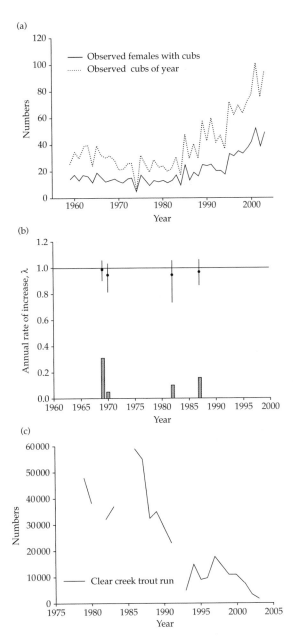

Figure 13.2 (a) The numbers of independent female grizzly bears and cubs counted in survey flights over the Greater Yellowstone ecosystem, 1958–2004. (b) Estimate of natural rate of population increase for grizzly bears using survival estimates based upon radiotelemetry data (after Dobson *et al.*, 1991). The upper points show the estimated annual rate of increase using demographic data collected at different time intervals, and the bars are the probability that the annual rate of increase is greater than unity. (c) Numbers of cutthroat trout counted in the Clear Creek trout stream in Yellowstone National Park, 1978–2003 (data from InterAgency Grizzly Bear Annual Reports; www.nrmsc.usgs.gov/research/igbst-home.htm).

more serendipitous survey data? Eventually researchers realized that very few of the females seen with cubs had radio collars; they lived in the more remote parts of Yellowstone and had very little contact with humans. In contrast, there was a high proportion of 'nuisance bears' in the radio-telemetry sample; these were bears that had wandered into human areas and had become acclimatized to the presence of humans. They suffer higher rates of mortality than the backwoods bears, largely because of road collisions and fatal interactions with hunters and property owners (Mattson et al., 1996; Dobson et al., 1997c; Pease and Mattson, 1999).

More recent work has shown that grizzly bears are intricately enmeshed with the Greater Yellowstone food-web. In spring a sub-population of bears focuses on catching cutthroat trout (*Oncorhynchus clarki*). Their ability to gain body fat after winter hibernation is highly dependent upon this food resource. Recent declines in Yellowstone's trout population have significantly reduced the fecundity of this section of the bear population (Figure 13.2c). Another crucial food resource for bears are the nuts of the white-bark pine (*Pinus albicaulis*), which the bears steal from red squirrels (*Tamiasciurus hudsonicus*) that cache them in convenient aggregations high on mountain slopes in the fall. Studies reveal that an abundance of white-bark pine nuts is crucial to grizzly over-winter survival (Mattson et al., 1992; Pease and Mattson, 1999). Unfortunately, Yellowstone's white-bark pines are host to a pine blister rust that attacks their cambium, leading to reductions in cone production and increasing their susceptibility to attack by pine bark beetles (Kendall and Roberts, 2001). Climate warming has increased the speed at which the beetles reproduce, so the synergistic interaction between blister rust and bark beetle has caused white bark pines to decline throughout Yellowstone. Their demise could reduce over-winter survival of grizzly bears and bring the bears back to the edge of extinction.

The crucial point is that understanding grizzly bear population viability requires models that go beyond estimating simple birth and death rates and that examine the bears' interactions with other species in the ecosystem. These models will by necessity be simplifications, but they will provide important conservation insights. If we based decisions regarding the health of the bear population on data obtained purely from demographic studies (even seemingly robust studies involving radio-telemetry), we would miss details that are crucial to successfully managing the bears over the long term.

Despite the insights that models can provide in cases like the grizzly bear, their use can be controversial. A bill to reform the Endangered Species Act that passed the US House of Representatives in September 2005 drew a distinction between empirical data and models, with the implication that the latter are somehow distinct from and inferior to the former. We presume this notion that data and models exist in separate realms stems from either ignorance about ecological models or a fear that the output of such models will be politically unpopular—or both. Yet without the use of models, it is all but impossible to make well-informed decisions about most aspects of endangered-species management.

13.3 Building a reserve network

Faced with limited resources, numerous species in need of assistance, and ongoing habitat loss, several methodologies have been developed to help determine global conservation priorities. Given that not all conservation organizations have precisely the same mission and objectives, it comes as no surprise that strategies for global prioritization differ (Redford et al., 2003). Conservation International, for example, focuses on both hotspots; regions that harbor large numbers of endemic species and have undergone substantial habitat loss, and additional, species-rich wilderness areas such as the Amazon (Myers et al., 2000; Mittermeier et al., 2003). The World Wildlife Fund targets 238 terrestrial ecoregions, identified via a set of metrics related to species richness, endemism, presence of rare ecological or evolutionary phenomena, and threats (Olson and Dinerstein, 1998). The recently created Alliance for Zero Extinction targets the rarest of the rare: its goal is to protect approximately 600 sites, each of which represents the last refuge of one or more of the world's mammals, birds, reptiles, amphibians, or conifers (Ricketts et al., 2005).

All of these strategies strive for efficiency, aiming to protect the largest number of conservation targets in the fewest sites or at the lowest cost (Possingham and Wilson, 2005). In the context of species conservation, that efficiency is largely a function of the degree to which species' distributions overlap: the greater the number of targeted species that co-occur, the smaller the amount of land that must be secured on their behalf. This rather obvious point has a number of important ramifications for conservation planning, stemming from the fact that species' distributions can be measured at various spatial scales, from a coarse-grained measurement of how species' ranges are distributed across space (extent of occurrence, *sensu* Gaston, 1991) to finer-grained measurement of how the populations of species are distributed within their respective ranges (area of occupancy, *sensu* Gaston, 1991). The degree of overlap (or lack thereof) at each scale fundamentally influences the efficiency of conservation.

At the broadest scale, that of whole ranges, there is often considerable overlap between species; this is demonstrated by the existence of centers of endemism for particular groups of organisms. The degree to which these centers of endemism overlap *between* groups is an issue of considerable importance to conservation practitioners; unfortunately it appears to vary geographically. Myers *et al.* (2000), for example, reported a high degree of congruence between endemic plants and endemic terrestrial vertebrates within hotspots such as Madagascar ($594\,000\,\text{km}^2$) and the tropical Andes ($1\,258\,000\,\text{km}^2$). On the other hand, the Cape Floristic Province ($74\,000\,\text{km}^2$) and the Mediterranean Basin ($2\,382\,000\,\text{km}^2$) each contained large numbers of endemic plants but relatively few endemic vertebrates.

Prioritization at the global scale can help efficiently allocate conservation resources by adding coherence to conservation efforts. Yet most conservation action necessarily takes place at a much finer scale (Dinerstein and Wikramanayake, 1993). Species, including threatened species, are concentrated in different areas within regions (Dobson *et al.*, 1997b). Many decisions about site protection and management must be made in the context of local conservation priorities for biodiversity targets and funding (Jepson, 2001), and global political agreement on any one comprehensive plan is unlikely. Moreover, to date, data necessary for the actual implementation of conservation at individual sites has been unavailable over a global extent. Thus, the development of global strategies over the past two decades has been accompanied by the parallel, but largely separate, development of theory and tools for the selection of networks of individual protected and managed areas.

13.3.1 Systematic reserve design

The problem of selecting sites within regions addresses the central issue of efficiency in conservation planning: select the set of sites that protects the greatest number of conservation elements (e.g. species, habitat types) for the lowest cost. Early approaches tackled this problem with stepwise algorithms (Kirkpatrick, 1983; Margules *et al.*, 1988). Later workers framed the question as an optimization problem (Cocks and Baird, 1989; Camm *et al.*, 1996). In its simplest form, the problem is one of identifying the smallest number of sites or 'minimum set' that includes all species on at least one site. Minimize

$$cost = \sum_{i \in I} x_i \qquad (13.2)$$

subject to

$$\sum_{i \in I} p_{ik} x_i \geq 1 \qquad \forall k \in K \qquad (13.3)$$

where i and I are, respectively, the index and set of sites, k and K are, respectively, the index and set of species, x_i is an indicator variable equal to 1 if and only if site i is included in the reserve network, and p_{ik} is an indicator variable equal to 1 if and only if species k is present in site i. Eqn 13.2 minimizes the number of sites in the reserve network, whereas eqn 13.3 ensures that the reserve network includes each species in at least one site. This formulation is a binary integer program which can be solved with optimization software.

Not surprisingly, these optimal formulations outperform a variety of heuristic methods (algorithms whose solutions are not provably optimal) in practice (Csuti *et al.*, 1997; Rodrigues and Gaston, 2002a), including, among others, various stepwise approaches. But several assumptions of these simple formulations—for example, that all sites have equal cost or that any one site is sufficient to protect a species—make them trivial for most conservation purposes. Fortunately, optimization methods can produce more relevant solutions by incorporating additional factors into the above models. Additional complexities include site-specific costs; weights so that some species are more 'valuable' than others, minimization of boundaries so that contiguous sites are preferred, and specifications of areas of suitable land required for each species rather than their simple presence or absence (Rodrigues *et al.*, 2000; Fischer and Church, 2005). These fuller descriptions of the desired properties of a reserve network can be much more difficult to optimize. Heuristic methods such as simulated annealing are potentially applicable to larger data-sets and problems of greater complexity than are optimal methods (Pressey *et al.*, 1996). However, identification of optimal solutions using mathematical programming remains the preferred method for problems of manageable size (Csuti *et al.*, 1997), since results of optimization methods are either provably optimal, or, if solutions are not found in a short-enough time frame, the distance from optimality can be quantified. (Although the notion of knowing how far one is from an unknown optimal solution is somewhat counterintuitive, this can be achieved by solving the problem to optimality under relaxed constraints; the distance to the relaxed optimum is then an upper bound on the degree of actual suboptimality.) Moreover, optimal techniques are increasingly applicable to a broader number of complex problems (e.g., see Rodrigues and Gaston, 2002a), and even when optimal solutions are not feasible within time constraints, best solutions found under truncated solution times (e.g., 2 min; Fischer and Church, 2005) have been shown to outperform simulated annealing solutions to the same site-selection problems.

13.3.2 Site-selection algorithms meet real-world complexities

The potential advantage of optimal sets of sites is straightforward: by definition, acquiring all sites in such a set is the most efficient means to satisfy conservation objectives subject to the constraints and criteria used. Traditionally, theoretical work on site selection assumes a static world: a reserve network is identified, and then all sites in that network are acquired for protection simultaneously. In practice, however, reserve networks are often acquired over a period of several years (Pressey and Taffs, 2001). During this time unanticipated complications can wreak havoc with

what had once been an optimal solution. Simply recomputing the optimal solution based on updated data each year cannot overcome these shortcomings (Turner and Wilcove, 2006).

If acquisition is not instantaneous and uncertainty exists with respect to when sites become available for acquisition, if unprotected sites become unsuitable over time, or if budgets fall short of those initially envisioned, even optimal portfolios may perform poorly (Faith *et al.*, 2003, Drechsler, 2005). Simulation of site acquisition under such uncertainty has revealed that optimal solutions to the minimum-set problem may be outperformed by several heuristic methods (Meir *et al.*, 2004). Among these is a 'greedy' algorithm which simply purchases those sites adding the most additional targets to the existing network at any given time. The problem is that, when many sites are available, the chances are good that most of the sites in a single optimal set can be acquired. But when site availability is limited, such an approach is handicapped by its narrow focus on a few key sites, many of which may never become available. Thus the simpler greedy algorithm, which considers sites outside of a fixed initial portfolio, is able to fare better, even under uncertainty. On the other hand, this potential benefit brings with it opportunity cost: when site availability is high, considering more sites (i.e. lower-quality sites) for acquisition may result in a reserve network more costly than necessary to meet targets. Furthermore, conservation planners do not always know in advance what site availability will be, making it difficult to identify a decision rule for reserve acquisition that will perform well in a given situation.

Adaptive decision rules (Turner and Wilcove, 2006) offer one approach to address these uncertainty issues. Adaptive decision rules combine the relative strengths of the minimum-set approach (optimum cost efficiency when sufficient sites are available) and heuristic rules (ability to create networks that meet more targets when site availability is lower). The best-of-both-worlds (BOB) rule is an adaptive rule that proceeds as follows. During each year, first compute the mean number of sites that must be acquired per year to exhaust the budget at the end of the multiyear acquisition period. Then acquire any available sites that appear in the current optimal solution. If the number thus acquired in a given year falls short of the mean number needed per year, use the greedy rule as a backup to make up the difference. This is repeated for each successive year.

The key innovation of BOB as a decision rule for acquisition is that it adapts to uncertainty or variation in availability. In years in which the optimal algorithm falls short due to low availability of sites in the optimal portfolio, BOB attempts to correct the shortfall immediately with a backup method (the more aggressive, greedy method). Adaptive rules such as BOB outperform non-adaptive rules such as the minimum-set approach or existing heuristic methods on a variety of data-sets and under a broad range of availability rates (Figure 13.3), degradation rates, and overall acquisition budgets (Turner and Wilcove, 2006).

Adaptive rules attempt to address real-world challenges (i.e. the dynamic, uncertain nature of reserve implementation) by linking the processes of conservation planning and reserve acquisition under a single model. They thus require tight coordination and feedback between biological planners (the prioritization phase) and acquisition personnel (the acquisition phase). Biological data and up-to-date information on the success or failure of site acquisitions, at a minimum, must be fed back into a single comprehensive model (BOB or something similar) for these approaches to work. These methods are new, and it remains to be seen to what extent this institutional coupling of biological prioritization and acquisition can work in practice.

An alternative approach is to recognize the more disjunct institutional structure (less-tightly coupled biological prioritization/acquisition teams or phases) that exists in many situations and ask,

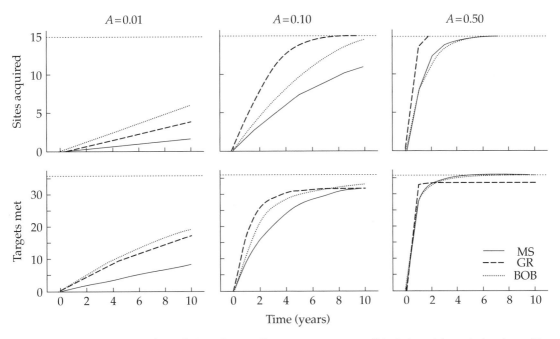

Figure 13.3 In reserve acquisition, adaptive decision rules generally meet more targets more efficiently than existing methods under conditions of uncertain or incomplete site availability. The annual probability of each site becoming available, A, strongly affects performance of decision rules over the course of reserve acquisition. Results are shown for 36 species of the Lake Wales Ridge ecosystem in Florida, USA, with targets of three sites/species and a budget of 15 sites. Upper and lower rows of graphs show the number of sites acquired and number of targets met, respectively, with horizontal lines indicating the initial minimum set size and maximum possible targets met, respectively. Acquisition-decision rules include an improved rule based on the optimal minimum-set (MS) algorithm, a greedy richness-based (GR) algorithm, and the best-of-both-worlds (BOB) adaptive rule.

what potential products from the prioritization phase will be most useful to the acquisition phase under the broadest range of realistic scenarios? The principle of irreplacibility, which values sites according to the likelihood that their protection will be required for the reserve network to meet or maximize conservation objectives, appears particularly well suited to this task. It may be used to create products that are robust to unknown conditions, yet may be tailored to those conditions which are known (e.g. total budget; Turner et al., 2006). However, one drawback with irreplacibility scores is that they are at present a means to rank sites but are not themselves a prescription for the identification of a full reserve network. Indeed, for this reason an irreplacibility-based acquisition rule failed to protect as many biodiversity targets in simulations as the adaptive rule BOB (Turner and Wilcove, 2006).

13.4 Habitat destruction

The need to create networks of reserves stems primarily from the rapid rate at which natural habitats are being converted to other uses. These habitats are most often converted from their pristine or near-pristine state into agricultural land, and then into either degraded land, if they fail to sustain agriculture, or into housing developments, golf courses, or cities (Turner et al., 1990; Tilman et al., 2001b). In some cases, conversion to agriculture (and pastureland) is at least partially reversible. In the eastern USA, for example, forests were converted to farmland in the eighteenth and nineteenth centuries. After many of these farms were abandoned in the late nineteenth and early twentieth centuries, forests regenerated over substantial areas.

13.4.1 A model of land-use change

Models with similar structure to those developed to examine the dynamics of infectious diseases and forest fires can provide insights into the dynamics of land-use change (Dobson et al., 1997a). In essence these models consider pristine land to be equivalent to a susceptible population of hosts, which is then colonized (or infected) by humans

who use the land for agriculture. The land is then either farmed in perpetuity or until the time when it is no longer productive, when it is abandoned and left to recover slowly through succession, until it is once again susceptible to colonization for agriculture. The relative duration of time it takes to recover and the duration of time it lasts under agriculture are equivalent to the durations of infectivity and resistance in standard SIR (susceptible, infectious, and resistant) infectious disease models; here they determine the equilibrium amount of land under agriculture, A, in recovery, U, pristine (and recovered), F, and the size of the human population supported by agriculture, P.

$$\frac{dF}{dt} = sU - dPF \tag{13.4}$$

$$\frac{dA}{dt} = dPF + bU - aA \tag{13.5}$$

$$\frac{dU}{dt} = aA - (b+s)U \tag{13.6}$$

$$\frac{dP}{dt} = rP\frac{A - hP}{A} \tag{13.7}$$

The model assumes that land is viable for agriculture for a period of time $1/a$ years. The land is then abandoned and slowly returns to the original habitat type by a successional process that takes $1/s$ years. Abandoned land may also be returned to agricultural use at a rate b. The model assumes a human population birth rate r (around 4%) and that each human requires h units of land to sustain them. Each of these parameters can be quantified for different habitats and historical levels of food productivity. The system settles asymptotically to the following equilibrium conditions:

$$A^* = hP^*; \quad U^* = \frac{aA^*}{(s+b)};$$

$$F^* = \frac{ah}{d}\left(\frac{s}{s+b}\right); \tag{13.8}$$

$$P^* = \frac{F_0 - \left(\left(\frac{ah}{d}\right)\left(\frac{s}{s+b}\right)\right)}{h\left(\frac{a}{s+b} + 1\right)}$$

where F_0 is the original extent of the forest (or savanna).

We can also derive an approximate expression for the initial rate of spread of agriculture, equivalent to the R_0 of epidemic models. For this system $R_0 = rdF_0/a$, suggesting that agricultural development will expand at a rate determined by human population growth, the efficiency of habitat conversion, and the duration of time over which the land supports agriculture.

This simple approach is insightful in that it suggests counterintuitive results. For example, the longer land is viable for agriculture, the less of it will remain in a pristine state (Figure 13.4). This occurs because large amounts of agricultural land support an increasing human population that constantly demands more land. When agricultural land is productive for only a short period of time, as is typical of slash and burn (swidden) systems, humans are constantly forced to move on, resulting in a mosaic of patches in different stages of succession. Echoes of this simple theoretical prediction are seen in the patterns of land-use change in Puerto Rico, the Philippines, and the USA (Figure 13.4). In particular, as land use tends to decline with altitude and the time to succession is positively correlated with duration of time for which land is used, then this can produce a

changing mosaic of land-use patterns across altitudinal locations and soil types.

The efficiency with which land is used for agricultural production has led to a heated debate in conservation biology, where again theoretical insights have proved important. There is instinctive gut reaction among many environmentalists that genetically modified food and big agriculture are bad for biodiversity. Certainly, the massive expansion of industrial agriculture in the twentieth century has converted many natural habitats into fields and orchards. (One of the first books to consider conservation as a topic for valid scientific study was Dudley Stamp's *New Naturalist* volume, which focused on how loss of hedgerows due to agricultural expansion was having major impacts on British farmland birds; Stamp, 1969). However, the Millennium Ecosystem Assessment (MEA) provides an important counter example. Land-use changes in the Far East over the last 40 years have been driven dramatically by the need to feed the area's rapidly growing human population. Industrialization, in turn, has hugely increased the region's wealth and food purchasing power. In the early 1960s about 35% of the available land was already converted to agricultural production; if

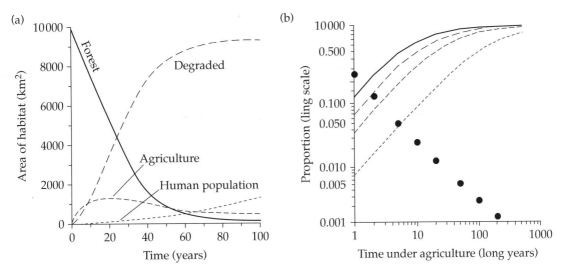

Figure 13.4 (a) Temporal change of land use based on a simple SIR forest fire model of land-use change. (b) Equilibrium proportions of land remaining as undisturbed forest (black dots) and as agricultural land (lines). The land retained under agriculture is depicted for four different rates of recovery of abandoned land: 5 years (solid line); and 10, 25, and 100 years (highest to lowest dashed lines). From Dobson *et al.* (1997a).

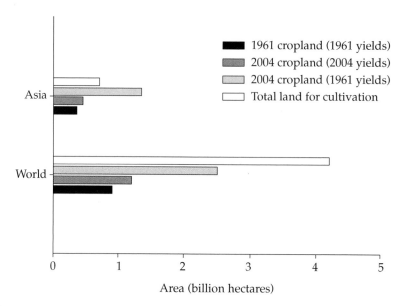

Figure 13.5 Land use and agricultural intensification. The graph shows the areas of land needed to feed the population of Asia (upper bars) and the world (lower bars) in 1961 and 2004. In each case the top bar gives the total land area available for cultivation (excluding mountains and cities), the lowest bar gives the total area of land under cultivation in 1961, the bar above this gives the land under cultivation in 2004, and the green bar gives the area of land needed to produce 2004 agricultural yield if agricultural efficiency had remained at 1961 levels (that is, with no Green Revolution). After Wood *et al.* (2006).

there had been no advances in agriculture, then all of the remaining land and a significant portion of Africa would be needed to feed the current population of the Far East (Figure 13.5).

However, increases in agricultural efficiency, particularly the development of high-yield varieties of rice and grains (see Chapter 12 in this volume; Conway, 1997), have allowed rates of land conversion to proceed at a much lower rate. In the terms of this model, high-yielding agriculture acts to decrease h, the amount of land needed for each human. Ironically this increased agricultural efficiency comes with an increased dependence upon water for irrigation, and this, in turn, may provide a further motivation to conserve habitats such as forests that store water and regulate its flow across the landscape. In the following section, which discusses models of ecosystem services, we describe how this model can be adapted to include a relationship representing support for the human population, P, from the pristine land, F (eqn 13.9; see below).

More general explorations of this effect have been examined in a series of models developed by

Green *et al.* (2005; but see Vandermeer and Perfecto, 1995, 2005). These models suggest that the best type of farming for the persistence of other forms of biodiversity is dependent both upon the demand for agricultural products and on how the population densities of sensitive species change with agricultural yield. If species are highly sensitive to reductions in pristine habitat, then intensive agriculture may be a better option, as it minimizes the land used for any level of agricultural productivity. In contrast, wildlife-friendly agriculture may require a much larger area of land under cultivation to achieve similar levels of food productivity, and this could be much more detrimental to species that are unable to use land that has been partially converted for agricultural use.

13.4.2 Habitat loss, species extinctions, and extinction debts

Conservation biologists have long wrestled with how to link loss of a habitat to loss of the species it contained. Initial work in this direction focused on using the species area curves of MacArthur and

Wilson (1963, 1967) to extrapolate from loss of habitat to loss of biodiversity (Diamond and May, 1981; Reid, 1992; Simberloff, 1992). This in turn led to the SLOSS (single large or several small) debate (Gilpin and Diamond, 1980; Higgs and Usher, 1980; Wilcove et al., 1986), which focused on ascertaining whether two small reserves were more effective than one large reserve in conserving a maximum number of species.

Whereas some uses of the species-area extrapolation have been successful in predicting the expected number of extinctions for birds in the eastern USA (Pimm and Askins, 1995) and in southeast Asia (Brooks et al., 1997), the approach has been less successful for other taxa (Simberloff, 1992). This may be because the basic species area curve assumes that both species and habitat loss occur at random, and when they are non-random and correlated then species may be lost at a faster rate than is predicted (Seabloom et al., 2002). There may also be a considerable lag before habitat loss leads to species loss from the landscape. Tilman and colleagues (Tilman et al., 1994) have suggested that because of this time lag the conversion of forests, savannas, and other natural habitats into agriculture creates an extinction debt. That is, in the absence of intervention, extinctions will continue to occur after habitat destruction has ceased. These extinctions are predicted to occur among the species persisting in the remaining, isolated patches of natural habitat in a fashion largely determined by the dispersal and competitive strategies of each species still present in the community (Nee and May, 1992). Two different management approaches can be adopted to reduce the rate of extinction: where possible patches in the landscape may be reconnected by protecting land that might serve as corridors for the dispersal of key species that would otherwise go extinct. Alternatively, habitat loss may be reduced by explicitly recognizing the dependence of the human economy upon the services provided by natural habitats.

13.5 Ecosystem services

A significant number of conservation studies over the last 10 years have focused on the importance of ecosystem services, defined as those goods and services that are provided by nature but not necessarily valued in the marketplace (Costanza et al., 1997; Daily, 1997; Daily et al., 2000; Balmford et al., 2002). Examples of such services include pollination of crops, production of water, climate amelioration, erosion control, and aesthetic enjoyment. The rest of this chapter focuses on some of those studies as we explore how ecosystem services can be monitored, how consideration of ecosystem services might be incorporated into models for land-use change, and how simple models can help us understand what patterns of change in ecosystem services we might expect as land use grows.

13.5.1 Monitoring economic goods and services provided by natural ecosystems

Ecologists and economists have spent much of the last decade wrestling with how to quantify the goods and services provided by natural ecosystems (Daily, 1997; Daily et al., 1997, 2000). At one extreme this discussion has focused on attempting to quantify the net annual economic benefit provided to the human economy by natural ecosystems (Costanza et al., 1997). However, such studies do not address a number of important practical issues. For example, few ecologists (or ecologically informed economists) would disagree with the assertion that the world's wetlands provide a large number of vital ecosystem services. But for a particular marsh at a particular site, what are the consequences to ecosystem services of filling 1, 5, or 10 acres? What is the impact of losing a particular species in that marsh (e.g. an endangered bird) or of a change in the composition of the dominant vegetation (for example, replacement of sedges by cattails)? Knowing the overall value of wetlands does little to inform these sorts of everyday decision. This has led to a heated discussion among ecologists regarding the dependence of ecosystem function upon species diversity (Tilman et al., 1997; Chapin et al., 2000; Kinzig et al., 2001; Loreau et al., 2001; Bond and Chase, 2002).

The recently completed Millennium Ecosystem Assessment has adopted a classification of ecosystem services that provides a useful way of framing a discussion about how we might measure changes in the rates at which they are delivered. The Millennium Ecosystem Assessment divides these services into supporting services and provisioning, regulating, and cultural services. Quantifying ecosystem services on a species-by-species basis is clearly an impossible task, particularly as many of the most important services are undertaken by microscopic species whose taxonomic status is unclear (Nee, 2004). Nevertheless, it is likely that different types of service will predominantly be undertaken by species on different trophic levels. For example, regulating services, such as climate regulation and water regulation and purification will be predominantly undertaken by interactions between species at the lowest trophic levels. In contrast, cultural services, such as recreation and tourism, as well as aesthetic and inspirational services, will often require ecosystems that contain a near-complete suite of large mammals, raptors, and other charismatic species; this requires that the upper trophic levels remain intact. Thus the trophic diversity of ecosystems can provide important information on their ability to provide different types of services. Ultimately, a key index of the health of an ecosystem might be some measure of trophic diversity, such as food-chain length, or the standard deviation of trophic position for all the species in a food-web of the region (Dobson, 2005). This index could provide a potentially important indication of the diversity of ecosystems services supplied by the system.

13.5.2 Models of ecosytems services and land-use change

Although hard to measure, ecosystem services are an important factor in considering the impact of land-use change. In this section we present a model that explores how consideration of the economic value of pristine land changes our expectation of the growth of converted land. Let us initially consider a simple modification of the model described above for the dynamics of land-use change (eqns 13.4–13.7). One way to incorporate ecosystem services would be to include an additional expression for the dynamics of an ecosystem service that is supplied by the forest to the converted land, for example, provision of water for crops and sanitation. Here we will assume that agriculture and human well-being are fundamentally dependent upon the presence of water (which is certainly true), but that the volume of water available is a function of the area of land still maintained as natural forest (or savanna). Madagascar's Ranomafana National Park provides a classic example of this effect. Although the park is justly famous for containing the world's highest diversity of lemur species, as well as numerous endemic bird and plant species, the main reason it was protected is because it contains a hydroelectric dam whose efficiency is determined by a steady stream of water from the forested watersheds that form the national park. The dam is not especially intrusive, but its turbines are dependent upon the forest retaining water and releasing it at a steady and continuous rate. This dam supplies most of the electrical energy for the major city of Antananarivo. Water from the park is also vital for the agriculture that covers the coastal plain between the park and the Indian Ocean. The potential of the forest to store and maintain a steady supply of water for electrical power and irrigation is a classic example of an ecosystem service in action. We can readily include this effect into our model of habitat conversion by assuming that the rate of food production on the land converted to agriculture is a simple Michaelis–Menton-type function of the area of land retained as pristine forest (e.g. $A \sim > AF/(F + F_{50})$, where F_{50} is the area of pristine land at which agricultural productivity drops to half its maximum level). The expression for human population growth is the only equation that needs be modified. It now becomes

$$\frac{\mathrm{d}P}{\mathrm{d}t} = \frac{rP(A - hP(1 + F_{50}/F))}{A} \tag{13.9}$$

This simple modification has a number of important consequences. Most notably, the system now settles to a new set of equilibria. The expressions for abandoned land remains unchanged from above, $U^* = \frac{aA^*}{(s+b)}$, but the agricultural land, forest,

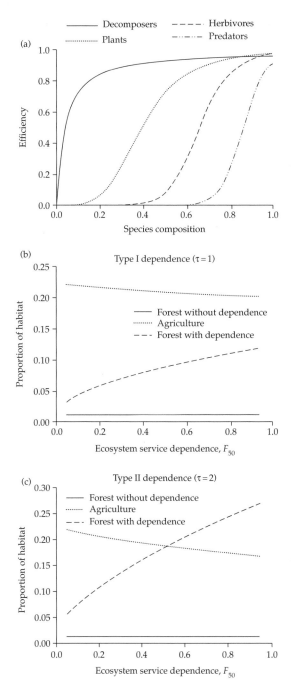

Figure 13.6 (a) The phenomenological relationship between the efficiency with which ecosystem services are undertaken and species composition (as a proportion of original intact community). The curves are drawn for services that are predominantly driven by species on specific trophic levels (e.g. ecotourism may be highly dependent on top carnivores; soil productivity and photosynthesis are dependent upon bacteria and plant diversity). They assume that a Michaelis–Menten type function can be defined for a 50% level of ecosystem service efficiency at some level of species diversity (after Dobson *et al.*, 2006). (b and c) Equilibrium proportion of land under agriculture and retained as 'natural habitat' when agricultural land has a dependence upon services supplied by the 'natural habitat'. In each case the *x* axis describes the position of the 50% ecosystem service efficiency inflexion point, the solid line indicates the area of pristine land in the absence of any recognition of dependence of agriculture on ecosystem services, the dashed line indicates the area of land retained in its pristine, forested form as the dependence of agricultural productivity increases, and the dotted line is the area under agriculture. In (b), we have assumed that the slope of the agricultural dependence on pristine land is unity ($\tau = 1$) at the inflexion point, in (c) we assume the slope is 2; thus (b) phenomologically corresponds to ecosystem services produced by species at the lowest trophic levels (e.g. bacteria, soil micro-organisms) and (c) corresponds to services produced by species at low to intermediate trophic levels (e.g. plants). As the slope of the dependence increases (corresponding to higher trophic levels) it is possible to get multiple-stable states (A.P. Dobson, unpublished work).

and human population settle to new levels:

$$A^* = hP^* \left(1 + \frac{F_{50}}{F^*}\right)$$

$$F^* = \tfrac{1}{2}c_1 \pm \tfrac{1}{2}\left(c_1^2 + 4c_1 F_{50}\right)^{1/2} \text{ where } c_1 = \frac{sah}{d(b+s)}$$

$$P^* = \frac{F^*(F_0 - F^*)}{h(F_{50} + F^*)\left(\frac{a+b+s}{b+s}\right)} \qquad (13.10)$$

The primary effect of including this phenomenological form of ecosystem service is to reduce the amount of habitat that is converted to agricultural land. This, in turn, reduces the size (and density) of the human population supported by the agricultural land; more explicitly it accurately recognizes that the size of the human population that can be supported is not solely dependent upon the land converted to agriculture but also upon the land remaining as forest; this slows future demands for habitat conversion (Figure 13.6). It is a relatively trivial exercise to modify eqn 13.9 so that the term in F_{50}/F is raised to the power of τ (Figure 13.6b and c illustrates the case for $\tau = 1$ and 2); this will correspond to increasingly strong (steep) dependence of agriculture on the remaining forest land. As the value of τ increases, the system now has the potential to settle to either of two (or more) alternative states (A.P. Dobson, unpublished work). These results echo the earlier work of May (1977c) and the more recent explorations of others (Carpenter and Cottingham, 1997; Scheffer et al., 2001). Plainly, the results are dependent upon the functional forms we have chosen to represent the dependence of agriculture on water and the efficiency with which increased agricultural production translates into increased human population or increased demand for resources. Notice also that the larger the dependence of agriculture on pristine habitat, the greater the proportion of forest that is retained. We suspect that these are all functions that could be measured for different crops, habitats, and human culture.

The main point of this exercise is to illustrate that once we recognize a dependence of the human economy on services provided by natural resources, we begin to see changes in the predicted rate at which we convert natural habitats. Notice too that an interesting psychological switch has occurred in the motivation to conserve wild areas such as Ranomafana. Initially the land was set aside for purely utilitarian reasons. Once the area's biological wealth was appreciated, this became an equally important justification for protecting the area. If the principal way to protect biological diversity is to set aside land in nature reserves, we will be more successful if we can identify and quantify the diversity of utilitarian services provided by different types of natural ecosystems in addition to their value as reserves for imperiled species.

13.5.3 How will the value of ecosystem services change as habitat is converted?

This initial consideration of ecosystem services assumes that the converted land is entirely dependent upon the non-converted land for services. In most cases, however, the converted land will not only provide new services of its own (e.g. farm crops, retail outlets, golf courses) but also will continue to provide a significant number of the services that the land provided prior to its conversion. The new set of plants will also photosynthesize and create and retain soil. (Indeed, some invasive plant species do so at a more efficient rate than the native species they displace.) We therefore need to develop frameworks that consider how different ecosystem services change as we modify the proportion of converted and pristine land in the landscape.

A handful of studies have examined the relationship between economic goods and services provided by natural systems and the services supplied in adjacent, or equivalent, modified systems (Peters et al., 1989; Bonnie et al., 2000; Kremen et al., 2000; Balmford et al., 2002). Cost-benefit analyses of these studies suggest that in most cases the net economic value of habitat that is totally converted declines by an average of around 50%, and the benefit/cost ratio of conserving the remaining unconverted habitats may be as high as 100:1 (Balmford et al., 2002).

The essentials of these phenomena have been examined in a general model of land-use change (Dobson, 2005). The principal objective of this terse model of the services provided by pristine and modified environments is to examine the underlying factors that confound our ability to detect changes in the rate at which ecosystem services are supplied.

Let us assume that the net production of ecosystem services can be characterized by the net relative value (NRV) of the goods and services produced by both the pristine and converted habitat. Let us then consider the situation where a proportion, p, of habitat has been converted from its pristine state into agriculture, mining, or some other modified use. We will define the current value of the total landscape as a simple sum of the converted and unconverted portions, relative to its initial value of unity in its pristine state when all it supplied were indirect services to the human economy. Doing so naively assumes we have some way of quantifying this value when, at present, all we really know is that we tend to undervalue it (Daily, 1997; Daily et al., 1997; Balmford et al., 2002).

The model is developed in three steps. The first is to define how the goods and services produced by the pristine land decline as the amount of pristine habitat shrinks. The second is to describe how the modified portion of the landscape produces goods and services. The total goods and services produced, NRV, is then the sum of these two parts.

Let s' be the value of the goods and services produced by a unit area of pristine land, a function of three parameters: p the proportion of land converted; ES_{50} the proportion of land converted at which ecosystem services produced decline to half of their maximum, pristine value; and τ, a shape parameter describing how sharply goods and services derived from pristine land fall as the amount of remaining pristine land shrinks. The first step in the model is to define s', goods and services provided by the remaining area of pristine land as:

Goods and services from a unit area of pristine land
$$= s' = ((1 - p)/p)^{\tau}/(ES_{50} + (1 - p)/p)^{\tau}) \quad (13.11)$$

In Figure 13.7 the solid lines show the quantity $(1 - p)s'$, which is the goods and services produced by all of the remaining pristine land.

The second stage in developing the model is to define how the converted land produces goods and service. These falls into two parts: those

produced by the converted land independently of the pristine land, s, and those produced by the converted land using services from the pristine land, sds'. Here d represents the degree of dependence of converted land on pristine land. For example, pollination of farm crops may be strongly dependent upon the diversity and abundance of pollinators in the remaining patches of natural habitat (Kremen et al., 2002; Ricketts et al., 2004).

Goods and services from a unit area of converted
$$\text{land} = s + sds' \quad (13.12)$$

In Figure 13.7 the dotted lines show the quantity $ps(1 + ds')$, which is the goods and services produced by all of the converted land.

The final step is to add these two terms together to give the NRV of the goods and services produced by the total habitat (converted and pristine):

$$NRV = (1 - p)s' + ps(1 + ds') \quad (13.13)$$

Figure 13.7 illustrates how this value, the net service supply rate, can change as a function of p, the amount of land converted. Notice in particular how an ecosystem where the modified land is heavily dependent on the pristine land and the pristine land fails to provide goods and services at high p leads to a situation where total goods and services are high for all low and intermediate values of land use, but fall precipitously as the proportion of land used approaches 1.

The key point is that if we are to monitor declines in ecosystem functions as changes in the net value of goods and services they supply, then we need to know more about the shapes of these dependence curves. This is principally because our ability to detect changes in the economic value of natural habitats will be a subtle function of the relative value of services supplied by the pristine and modified habitats as well as the dependence of the services provided by the modified landscapes on the presence of unmodified habitat.

A handful of simple messages emerge from the model. We usually undervalue the economic services supplied by the natural environment as we are able to put a quantitative economic value on only a subset of the services provided by the natural habitat (Balmford et al., 2002). If the value

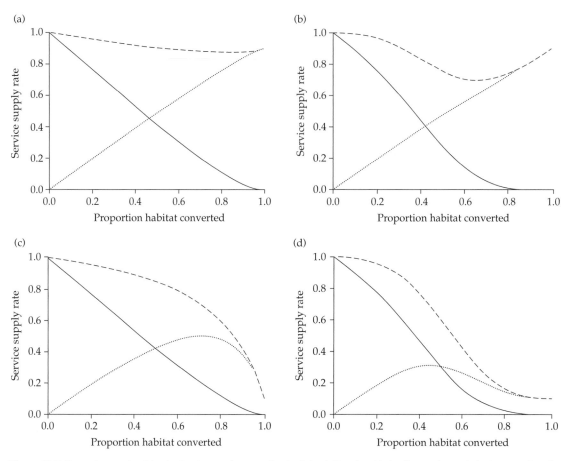

Figure 13.7 Net services produced by simple mixtures of converted and pristine habitat (y axis); in all cases the x axis is the proportion of habitat converted from pristine habitat to agricultural land, the solid line illustrates the declining rate of production of ecosystem services (to converted habitat) as pristine land is converted, the dotted line is the value of services produced in the converted (agricultural) habitat, and the broken line is the sum of the services produced in the mixture of two habitats at this level of conversion (eqn 13.10 in the text). We have illustrated four different scenarios that reflect all four combinations of high/low dependence and fast/slow rates of ecosystem service decline. (a) Weak dependence of agriculture on pristine land ($d = 0.1$) and services decline slowly ($ES_{50} = 0.5$) as habitat is lost. (b) Weak dependence of agriculture on pristine land ($d = 0.1$) and services decline rapidly ($ES_{50} = 0.2$) as habitat is lost. (c) Strong dependence of agriculture on pristine land ($d = 0.9$) and services decline slowly ($ES_{50} = 0.5$) as habitat is lost. (d) Strong dependence of agriculture on pristine land ($d = 0.9$) and services decline rapidly ($ES_{50} = 0.2$) as habitat is lost.

of the services provided by the modified habitat are assumed to be similar to or less than those provided by the pristine habitat, then we will be able to detect changes in the value of these services only when the modified habitat's value is highly dependent upon the area of pristine habitat. When the services provided by the modified habitat are largely independent of the pristine habitat, the rate of change of land value may be too shallow to be detected as more of the pristine habitat is

degraded or converted. In many cases, the initial decline in land value will be reversed as the new land use comes to dominate the landscape. Unfortunately, mistakes are costly; doubly so as the cost of restoration may take many years to be recovered. If developers discount the future, services supplied by pristine habitats may never be recovered once conversion has proceeded beyond a critical economic threshold where the cost of restoration exceeds their future discounted potential value.

13.6 Conclusions

Our goal in writing this chapter has been to illustrate and suggest emerging areas in conservation biology where theory can provide valuable insights into real-world problems. The problems we explored ranged from the very specific (e.g. is the grizzly bear population around Yellowstone National Park in danger of extinction?) to the very general (e.g. what is the relationship between land-use change and the loss of ecosystem services?). This is significant, for it indicates the breadth of opportunities awaiting theoreticians interested in tackling applied problems. Indeed, we note with pleasure that a large fraction of the students applying to graduate programs in ecology today do so with an avowed interest in conservation biology.

It is also worth emphasizing that many of the most interesting examples discussed in this chapter involve the integration of the social sciences, especially economics, into solutions to ecological problems. Examples include the incorporation of cost constraints into the selection of reserve networks (a problem that also requires one to delve into operations research) or linking rates of deforestation to the values of ecosystem services produced by pristine and disturbed lands. Economic considerations have been a part of some ecological models (e.g. fisheries) for decades, but on the whole, the social sciences have yet to be incorporated into much of the theoretical research now underway in conservation biology. We predict (and hope!) this will change, given that a viable solution to an environmental problem must make sense not only from a scientific perspective but also from a socio-economic one.

There is no reason why advances in applied fields cannot yield insights relevant to more academic disciplines, and this is certainly true in the case of conservation biology. The population viability models used to guide policies for imperiled species continue to shed light on the structure and functioning of metapopulations. Conversely, metapopulation theory has been enormously influential in conservation biology. Thus, one is tempted to conclude that the distinctions between theory and practice, or pure and applied, are illusory to a large degree. For aspiring theoreticians, we offer this advice: Pay attention to applied problems, for they offer plenty of theoretical challenges as well as practical benefits. And for conservation practitioners, we suggest that theory (in its many forms) will play an increasingly valuable role in the search for solutions to real-world problems.

Of course, the conservation of biodiversity will depend ultimately upon the preservation of natural habitats, the restoration of degraded ones, and the discovery of new ways to maintain important ecosystem services on lands largely devoted to activities such as farming and energy generation. Our major worry, therefore, is that whatever scientific progress we are making occurs against a backdrop of a worldwide series of uncontrolled experiments, involving the destruction of wildlands, extinction of species, changes in nutrient cycles, and even changes in the Earth's climate. If we may offer a final, whispered word of advice to the upcoming generation of theoretical ecologists and conservation biologists, it is simply this: hurry!

CHAPTER 14

Climate change and conservation biology

Jeremy T. Kerr and Heather M. Kharouba

14.1 Introduction

It is increasingly recognized that, as a result of ever-growing atmospheric inputs of greenhouse gases like carbon dioxide from the burning of fossil fuels, the climate is changing regionally and globally. This has been affirmed, in light of increasing scientific understanding, in the latest report of the Intergovernmental Panel on Climate Change (IPCC) in 2001, by the US National Academy of Sciences in its 2001 report, and most recently by a statement from the Science Academies of all G8 countries, along with China, India, and Brazil. This latter statement calls on the G8 nations to 'Identify cost-effective steps that can be taken now to contribute to substantial and long-term reduction in net global greenhouse gas emission [and to] recognize that delayed action will increase the risk of adverse environmental effects and will likely incur a greater cost'.

Global warming caused by elevated greenhouse gas levels is expressed with long time lags, which can be difficult to appreciate by those unfamiliar with physical systems. Once in the atmosphere, the characteristic residence time of a carbon dioxide molecule is a century. And the time taken for the ocean's expansion to come to equilibrium with a given level of greenhouse warming is several centuries. If current trends continue, by around 2050 atmospheric carbon dioxide levels will have reached more than 500 parts per million, which is nearly double pre-industrial levels. The last time our planet experienced levels this high was some 20–40 million years ago, when sea levels were around 100 m higher than today. It can also be difficult to relate intuitively to the seriousness of

the roughly 0.7 °C average warming of the Earth's surface over the past century. And the warning by the IPCC in its 2001 report, that global warming would be in the range of 1.4–5.8 °C by the end of this century, may also seem unalarming when we experience such temperature swings from one day to the next. There is, however, a huge difference between daily fluctuations, and global averages sustained year on year; the difference in average global temperature between today and the last ice age is only around 5 °C.

There remain, of course, uncertainties about many of the details of climate change. In particular, we are unsure of the timescales for some important nonlinear processes. As the polar ice caps melt, the surface reflectivity declines, feeding forward to cause greater warming and faster melting. But the timescale for the ice-cap to disappear entirely—decades or centuries?—is unclear. As permafrost thaws, large amounts of methane gas are released, further increasing global warming (methane is a more efficient greenhouse gas than carbon dioxide). Increased precipitation in the North Atlantic region and rapidly melting glacial ice in Greenland reduces the salinity of ocean water at the surface; this reduces the water's density and it sinks more slowly. Such changes in marine salt balance have, in the past, modified the fluid-dynamical processes which ultimately drive the Gulf Stream, turning it off over decadal timescales. Similarly pressing questions—especially how much? and how fast?—can be asked of sea level rise. For a survey of other nonlinear effects with potentially dangerously large impacts see Schellnhuber *et al.* (2006).

Climate change is responsible for significant human mortality, partly because of recent and unprecedented heat waves. A recent Royal Society report addresses the interplay between climate change and crop production, unhappily emphasizing that 'Africa is consistently predicted to be among the worst hit areas across a range of future climate change scenarios', and also noting that present drought conditions across large areas of sub-Saharan Africa are almost surely associated with increases in surface temperatures in the Indian Ocean caused by global warming. World Health Organization statistics indicate that, conservatively, more than 10^5 people die annually from anthropogenic climate change (Patz *et al.*, 2005). This issue has received comparatively little attention, perhaps because mortality from anthropogenic climate change is relatively diffuse.

It is difficult to think of any characteristic of the environment that affects the living world so pervasively as climate. Because of its fundamental significance to so many biological phenomena, climate change presents a formidable array of questions to ecologists and evolutionary biologists. Some of these questions are already subjects of long-standing empirical and theoretical debate. How does climate relate to present-day spatial variation in species richness? Will future climate change impacts on species be predictable using observations of past changes? Are species' ranges determined by climate or biotic interactions and by which mechanisms will climate change force species to shift? Has global change, of which climate change is a critical ingredient, committed the world to a new mass extinction? This chapter briefly reviews the mechanisms that may govern how species and species assemblages are affected by climate and thence assesses some of the implications of climate change for the future of biological diversity in the twenty-first century.

14.2 Climate effects on species distributions

Climatic tolerance is known to affect species' geographical distributions (Woodward, 1987; Root, 1988a; Parmesan *et al.*, 2005 and references therein). Fundamentally, climatic effects on metabolic rates contribute strongly to whether a species is likely to persist, go extinct, or shift from a particular area. For many species, potential metabolic constraints on species distribution can be reduced to a simple question: can the species elevate its metabolism to a sufficiently high level to sustain itself? However, climate impacts on species distributions are not limited to effects on metabolic rates. Freeze tolerance or avoidance present additional mechanisms limiting where species are found and how they will likely respond to shifting climatic conditions. Climatic constraints clearly include more than just the changes in mean annual temperature that are most commonly reported in widely disseminated scientific consensus documents about climate change. The extremes of climate sometimes play a more important role in shaping species distributions than mean annual conditions (e.g. Kukal *et al.*, 1991). For instance, minimum winter temperatures in many areas have changed far more than mean annual temperature (IPCC, 2001). Such changes have reduced sea ice coverage around Hudson Bay, reducing polar bear populations that rely on the ice to reach some of their prey populations (Derocher *et al.*, 2004). In this instance, population reductions follow from reduced resource availability for the species, which is ultimately a result of warming minimum winter conditions.

Climatic effects on mammal species range limits are likely to be determined, in part, by the temperature dependence of metabolic rates. The ecological and evolutionary implications of that link form the basis for a broad range of related research that collectively falls under the umbrella of metabolic theory (e.g. Allen *et al.*, 2002; Brown *et al.*, 2004; but see Algar *et al.*, 2006; Muller-Landau, 2006). Independent field evidence collected from mammal (and bird) populations demonstrates that the lower limits of metabolic rate increase toward the poles, suggesting that metabolic rates must indeed be higher for populations to persist in those environments and that fewer metabolic strategies exist that would allow an organism to sustain itself as temperature declines (see Kleidon and Mooney, 2000; Anderson and Jetz, 2005). However, Humphries *et al.* (2005) review evidence that niche space in colder environments may be broader than previously thought. For at least one species, *Tamiasciurus hudsonicus* (red squirrels), field metabolic rate actually declines as temperature drops. Humphries *et al.* (2005) argue that this finding, although isolated, may be less unusual than suggested by Anderson and Jetz (2005) because so few measurements of metabolic rate, and particularly field metabolic rate, have been made in very cold environments. The implications of these findings for the effects of changing climates on species ranges are variable, depending on how metabolic rate varies with temperature, but it seems clear that predicting individual species' responses will often be difficult.

Mechanisms determining how temperature might limit species' northern range limits are still uncertain. Using data on winter distributions, Root (1988a) demonstrated that the northern range limits for many North American bird species were likely affected by minimum winter temperatures or annual frost-free period. The former factor most likely affects range limits through climatic tolerance (i.e. increased mortality in response to extremely low temperatures) while the latter may imply a stronger dependence on resource availability (longer frost-free periods allow for greater resource acquisition). Root (1988b) suggested that minimum winter temperatures affect bird range limits through their effects on birds' energy expenditures. At their northern range limits, the basal metabolic rates of species limited by minimum winter temperatures rise to approximately 2.5 times the rates observed further south. Areas where winter cold would require an even higher basal metabolic rate to avoid freezing would exceed the tolerance of these species. Climatic effects on bird species range limits may also act indirectly by altering resource availability (e.g. Repasky, 1991; Gross and Price, 2000). Whatever mechanism governs this relationship, one of the results is a very strong relationship between species richness and temperature (Figure 14.1).

The distributions of mammal species have long been known to relate to aspects of climate. For instance, Bergmann's rule states that spatial variability in body size is related to climate, with larger-bodied organisms occurring proportionately more frequently in colder areas than smaller-bodied

organisms. Several mechanisms for such gradients are possible, including: the body-size dependence of post-glacial recolonization; the random recolonization of cold areas by large-bodied ancestors of present-day species that then diversified *in situ*; that large-bodied organisms tolerate cold more effectively because of their smaller surface area/volume ratios, leading to far better heat conservation than small-bodied organisms; the converse, that losing excess heat is easier for small-bodied organisms in warm areas; or that large body size leads to improved starvation resistance in cold areas where there may be long periods of low resource availability (Blackburn and Hawkins, 2004). Blackburn and Hawkins' (2004) tests of these mechanisms for Bergmann's rule among mammals in areas of North America that were covered with ice during the last glacial maximum find support for simple cold tolerance and, to a lesser extent, starvation resistance, based on measurements of average annual temperature (for cold tolerance) and coarse-resolution Normalized Difference Vegetation Index (NDVI) data composited over the study region (see Kerr and Ostrovsky, 2003 for discussion of ecological applications of satellite data). Inability to tolerate freezing conditions may also influence mammal species richness across the Americas (Figure 14.1a): the best predictor of species richness across this region is minimum annual temperature and this relationship has a pronounced breakpoint at 0 °C (Figure 14.1b). From this observation, a second prediction can be derived, namely that body size should also vary with minimum temperature, a minor modification of Bergmann's rule. The minimum temperature/mean body-size relationship supports the prediction (Figure 14.1c). Determining the mechanisms—perhaps cold tolerance, starvation resistance, or some combination of these factors—behind these relationships will improve the odds of using such results to predict how species richness will track shifting climatic conditions.

Climatic change is likely to affect ectotherms particularly quickly, given their well-documented and multi-faceted dependence on environmental temperature (e.g. Lederhouse *et al.*, 1995; Blouin-Demers and Weatherhead, 2001). For instance,

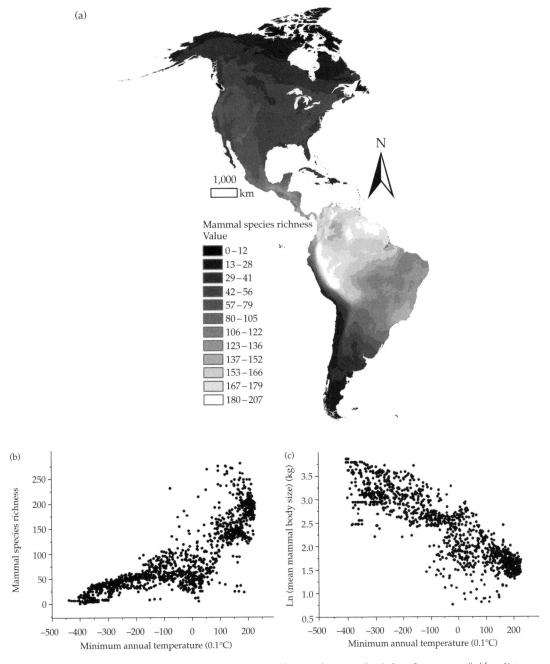

Figure 14.1 (a) Geographical patterns of continental mammal species richness in the western hemisphere. Data are compiled from Natureserve range maps for all mammal species (1755 species). (b) A graph showing the relationship between mammal species richness per quadrat and the strongest predictor of that variable, minimum annual temperature, throughout North and South America. There are 1255 equal-area quadrats covering the Americas. There is a breakpoint in this graph at 0 °C, suggesting that tolerance to freezing conditions may contribute mechanistically to the distribution of mammals in this area. (c) A secondary prediction of the tolerance mechanism is that body size should be inversely related to minimum annual temperature because large-bodied organisms lose heat in cold conditions far slower than small-bodied organisms. The correlation presented in panel b predicts Bergmann's rule, which is shown here for all mammals (1755 species) across North and South America, where each data point represents the mean body size of all mammals found in that quadrat ($n = 1255$).

spatial variation in species richness for many ectotherm taxa, including blister beetles (Kerr and Packer, 1999), tiger beetles (Kerr and Currie, 1999), ants (Kaspari *et al.*, 2004), butterflies (Kerr *et al.*, 1998, 2001; Turner *et al.*, 1987; Algar *et al.*, 2006), reptiles (Currie, 1991), amphibians (Currie, 1991), and fish (Kerr and Currie, 1999), relates strongly to climate and particularly temperature. Although species' responses to shifting climatic conditions are likely to be individualistic, the aggregate effect of these individual species-climate interactions is the strong relationship observed between species richness and climate (particularly temperature but also precipitation) across most of the world (Hawkins *et al.*, 2003; Currie *et al.*, 2004). Changes in climatic conditions are likely to cause widespread alteration in the distribution of species and, as a result, change patterns of species richness.

14.3 Climate change effects in mountainous regions

Climate change, particularly temperature increase, is expected to force many species inhabiting mountainous areas to shift toward higher elevations to track climatically suitable areas (Peters and Darling, 1985). In principal, it should be easier for species to track shifting climate along elevation gradients than through the far greater distances required along latitudinal gradients. After all, according to Hopkins' Bioclimatic Law, a 1 °C cooling is about the same as traveling 100 km toward the nearest pole but the same degree of cooling is achieved with perhaps a 130 m elevation increase (Hopkins, 1920). However, mountains become smaller toward their peaks, so species that shift upward with warming temperatures may suffer serious habitat loss. Thus, it may be easier to track shifting climate up mountainsides, but doing so has an inexorable cost in terms of habitat availability.

One of the first studies to detect climate-driven elevation shifts considered alpine plants in Austria (Grabherr *et al.*, 1994). Observed warming in the study region had been about 0.7 K in the 70–90 years prior to a series of field surveys conducted in 1992. Based on lapse rates with altitude, migration rates of roughly 1 m/year would have been

required for alpine plants to remain within the same climatic envelope over this time period. The observed rates of elevation shift, derived from detailed historical records of species locations, were closer to 0.1 m/year and even the most rapid shifts upward in elevation among these plants were only about 0.4 m/year. Hence, these species are lagging progressively farther behind the expanding edge of their climatic envelope and it is reasonable to expect an increase in extinction rates from these effects. This study demonstrates that even along relatively short elevation gradients, it is likely that many species will not be able to shift sufficiently rapidly to track the climatic zones they have most recently inhabited.

Evidence pointing to climate-induced elevation shifts among a variety of taxa has accumulated rapidly in the past decade. Parmesan (1996) discovered that bay checkerspot (*Euphydryas editha*) butterfly populations in western North America were less likely to go extinct at high elevations but also detected no retraction of the butterfly's population from the lowest elevation sites. A number of studies confirm part of the trend demonstrated by this early work: many species occurring in mountainous areas are shifting toward higher elevations. However some are also retracting from their lower elevation limits. Konvicka *et al.* (2003) found evidence suggesting that 12–15 of the 119 butterfly species in the Czech Republic were shifting upward in elevation. Some of these butterfly species expanded their ranges upward without losing low-elevation populations. The remainder have lost low elevation populations so now have smaller ranges because of reduced habitat availability towards higher elevations. Hill *et al.* (2002) showed that butterflies with southern range limits in Britain were more likely to lose populations at low elevations than species that reached their northern range limits in Britain. Similarly, Wilson *et al.* (2005) have documented the retraction of mountain-dwelling butterfly species' ranges toward higher elevations in Spain. These population losses led to a mean loss of a third of the regional range of these species with projections of perhaps 80% reductions in ranges expected in the next century. It is interesting that climate appears to strongly affect the location of range limits for species at the warm edge of their ranges in these studies, given MacArthur's (1972) hypothesis that the warm edge of species' ranges would most likely be determined by biotic interactions.

14.4 Past responses as a guide for future climate-change effects

In the search for better scientific tools to predict biotic responses to anthropogenic climate change, there is probably no clearer guide than observations of species' past responses to changing climate. Even though such climate changes were usually relatively gradual (with some exceptions; see Davis and Shaw, 2001), observations of past climate change have the potential to greatly enhance predictions of its future effects on species. In particular, discoveries of mechanisms forcing species to respond to shifting climatic conditions, the transitory nature of biotic communities, and species' individualistic responses to climate change were made using historical or paleoecological data-sets (reviewed in Graham et al., 1996 and Overpeck et al., 2005). It is essential to note that, although past climatic changes are similar to recent anthropogenic climate change, they are not identical. First, current climate change will likely lead to higher mean global temperatures than experienced at any time in the rest of the Quaternary period (the past 2 million years; Crowley, 1990; Overpeck et al., 2005). Many species that exist today have never encountered the climates likely to be observed in the next century. Second, changes will probably occur at least an order of magnitude faster than during the best-studied periods of prehistoric climate change (Webb, 1992; Malcolm et al., 2002). Last, and perhaps most important, are potential interactions between rapid climate change and the widespread conversion of natural environments to human land uses, which already presents the leading threat to biodiversity in North America (Dobson et al., 1997b; Kerr and Cihlar, 2004; Kerr and Deguise, 2004). There is simply no precedent in historical terms for this sort of interaction.

Reconstructing the relationship between historical climate and species distribution relies on fossil evidence of one form or another. In particular, microfossils that accumulate in lacustrine and marine sediments (Overpeck et al., 2005) provide the foundations of paleoclimate reconstruction. Pollen diagrams, pollen records, macrofossils, and DNA recovered from fossil pollen provide evidence of how plants have responded to past climate changes (Davis and Shaw, 2001). Particularly promising is the use of ancient DNA, which can be extracted from even badly degraded specimens and used to reconstruct life-history characteristics and environmental interactions for some species (e.g. Willerslev et al., 2003), including mechanisms of their decline or extinction (e.g. Shapiro et al., 2004).

Most evidence derived from fossil sources suggests that plants and animals have tracked past climatic changes individualistically. This result is now firmly established (Graham and Grimm, 1990; Graham, 1992; Graham et al., 1996; Davis and Shaw, 2001; Lyons, 2003) and supports the argument that biotic communities are contingent and that many species associations are transitory (Webb, 1992). Communities are not, for the most part, assembled from tightly linked components, nor do their parts form a highly coevolved assemblage (see Graham et al., 1996; Lyons, 2003). Climate change is likely to divide these assemblages into their constituent species, each of which responds according to its unique characteristics. To the extent that species respond to climate change individualistically, modern communities need not be, and often are not, analogous to past communities, nor will future biotic communities necessarily resemble their modern counterparts (Lyons, 2003). Evidence from the FAUNMAP database demonstrates that mammal species responded to past environmental changes by shifting their ranges individualistically (Graham et al., 1996; Lyons, 2003). Some species significantly expanded their range while others did not (Lyons, 2003). As with the purely natural climate changes of the past, anthropogenic climate change will have winners and losers in terms of species' responses, although the far more rapid pace of climate change in the present day could plausibly tip the balance toward species' declines.

Observations of rates at which trees have shifted in the past may have been overestimated, suggesting that species may have much more trouble tracking shifting climatic conditions than previously believed (Pearson, 2006). A classic study (Skellam, 1951) demonstrates that, even over the several thousands of years since the end of the last glacial period, oak-tree dispersal rates were far too low to have allowed them to recolonize Britain and reach their present-day northern boundary there. This was elegantly demonstrated by the development of an analytical diffusion process, seeded with liberal assumptions that improved acorn dispersal distance and frequency. Despite these, Skellam concludes that oak dispersal must have been considerably accelerated by animals, such as rooks. More recently, McLachlan *et al.* (2005) used chloroplast DNA surveys to show that past migration rates of temperate trees were significantly slower (<100 m/year) than originally thought (100–1000 m/year). These new estimates of past migration rates are up to an order of magnitude slower than previously believed. However, the molecular evidence presented by McLachlan *et al.* (2005) also suggests that recolonization following glacial retreat was greatly facilitated by the presence of small refugia where pocket populations had persisted; a possibility also raised by Skellam (1951) in his much earlier theoretical contribution. It is a positive sign that tree species are able to persist in these refugia for significant periods and then successfully re-colonize broad areas (McLachlan *et al.*, 2005; Pearson, 2005) but perhaps a dangerous sign that, should refugia not be viable under circumstances created through human activities, dispersal rates for many species are probably too slow to keep them within suitable climatic zones. Populations will then need to rely on long-distance dispersal events, which are clearly rarer, at least for trees, than previously thought.

Although the fossil record does demonstrate that trees and mammals have shifted large distances in the past, often successfully responding to relatively gradual, natural climate changes, there remains significant risk that these taxa will not be able to respond successfully to anthropogenic climate change (Malcolm *et al.*, 2002; Pearson, 2005). Recent evidence regarding megafaunal extinctions following the end of the last glacial period suggests that even reasonably gradual climatic change can cause extinctions. For instance, Guthrie (2006) provides evidence from macrofossil records in the Yukon Territory and Alaska of some large mammal species declines, such as of horses (*Equus* spp.) and mammoths (*Mammuthus* spp.). These results suggest that these declines were ultimately driven by natural, climate-induced changes in vegetation, particularly forage quality, at the Pleistocene–Holocene transition between 11 500 and 13 500 years before present. Guthrie (2003) also notes that horse body size declined substantially before the arrival of humans and that ecological change, rather than overkill by newly arrived human hunters, is likely to have precipitated horse declines in northwestern North America. Although these results do not preclude the possibility that human hunting pressures delivered the *coup de grace* for some large mammals at the Pleistocene–Holocene transition, they do demonstrate that climate-induced changes are likely to have pushed some species to the brink of extinction.

14.5 Evolutionary responses to climate change

Evolutionary responses to altered environment are possible and may occur alongside range shifts (Thomas, 2005). Expanding range margins, for plants at least, involves selection for phenotypes that tolerate local conditions. If the abundant centre hypothesis is correct (see Brown, 1984; Lawton, 1993), populations near the edge of a species' range are expected to be smaller, perhaps maintained through immigration from the core of species' ranges, and to occupy a narrower range of habitats than larger, core populations. In a recent review of tests of these patterns, however, Sagarin and Gaines (2002) point out that only about 39% of reported ranges follow the expected pattern, a result that is probably inflated by under-reporting of negative results (the so-called file-drawer problem; Csada *et al.*, 1996). These results are underscored by new work on eastern North American tree populations (Murphy *et al.*, 2006), which rarely show the pattern expected from the abundant centre hypothesis. Edge populations are often larger than expected from the abundant centre hypothesis (e.g. Herlihy and Eckert, 2005), and possibly have reproductive strategies that differ from core populations. This suggests that the potential expansion of range margins for many species will not by hampered by gene flow from core areas, contrary to expectations (see Kirkpatrick and Barton, 1997, who also note that small changes in patterns of gene flow could plausibly be caused by climate change and can cause peripheral populations to become demographic sinks), and that core populations reduce the fitness of edge populations confronted by distinctive environmental conditions.

Even though climate change will almost certainly act as a strong selection force for many species, evidence from the geological record suggests that plants and animals have more commonly shifted their distributions instead of evolving new adaptations *in situ* (Noss, 2001; Thomas, 2005). However, most of the work on species' responses to climate change during the Pleistocene has focused on species with dramatic range shifts and not species that shifted little or not at all (Lyons, 2003). Evolutionary responses to climate change, not to mention the interaction between climate change and other forms of environmental modification, remain poorly understood (Hughes *et al.*, 2003). Many species are likely to have sufficient phenotypic plasticity to avoid shifting in response to climate change (Parmesan and Galbraith, 2004). This provides a potential explanation for the large numbers of species that remained in place during periods of climate change in the geological record (then again, poor data regarding historical species ranges may also partly explain these observations). It is not yet clear what proportion of species may remain relatively unaffected by climate change because of their innate phenotypic plasticity.

The extent to which species are at equilibrium with climate strongly affects our understanding of present-day range limits and our ability to predict future responses of species to climate change. Following Araujo and Pearson (2005), species can

be considered to be at equilibrium with climate if they maintain a presence in all climatically suitable, contiguous areas and are absent from unsuitable ones. Equilibrium between climate and species ranges need not imply that ranges are static but that climate does not cause large fluctuations of range boundaries. This definition avoids circularity if the current climatic boundaries of a species are not used to define the species' tolerance. The exact locations of range boundaries are known to be extremely dynamic (Brown *et al.*, 1996), making it somewhat more difficult to define climatic tolerance precisely. However, because predictions of future distributions of species often draw on purely spatial, present-day data, the degree to which species distributions are at equilibrium with current climate has important implications for species–climate 'envelope' modeling, which typically assumes equilibrium between species distributions and climate (Araujo *et al.*, 2005). Clearly, the capacity to track shifting climate conditions, and to maintain approximate equilibrial range boundaries, depends strongly on dispersal ability. Assemblages of plants and breeding birds have been found to be closer to equilibrium than reptile and amphibian assemblages (Araujo and Pearson, 2005), and tree distributions in Europe deviate significantly from equilibrium conditions (Svenning and Skov, 2004).

Many studies have modeled future species distributions following prolonged climate change (e.g. Illoldi-Rangel *et al.*, 2004; Peterson *et al.*, 2004). The reliability of these models is still debatable (Hampe and Petit, 2005; Carmel and Flather, 2006). Many of these models rely on bioclimatic envelope modeling or statistically sophisticated analogs (see Thuiller *et al.*, 2004). Most techniques neglect the potential effects of biotic interactions on range dynamics, assume that species are evolutionarily homogeneous and unchangeable entities across their range, and do not consider dispersal capacity, which has the potential to constrain future migrations of species and restrict species to habitats that are predicted to remain suitable following climate change (e.g. Hampe and Petit, 2005; Carmel and Flather, 2006; Crozier and Dwyer, 2006). At broad scales, multiple environmental factors covary and thus may limit the predictive value of these models (Currie *et al.*, 2004; Crozier and Dwyer, 2006). Moreover, the assumption that present-day spatial patterns will translate to temporal responses to climate change, particularly since the rate and magnitude of climate change exceeds levels within the unique evolutionary history of many species, is often implicit, untestable, and possibly unreliable in these studies (see White and Kerr, 2006). One way around this issue might be to model future responses to anthropogenic climate change based on observations of responses over the twentieth century, although few data-sets are sufficiently extensive and intensive to provide such power (but see Kerr *et al.*, 2001; White and Kerr, 2006). Predicting how species' distributions will change in the next century is clearly an inexact science that requires a number of assumptions. These are not unreasonable. To a large extent, these studies (particularly when they include error estimates, as in Thomas *et al.*, 2004) demonstrate that biotic responses to climate change will be very large and variable, but with potentially massive negative consequences.

14.6 Land-use and climate-change interactions

Although there is little doubt that both climate and land-use changes affect biodiversity, and will continue to do so into the future (Sala *et al.*, 2000), both of these factors have already had widespread impacts. Effects specific to climate are more difficult to detect than the effects of land-use change, which can be dramatic and obvious. To date, loss of habitat due to conversion to human land use almost certainly is the leading cause of extinction for most mainland assemblages. Land-use change, which encompasses many kinds of land-use/land-cover conversions (e.g. urbanization, agricultural expansion, or contraction), has accelerated extinction rates by perhaps three orders of magnitude (e.g. May *et al.*, 1995). Because interactions between land use and climate changes are already evident, a few comments specific to land use are justified, not least because much of the evidence to date on recent climate-change impacts on species distributions has been collected from areas where human land uses are spatially coincident.

Since World War II, agricultural land uses have intensified in much of North America and Western Europe (Freemark and Boutin, 1995; Benton *et al.*, 2002). To increase yields, agriculture now relies relatively heavily on massive pesticide and fertilizer use, mechanization, and irrigation, leading to the widespread loss and degradation of field boundary features (Freemark and Boutin, 1995; Longley and Sotherton, 1997). The ecological outcome has been the simplification of agricultural systems, a reduction of natural enemy diversity (Freemark and Boutin, 1995; Wilby and Thomas, 2002), the loss of natural habitat (Burel *et al.*, 1998; Benton *et al.*, 2002), the fragmentation of landscapes, and the degradation of remaining habitats (Burel *et al.*, 1998). The net result of agricultural development is a reduction in the capacity of the environment to support biodiversity (Gaston *et al.*, 2003).

Habitat loss to agriculture is a primary cause of species endangerment in the USA (Czeck *et al.*, 1997; Dobson *et al.*, 1997b) and is also the best predictor of numbers of endangered species across Canada (Kerr and Cihlar, 2003, 2004; Kerr and Deguise, 2004). Moreover, agricultural activities prevent endangered species recovery in Canada by making habitat loss permanent (Kerr and Deguise, 2004), and hinder the establishment of effective protected areas networks (Deguise and Kerr, 2006). A consequence of these observations is that conservation beyond the boundaries of traditional protected areas will be essential. In that vein, Leopold's (1948) famous call for a new land ethic was prescient.

The potential interaction between land-use and climate changes seems poised to accelerate extinction rates considerably (Thomas *et al.*, 2004, see also Buckley and Roughgarden, 2004; Harte *et al.*, 2004; Ladle *et al.*, 2004; Thuiller *et al.*, 2004), especially if they act synergistically (Harte *et al.*, 1992; Myers, 1992; Thomas *et al.*, 2004). Following the last glacial retreat, species tracked shifting climatic conditions across landscapes that presented relatively few natural barriers (Collingham and Huntley, 2000). However, recent anthropogenic habitat modifications have led to widespread habitat losses and fragmentation, and consequently have generated barriers to movement that may prove insurmountable for many species (Dennis and Shreeve, 1991; Collingham and Huntley, 2000; Hill *et al.*, 2001; IPCC, 2001). In human-modified landscapes, many areas that will become climatically suitable given expected climatic changes may be remote from current distributions and beyond the dispersal capacity of many species (Hill *et al.*, 1999; Walther *et al.*, 2002; Pearson and Dawson, 2003). This could confine species that are not adapted to agricultural environments to unsuitable, smaller areas as climate causes their ranges to retract, thus increasing their likelihood of extinction (Peters, 1992; IPCC, 2001; Parmesan and Galbraith, 2004). Evidence from Britain over the past century reveals that lack of habitat suitability is already slowing the movement of butterflies ranges northwards (Hill *et al.*, 1999). In Canada, butterfly species are responding positively to climate change but negatively to land-use change (Figure 14.2). Populations may also be negatively affected by changes to a variety of important biotic interactions, such as increased predation, competition, and transmission of disease (IPCC, 2001). This last factor, which can be affected by climate, has been implicated as a cause of widespread amphibian decline (Pounds *et al.*, 2006). In short, many species may fail to track shifting climatic conditions because of extensive and intensive land-use change (Hill *et al.*, 1999; Warren *et al.*, 2001).

Although climate changes to date remain small, habitat availability is preventing some butterfly species from tracking shifting climatic conditions successfully Hill *et al.* (1999); Parmesan *et al.* (1999). Warren *et al.* (2001) also report that climate and land use changes are causing the distributions of habitat specialist butterflies to decline more than those of habitat generalists. Changes in climate and habitat act as opposing forces near these species' range margins: the negative impacts of land use changes along range margins overwhelm the relaxation of climatic constraints along those margins (Warren *et al.* 2001). At the scale of entire species' ranges, climate change and habitat loss may interact to increase extinction risk (Travis 2003). For example, patch occupancy models demonstrate that extinction becomes very likely below a species-specific threshold of habitat

availability (Travis 2003) and that this threshold may be higher during periods of climate change. That the interaction of climate and land use changes have already exerted significant effects makes predictions of impending increases in extinction rates (Sala *et al.* 2000) much more plausible.

14.7 Autecological characteristics and species' potential responses to climate change

Since landscapes are increasingly becoming modified by humans, certain autecological characteristics are expected to affect an individual species' potential to respond successfully to climate change (O'Grady *et al.*, 2004; Kotiaho *et al.*, 2005; Thuiller *et al.*, 2005). These include characteristics that have been linked to extinction risk, such as local abundance, range size, range fragmentation, body size, and life-history specialization (O'Grady *et al.*, 2004). Several factors, such as low dispersal ability and obligate mutualisms, may cause species to respond slowly to climate change. Such lagged responses may result from limited dispersal potential that delays poleward/upward colonization (Warren *et al.*, 2001) or when a necessary resource, such as an obligate food plant, itself responds slowly to change (Parmesan *et al.*, 1999).

During past climate changes, dispersal rates were crucial to a species' ability to colonize suitable habitat (Peters and Darling, 1985; Thomas, 2005). However, accelerating climate change will likely require much higher migration rates than those observed postglacially (Malcolm *et al.*, 2002; Pearson, 2005). The ability of a species to migrate at a sufficient rate to keep up with the changing climate will be dependent on the dispersal characteristics of individual species and the natural (e. g. mountain ranges) and human-modified (e.g. habitat fragmentation) features of the landscape over which dispersal is occurring (Peters, 1992; Pearson and Dawson, 2003; Thomas, 2005). Colonization ability at the expanding range margin can also be affected by flight capacity and the species' habitat preferences (Thomas, 2005). However, for most species a lag will be expected before extensive colonization is possible so their ranges may decline initially (Peters, 1992). It is not clear how

(a)

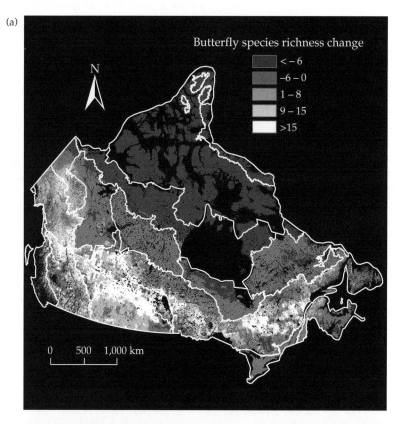

(b)

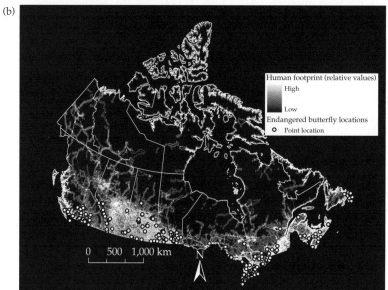

Figure 14.2 (a) Relationship between climate change and butterfly species richness change in Canada. Shades at the top of the scale represent areas where butterfly species richness has decreased and shades at the bottom represent areas where butterfly species richness has increased. (b) Relationship between areas of high human footprint and threatened butterfly species in Canada. Dots show observation points for butterflies listed under Canada's Species At Risk Act including those that have recently been extirpated from Canada. These observation locales are overlaid on a map of human footprint (Sanderson *et al.* 2002), showing gradients of the intensity of human activities (where ▨ is the lowest and ▩ the highest intensity). In both panels white lines show the boundaries of the 15 ecozones in Canada.

long such a lag may be. It seems safe to say that highly vagile species will have relatively little difficulty tracking shifting climatic conditions, whereas the survival of species with low dispersal rates will depend more on the availability of suitable habitats (Pearson and Dawson, 2003; Meynecke, 2004). Evidence suggests that range-restricted species tend also to have poorer dispersal abilities (Gaston, 1994). Habitat specialists with low colonization ability and poor dispersal are likely to be most prone to extinction during climate change and habitat loss (Travis, 2003).

Dispersal ability will strongly affect species' ability to track rapidly shifting climatic conditions and, thus, odds of survival in recently fragmented landscapes (Thomas, 2000; Williams *et al.*, 2005). Modern fragmented landscapes and rapidly changing climatic conditions favour well-developed dispersal abilities and opportunistic species (Gaston, 1994; Malcolm *et al.*, 2002). A recent study measured the migration distance of North American tree species and found that fragmented landscapes limited their dispersal (Iverson *et al.*, 2004). Moreover, Turin and den Boer (1988) found that British

plant species with effective dispersal abilities were generally increasing in range size, while those with poor dispersal abilities were decreasing. Therefore, a possible selection pressure may exist for species to disperse larger distances. In the case of butterflies, which have at least one generation per year, it is reasonable to predict that global change-induced selection pressure for better dispersal may have caused some species to increase their flight capacity over the past century.

Interspecific interactions will likely be altered by climate change and will play a major role in determining new species distributions (Peters, 1992). More specifically, it is likely that mutualisms will significantly influence the ability of species to disperse (Harte *et al.*, 1992). Although many species will be physically capable of migrating quickly enough to track a changing environment, the distribution of some species is limited by specific plant distributions or suitable habitats (Peters, 1992). Therefore, dispersal rates may be largely determined by those of co-occurring plants and habitat. For example, the dependence of many butterfly species in northern Europe and Canada

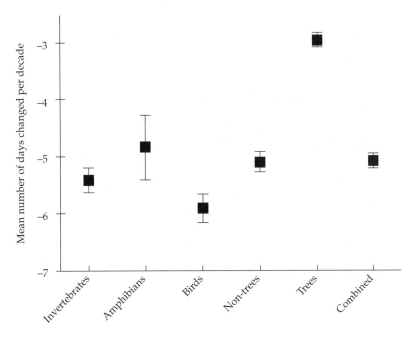

Figure 14.3 Phenological shift among different species groups, measured in the number of days (±S.E.M.). All assemblages show earlier phenologies, ranging between about 3 days earlier (for trees) to nearly 6 days earlier (for plants, excluding trees). Taken from Root *et al.* (2003).

on their larval food plants may influence their response to climate change as the range limits of plants are likely to be slower than those of butterflies (Grabherr et al., 1994; Parmesan et al., 1999; Peterson et al., 2004). Plant species, in turn, frequently depend on soil organisms that may disperse even more slowly (Harte et al., 1992). Larval stages of many butterflies also have obligate mutualisms with ants to protect them against predation and parasitism (Eastwood and Fraser, 1999), so it is likely that these species' ranges will be tied to those of their mutualist partners (Dennis and Shreeve, 1991).

Substantial evidence demonstrates that global warming has already had significant impacts on the phenology of both plants and animals (Walther et al., 2002; Root et al., 2003; Root and Hughes, 2005; Figure 14.3). Temperature and precipitation both affect phenological development among plants and animals, including the timing of migration (among animals) or seed production (among plants). Differential phenological shifts in response to climate change could prove extremely disruptive (Root and Hughes, 2005). It is easy to imagine that such different phenological responses could disrupt obligate mutualisms, potentially leading to the extinction of one or both partners. Anthropogenic climate change has also affected the body size of a number of mammal species, perhaps by altering their growth rates (see review by Millien et al., 2006). Autecological observations will be essential to untangle the array of biotic interactions—and perhaps identify the ones that matter most—affecting how species and communities respond to climate change.

14.8 Protected areas and conservation in a changing climate

Climate and land-use changes are likely going to force many species to shift beyond the boundaries of existing protected areas. The addition of climate change to the other stressors already affecting habitats and their constituent species presents a major challenge for the future conservation of biodiversity (Parmesan and Galbraith, 2004). Protected areas have historically been at the centre of conservation strategies and are widely viewed as a

particularly effective tool for conserving biodiversity, especially when managed well (Lawler et al., 2003; Chape et al., 2005; Hannah et al., 2005). Protected areas will remain an important conservation tool in the future since the area under protection continues to increase worldwide while remaining undisturbed habitat continues to decline (Sanderson et al., 2002).

Protected areas are fixed in place while species ranges are dynamic, leading to the real possibility that global change will alter species compositions in those areas (Scott et al., 2002; Burns et al., 2003; Figure 14.4). Climate change impacts on the effectiveness of parks will vary regionally (Suffling and Scott, 2002), complicating efforts to rely on them to protect biodiversity. New protected areas may need to be planned for areas that are currently unsuitable for target species (Williams et al., 2005), underscoring the need for accurate prediction of how species and communities will respond to climate change. Furthermore, many protected areas have been established or managed to conserve 'representative' ecosystems or complements of species that may be reshuffled in the future (Hannah et al., 2002b), reducing the effectiveness of reserve-selection procedures based only on present-day patterns. Land-use changes around protected areas (e.g. Wiersma and Nudds, 2001) may also expand the threat posed by climate change (Pyke et al., 2005). These challenges loom large on the horizon for conservation strategies based on protected areas.

Conserving biodiversity in a dynamic world will require new strategies that go beyond static reserve-selection methods (Pyke et al, 2005). Conservation planning must explicitly account for the individualistic species responses to climate change as well as current patterns (Hannah et al., 2002a). A shift from reactive to proactive conservation approaches is necessary to assess how, where, and when future threats will affect species persistence (Bomhard et al., 2005). Conservation strategies should not neglect populations at the trailing edge of species' ranges either: these may have distinctive ecological features and are sometimes disproportionately important for the species' survival and potential to retain an evolutionary response to change (Hampe and Petit, 2005).

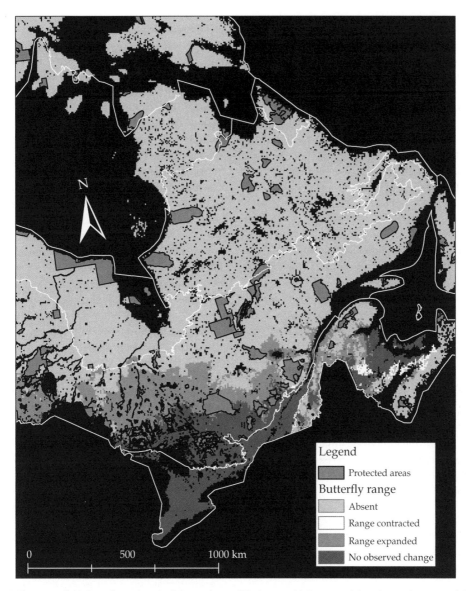

Figure 14.4 The range of this butterfly species, the little wood satyr (*Megisto cymela*), has expanded north over the course of the twentieth century and moved into new protected areas. Protected areas are dark outlined in black. The changes in butterfly range are indicated by different shading and ecozones (similar to biomes) are outlined in white. This species, along with many others, is tracking rapidly warming winter and growing season temperatures. Data adapted from White and Kerr (2006).

Although the respective effects of land-use and climate change differ, the best conservation strategies to address these aspects of global change converge. Land-use conversion to agriculture is now widespread, largely irreversible, and likely to prevent much expansion of current protected areas within affected lands (Deguise and Kerr, 2006). Complementing traditional conservation focus on protected areas, maintenance and restoration of natural and semi-natural areas within human-dominated landscapes can significantly increase the diversity that can persist within a landscape (e.g. Ricketts, 2001; Hannah *et al.*, 2002a; Da Fonseca *et al.*, 2005). Increasing emphasis on the matrix, not just the habitat islands within it, will also increase landscape connectivity with obvious benefits for species that must shift in response to climate change. Much can be learned from natural corridors that were conduits for species dispersal during past climate changes (Delcourt and Delcourt, 1984). Maintenance of landscape linkages running parallel to existing climate gradients will improve landscape connectivity that will assist biodiversity conservation efforts (Hunter *et al.*, 1988; Hobbs and Hopkins, 1991; Noss, 2001), regardless of climate change. Finally, relatively low-intensity (see Kerr and Cihlar, 2004) or mixed-use agricultural landscapes (e.g. agroforestry; Hannah *et al.*, 2002b) can maintain increased habitat and resource availability (Benton *et al.*, 2002) that make dispersal through such landscapes more likely (see Parmesan and Galbraith, 2004; Hannah and Hansen, 2005).

14.9 Conclusion

Because climate strongly affects where species are found as well as their phenology, the significant effects of the relatively modest climate change that has already been observed can be expected to accelerate in the near future. Further theoretical and empirical developments are essential to predict how species will respond to the global changes anticipated during the twenty-first century. Macroecology and theoretical ecology provide an essential framework for such research but contributions to conservation in the face of the massive threat posed by anthropogenic climate change will be strongest when they can provide credible mechanisms, preferably linked to species' autoecological characteristics.

Although climate change presents a serious threat to the human enterprise, its indirect effects on biodiversity will also be profound and are likely to accelerate extinction rates dangerously (for discussion of extinction rates, see Lawton and May, 1995). Many human activities rely on the stable provision of irreplaceable ecosystem services, which are syntheses of myriad species' interactions. Sensible interpretation of the precautionary principle suggests we must avoid disrupting such interactions. Strategies to accomplish such an ambitious goal, such as maintaining or restoring habitats in human-dominated landscapes, are eminently justifiable from a scientific perspective and should be implemented immediately. These will improve the outlook for biological diversity—and the human enterprise that rests upon the services which diversity provides—whether the principal threat comes directly from climate change, land-use change, or their interaction.

CHAPTER 15

Unanswered questions and why they matter

Robert M. May

The earlier chapters in this book could be thought of as travel notes from an intellectual journey across the landscape of ecological science. In particular, the previous five chapters, 10–14, implicitly or explicitly indicate some of the unintended consequences of the growth in numbers of people and in their environmental impacts. In this final chapter, I begin with a survey of some quantitative measures of the scale of human impacts. Emphasizing the many lamentable uncertainties in our knowledge base, I focus especially on the rising rates of extinction of plant and other animal species.

Why should we care about such impoverishment of our planet's biological diversity? I outline three kinds of possible reasons, under the headings of narrowly utilitarian, broadly utilitarian, and ethical. Each of these is then discussed, with emphasis on ways in which current lack of knowledge—lack of data and/or lack of theoretical understanding—is a handicap. In places, this carries the discussion into areas not commonly found in ecology texts (ethical, economic, and political questions, for instance). In other places, there is the more familiar exhortation for more research on this or that topic.

15.1 The growth of human populations

Contrary to some impressions, human population growth has been far from simply exponential. Broadly speaking, humans have been around for a couple of hundred thousand years (Deevey, 1960; Cohen, 1995). For essentially all this time, they were small bands of hunter-gatherers, with the total human population being variously estimated at around 5–20 million people.

With the benefits of the invention of agriculture, roughly simultaneously in various parts of the world around 10 000 years ago, things started to change. Denser aggregations of people became possible, and villages began their journey to cities. Following the advent of this agricultural revolution, human populations arguably grew more rapidly in the first 5000 years than in the more recent 5000, up to the beginning of the Scientific-Industrial Revolution around the 1600s. This relative slowing of population growth is almost surely associated with infectious diseases which were not sustainable at the low population densities associated with hunter-gatherers. Reader (2004) summarizes it well: 'Bacterial and viral diseases are the price humanity has paid to live in large and densely populated cities. Virtually all the familiar infectious diseases have evolved only since the advent of agriculture, permanent settlement and the growth of cities. Most were transferred to humans from animals—especially domestic animals. Measles, for instance, is akin to rinderpest in cattle; influenza came from pigs; smallpox is related to cowpox. Humans share 296 diseases with domestic animals.'

The next big upsurge in population growth resulted from agricultural and technological advances spurred by the Scientific-Industrial Revolution over the past few centuries. Human numbers reached their first billion around 1830. It took a century to double that.

Then arrived the third revolutionary upsurge, driven this time by advances in medical science—not

CAT scanners, but understanding of the transmission dynamics of infectious diseases, coupled with better hygiene and better nutrition (albeit still inequitably distributed). The next doubling took 40 years, to four billion around 1970. In 2006 we are about 6.5 billion.

Associated with population growth are great changes in patterns of urbanization. In 1700, about 10% of the world's population lived in cities. By 1900 it was 25%. Some time in 2006 (or maybe 2005 or 2007; our knowledge is not that precise) a child will be born marking a hinge in history where, for the first time, more people will live in cities than in rural areas.

The present marks another tipping point in that overall fertility rates have just dropped below replacement levels: globally, although still with big regional variations, the average woman is producing less than one female offspring. However, the pyramidal shape of age profiles in most parts of the world gives momentum to population growth, such that even if fertility rates remain below unity the human population will continue to grow throughout the present century, reaching 9 billion around 2050. And essentially all these added people will live in cities, taking the urban fraction in 2050 to 67%.

Walt Whitman once evoked the feelings we have about the vast numbers of humans to have lived before us: 'row upon row, rise the phantoms behind us'. But estimates of the total numbers of *Homo sapiens* ever to have lived run around 80 billion or so. So if they were regimented in tidy columns behind those alive today, each of us, looking over a shoulder, would see only a dozen or so phantoms. I find this a startling thought. It certainly underlines the singularity of our time.

15.2 The scale of human impacts

Our predecessors, even in hunter-gatherer times, had impacts on their environments. There are still debates about the extent to which early humans, as distinct from changing climate, contributed to the extinction of the Pleistocene megafauna in Europe, Africa, the Americas, and Australia, but there is no doubt that humans extinguished roughly half the bird species in Hawaii and New Zealand when they arrived less than 1000 years ago.

Significant though they were, these earlier impacts were not on the literally global scale of today's. It is estimated that humans now take to their own use, directly or indirectly, between 25 and 50% of all net terrestrial primary productivity (the commonly quoted figure is 40%; see Vitousek et al., 1986; Daily, 1997). Perhaps even more striking, it has been estimated that more than half of all the atoms of nitrogen, and also of phosphorous, incorporated into green plants today come from artificial fertilizers (produced with fossil-fuel energy subsidies) rather than the natural biogeochemical cycles which built, and which struggle to maintain, the biosphere. These estimates are necessarily imprecise, but they accord with a very recent study, using satellite imagery, which found 40% of the Earth's land surface being modified by human use, mainly for agriculture.

The Worldwide Fund for Nature (WWF, 2004) has presented estimates, country by country, of humanity's ecological footprint (EF) at current levels of consumption. The EF for a given country is defined as the biologically productive area required to produce the food and wood people consume, to give room for infrastructure, and to absorb the carbon dioxide emitted from burning fossil fuels. Thus estimated, the EF is expressed in 'area units'. Any such estimate is necessarily imprecise (the carbon dioxide bit arguably more so than other components), but on the other hand they are conservative in that other factors, such as requirements for natural ecosystem services to handle pollutants, are excluded. Having estimated individual countries' EF, the WWF adds them up to get the overall global EF shown in Figure 15.1. The observed increase over time derives partly from population growth, and partly from increases in the average footprint per person.

The WWF also estimate the total EF that individual countries, and thence the planet, could satisfy sustainably (the biological capacity, BC). Here the figures depend, to a degree, on assumptions about the footprints of future crops and energy sources. Figure 15.1 suggests we passed the point where humanity's actual EF exceeds the sustainable level—a milestone of milestones—around two decades ago. I again emphasize the ineluctable uncertainties in any such estimates

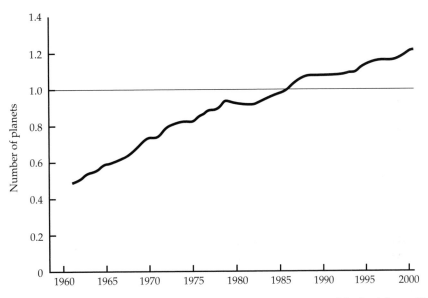

Figure 15.1 An estimate of the total ecological footprint of the human population, 1960–2001, as defined and discussed in the text. The straight line shows our planet's estimated biological capacity (BC)—the total ecological footprint available on a sustainable basis. The estimated actual total ecological footprint is expressed as a ratio to this baseline of 'one planet, used sustainably'. From WWF (2004).

of footprints. Even so, I believe Figure 15.1 is indicative.

The most unambiguous sign that human activities now are on a scale that rivals natural processes is, of course, climate change. The evidence for this, including the statement from the Science Academies of all G8 countries along with China, India, and Brazil, was set out at the start of the previous chapter.

15.3 Numbers of species

The remainder of this chapter will focus not on humans, but on consequences of our actions for the other species we share the planet with.

Seen through a wider-angle lens, the impending diminution of the Earth's diversity of plant and animal species could be an even greater threat than climate change. Unfortunately, analysis of the causes and consequences of accelerating extinction rates is impeded by the rudimentary state of our knowledge, which in turn derives more from past intellectual fashions than dispassionate assessment of scientific priorities. It is worth reflecting that Newton's Laws of Motion, and consequent

explanation of how planetary motions derive from the inverse-square gravitational attraction of the sun, came a full century before Linnaeus began the task of codifying living things (the date of the canonical tenth edition of *De Rerum Naturae* being 1758). This legacy of the lag between Linnaeus and Newton lingers today.

In a review of Terborgh's (1999) *Requiem for Nature*, McKibben (1999) has commented bitterly on these vagaries of human fashions and concerns: 'You can follow the changes in the value of the Japanese yen second by second from your desktop; reporters by the dozen struggle valiantly to explain the particulars of Microsoft's antitrust defense. But who can tell whether the tropical forest is disappearing more or less speedily than it was in the late 1980s when every singer worth her faded jeans was cutting a CD in its defense? This question is surely worth attention, since the equatorial jungles contain more examples of creation's fabulous imagination than any other ecosystem, and since its trees are a key part of the earth's system for cleansing excess carbon dioxide from the atmosphere. Perhaps you have a dim sense that some agreements have been signed to protect the rainforests, some programs put in place. But are they working? What strategies make the most sense to preserve what's left? Far more money and attention is devoted to, say, searching for and describing the possible remains of microbial life in the dust of Mars.'

So what is the current state of knowledge about the planet's organisms other than humans?

Currently around 1.6 million distinct species of plants and eukaryotic animals have been named and recorded (Hammond, 1995; May, 1999). Even this number—analogous to the number of books in the British Library or the US Library of Congress, which are precisely known—is uncertain to within around 10%, because the majority of species are invertebrate animals of one kind or another, for most of which the records are still on file cards in separate museums and other institutions. The consequence is a synonymy problem: the same species being separately identified and differently named in two or more places. Given, for example, that some 40% of all named beetle species are estimated to be known from only one geographic site (and sometimes from a single specimen), and that intercollated databases do not exist for many groups, this synonymy problem should not surprise us. Of course, a lot of taxonomic effort has gone, and today goes, into sorting out synonymies, especially in better-studied groups. But even as old synonyms are resolved, new ones are being added. In a seminal study, Solow et al. (1995) combined theoretical and empirical work to make a start on estimating true synonymy rates, with the aim of getting a better idea of just how many distinct species have been named and recorded. For thrip species recorded since 1901, Solow et al. found a directly observed synonymy rate of 22%, but estimated the true rate of synonymy in this group to be around 39%. Subsequent studies suggest comparable, although usually somewhat smaller, numbers for other groups (May, 1999). Overall, it could be argued that a discount factor of something like 20% might be applied to existing lists of known species. This seems to me to be an important area where more theoretical and empirical work is warranted.

Currently, new species are being identified at a rate of around 15 000–20 000 a year, while at the same time earlier synonyms are being resolved at around 3000–5000 each year, for a net addition of very roughly 15 000 species each year.

So much for what is known. But how many species may there be in total on Earth today? Recent estimates lie in the range 5 million–15 million (Hammond, 1995; May, 1999). Lower numbers, and also much higher ones, also have their advocates. Even if we take a low estimate of 3 million still to be identified, at the current rates just noted the job would take 200 years. Organizing better databases, and using molecular information about newly discovered species' genomes ('barcoding life'), promises to speed up this distressingly slow task (Godfray and Knapp, 2004; Savolainen et al., 2005). Even so, the craft of collecting material in the field will remain a seriously rate-limiting step.

15.4 Extinction rates

If we do not know how many species have been identified—much less their functional roles in ecosystems—to within 10%, nor the overall species total to within an order of magnitude, we clearly cannot say much about how many species are likely to become extinct this century. We can note that the IUCN Red Data Books in 2004, using specific and sensible criteria, estimate 20% of recorded mammal species are threatened with extinction, and likewise 12% of birds, 4% of reptiles, 31% of amphibians, 3% of fish, and 31% of the 980 known species of gymnosperms (IUCN, 2004). However, when these figures are re-expressed in terms of the number of species whose status has been evaluated (as distinct from dividing the number known to be threatened by the total number known—however slightly—to science), the corresponding numbers are 23, 12, 61, 31, 26, and 34% respectively. This says a lot about how much attention reptiles and fish have received.

The corresponding figures for the majority of plant species, dicotyledons and monocotyledons, are respectively 4 and 1% of those known, and 74 and 68% of those evaluated. Most telling are the two numbers for the most numerous group, insects: 0.06% of all known species are threatened, compared with 73% of those actually evaluated. The same pattern holds true for other invertebrate groups. For these small things, which arguably run the world, we know too little to make any rough estimate of the proportions that have either become extinct, or are threatened with it.

These disparities in our knowledge about different groups reflect differential attention from the research community. Rough estimates (and it would be good to have better ones) suggest the taxonomic workforce is roughly evenly divided between vertebrate, plant, and invertebrate species. Given that plant species are roughly 10 times as numerous as vertebrate ones, and invertebrates—by conservative estimates—at least 100 times more numerous, this reflects a gross mismatch between workforce and the task at hand (Gaston and May, 1992). Things get worse when we look at the conservation biology literature: a study of papers in the two leading conservation research journals from 1987 to 2001 showed roughly 70% dealing with vertebrates, 20% with plants, and 10% with invertebrates (of which half were butterflies and moths, enjoying the status of a kind of honorary bird; Clark and May, 2002). And when we turn to conservation-oriented non-governmental organizations, we find an even greater preponderance of attention given not just to vertebrate species, but to the roughly one-third that are birds and mammals.

Perhaps surprisingly, we can nevertheless say some relatively precise things about current and likely future *rates* of extinction in relation to the average rates seen over the roughly 550-million-year sweep of the fossil record (May *et al.*, 1995; May, 1999). For bird and mammal species (a total of approximately 14 000), there has been an average of about one certified extinction per year over the past century. This is a very conservative estimate of the true extinction rate, because many species receive little attention even in this unusually well-studied group. Such a rate, if continued, translates into an average 'species life expectancy' of the order of 10 000 years. By contrast, the average life expectancy—from origination to extinction—of a species in the fossil record lies in the general range 1–10 million years, albeit with great variation both within and among groups (May, 1999).

So, if birds and mammals are typical—and there is no good reason to assume they are not—extinction rates in the twentieth century were higher, by a factor of 100–1000, than the fossil record's average background rates. And four different lines of argument suggest a further 10-fold speeding up over the coming century (May *et al.*, 1995). Such an acceleration in extinction rates is of the magnitude which characterized the Big Five mass extinction events in the fossil record (Raup, 1998; Sepkoski, 1992). These Big Five are used to mark changes from one geological epoch to the next. Although there is much need for further work to refine estimates of this kind, it does seem likely that we are standing on the breaking tip of a sixth great wave of mass extinctions.

The crucial difference between the impending Sixth Wave of mass extinction and the previous Big Five is that the earlier ones stemmed from external environmental events. The sixth, set to unfold over the next several centuries—seemingly long to us, but a blink of the eye in geological terms—derives directly from human impacts.

15.4.1 Why should we care about extinctions?

What fraction of all eukaryotic species ever to have lived are alive on earth today? Following Raup (1998), Sepkoski (1992), and others, I can give a rough answer to this question. We saw above that the rough life expectancy of a species in the fossil record was typically a few million years. Juxtaposing this species lifetime against the roughly 550-million-year sweep of the fossil record leads to an estimate that the species extant at any one time represent roughly 1% of the total ever to have lived. The history of life on Earth, however, has been one of very approximately linear increase in diversification, so we might guess at approximately 2% for the proportion alive today.

Conversely, given that extinction has already been the fate of 98%, and possibly more, of all eukaryotes, why should the impending Sixth Mass Extinction concern us? I think the reasons can be brigaded under three broad headings, each posing an agenda for research.

15.5 Narrowly utilitarian considerations

Some would argue that known and yet-undiscovered species represent a precious resource of genetic novelties. These may be the raw stuff of tomorrow's biotech revolution, producing new pharmaceutical products, new foodstuffs, and other products for the global economy. Let us not burn the books before we have read them.

Whilst I am sympathetic to such attempts to move biological diversity into the ambit of conventional economics, in order to motivate political concern, I am sceptical of this argument. I think it more likely, with the pace of advances in understanding the molecular machinery of living things, that tomorrow's medicines will be designed from the molecules up, rather than emerge from high-tech bioprospecting.

There are, however, additional reasons for seeking a better understanding of the species richness and taxonomic details of neglected biota, especially invertebrates. For one thing, most of the benefits of modern medicine are oriented to the developed world (Kremer and Glennerster, 2004); the protozoan and helminth parasites which cause mortality and morbidity in developing countries deserve more taxonomic attention, both to themselves and to their vectors and (in some instances) non-human animal reservoirs. The transmutation of what was once traditional and local bushmeat consumption into a full-blown and indeed globalizing industry underlines this point in a different way (Bell *et al.*, 2005). HIV-1, HIV-2, and SARS are three viruses which made it into human populations this way. How many are yet to come?

On a different tack, if the tree of life is to be pruned by the anthropogenic extinction episode currently underway, would it not be a good idea deliberately to try to preserve the maximum possible amount of independent evolutionary history (IEH), rather than the cuddliest species? And, if so, how do we go about defining IEH? If we can preserve, for example, 50% of species in an assemblage, how do we quantify the difference between random and IEH-optimal preservation? Obviously, the discussion here overlaps with that under the later heading Ethical considerations.

The idea that IEH, or something equivalent, be used as an objective criterion in setting conservation priorities was originated by Vane-Wright *et al.* (1991) and characterized as the 'calculus of biodiversity' by May (May, 1990b). But how, precisely, are we to define IEH? Ideally, it might be best to assess it at the most fundamental, genetic level by some measure of information that is molecularly coded in DNA. At present, it seems more practical to work with information from the structure of phylogenetic trees: if k species from a total of n are saved, the fraction of IEH preserved can be measured by the overall branch lengths kept, as a ratio to the total branch lengths of the original tree. If we knew the tree structure, we would maximize IEH by first identifying the $k-1$ lowest branching points (nodes) in the tree, counting from the root. These define k clades, and we next select any one species from each clade. This algorithm works independent of whether we know the actual lengths of the branches or merely the branching order of the nodes (although, of course, firmer estimates of the fraction saved can be made in the former case). It is assumed here that all branch tips are equidistant from the root; more details about molecular evolution could give a picture in which such lengths varied, although the basic ideas would remain much the same (Faith, 1992; Nee and May, 1997). Other schemes for implementing such a calculus of biodiversity have been proposed, which differ in detail, but not in essentials, from the above (for recent reviews see Rodrigues and Gaston, 2002b; Pavoine *et al.*, 2005).

It turns out that, in general, simulations suggest a surprisingly large amount of IEH is preserved even when a large fraction of species is lost. Thus 80% of the underlying tree of life can survive even when 95% of species are lost. Furthermore, algorithms which maximize the preservation of IEH are generally not very much better than choosing the same number of survivors at random (Nee and May, 1997). In accord with intuition, the differences between the two extremes of optimal and random choices are more pronounced for trees with comb-shaped topology (non-bifurcating branches off a main stem) than for bush-shaped ones (new nodes equally likely on any branch).

Studies, such as that by Proches *et al.* (2006) on angiosperm assemblages from four different vegetation types in South Africa, indicate that assessments of IEH along the above lines do indeed reflect the known site-specific evolutionary history of South African flora 'remarkably well'. I think such studies of IEH in relation to 'what is there today' are reassuring, in the sense that they suggest these measures of IEH can indeed be a useful guide to the 'agonies of choice' (Vane-Wright *et al.*, 1991) forced upon us as we prune the tree of life.

Returning to the narrowly utilitarian theme of this section, it must be recognized that conservation concern will often be focused on an individual species for a particular reason—be it sentimental, or practical for its contribution to tourist revenue or other economic considerations, or both—regardless of the species' evolutionary significance. I will conclude by returning to these tensions among political, economic, social, or other constraining realities and the aspiration of optimizing the preservation of our evolutionary heritage.

15.6 Broadly utilitarian considerations

A more broadly utilitarian argument for concern about loss of biological diversity is that—as seen in Chapters 8, 9 and 13—we do not yet know enough about the structure and function of ecosystems to be able to predict how much disturbance and species loss they can undergo yet still deliver ecosystem services upon which we depend.

We have just seen how poor our knowledge is simply about how many species of animals, plants, and microbes are present on Earth today. Additionally, for most of those species which have been named and recorded—the majority of which are invertebrates—we know little or nothing about the roles they play in maintaining the ecosystems of which they are part. One estimate is that we have information about the behaviour and ecology of fewer than 5% of all identified animal species (Raven, 2004). So it is not surprising that we are not yet very good at predicting the effects upon local or regional ecosystems of the loss of species as a consequence of habitat disturbance, or over-exploitation, or introduction of alien species, or combinations of such perturbations.

The United Nations-sponsored Millennium Ecosystem Assessment, which involved some 1360 scientists from 95 countries and whose first global assessment of the world's ecosystems was published in 2005, represents a truly major effort to get to grips with these uncertainties (Millennium Ecosystem Assessment, 2005). It provides a comprehensive appraisal of the condition of, and trends in, the world's ecosystems. Ecosystem services are the benefit provided to humans as a result of species' interactions within the system. Some of these services are local (e.g. provision of pollinators for crops), others regional (e.g. flood control or water purification), and yet others global (e.g. climate regulation). In its massive report the Millennium Ecosystem Assessment identifies 24 categories of such ecosystem services, broadly grouped under three headings: provisioning, regulating, and cultural.

Table 15.1 summarizes these 24 categories of service, along with indications of whether the service is being enhanced or degraded. For provisioning services, enhancement is defined to mean increased production of the service through changes in area over which the service is provided (e.g. spread of agriculture or increased production per unit area). The production is judged to be degraded if the current use exceeds sustainable levels. For regulating services, enhancement refers to a change in the service that leads to greater benefits for people (e.g. the service of disease regulation could be improved by eradication of a vector known to transmit a disease to people). Degradation of regulating services means a reduction in the benefits obtained from the service, either through a change in the service (e.g. mangrove loss reducing the storm protection benefits of an ecosystem) or through human pressures on the service exceeding its limits (e.g. excessive pollution exceeding the capability of ecosystems to maintain water quality). For cultural services, degradation refers to a change in the ecosystem features that decreases the cultural (recreational, aesthetic, spiritual, etc.) benefits provided by the ecosystem.

Note that of the 24 categories of ecosystem services examined by the Millennium Ecosystem Assessment, 15—roughly two-thirds—are being degraded or used unsustainably. Whilst 15 have thus suffered, only four have been enhanced in the

Table 15.1 Global status of ecosystem services (Millennium Ecosystem Assessment, 2005).

Service	Status	Notes
Provisioning services		
Food		
crops	+	Substantial production increase
livestock	+	Substantial production increase
capture fisheries	−	Declining production due to overharvest
aquaculture	+	Substantial production increase
wild foods	−	Declining production
Fibre		
timber	+ / −	Forest loss in some regions, growth in others
cotton, hemp, silk	+ / −	Declining production of some fibres, growth in others
wood fuel	−	Declining production
Genetic resources	−	Lost through extinction and crop genetic resource loss
Biochemicals, natural medicines, pharmaceuticals	−	Lost through extinction, overharvest
Fresh water	−	Unsustainable use for drinking, industry, and irrigation; amount of hydro energy unchanged, but dams increase ability to use that energy
Regulating services		
Air-quality regulation	−	Decline in ability of atmosphere to cleanse itself
Climate regulation		
global	+	Net source of carbon sequestration since mid-19th century
regional and local	−	Preponderance of negative impacts
Water regulation	+ / −	Varies depending on ecosystem change and location
Erosion regulation	−	Increased soil degradation
Water purification and waste treatment	−	Declining water quality
Disease regulation	+ / −	Varies depending on ecosystem change
Pest regulation	−	Natural control degraded through pesticide use
Pollination	− *	Apparent global decline in abundance of pollinators
Natural hazard regulation	−	Loss of natural buffers (wetlands, mangroves)
Cultural services		
Spiritual and religious values	−	Rapid decline in sacred groves and species
Aesthetic values	−	Decline in quantity and quality of natural lands
Recreation and ecotourism	+ / −	More areas accessible but many degraded

+, enhanced; −, degraded, in the senses defined in the main text. *The evaluation here is of 'low to medium certainty'; all other trends are 'medium to high certainty'.

past 50 years, of which three involve food production: crops, livestock, and aquaculture. The status of the remaining five is equivocal, as indicated in the table's notes.

The way economists conventionally calculate gross domestic product (GDP) takes little or no account of the role of ecosystem services. Thus an oil tanker going aground, and wreaking havoc on the region's biota, will typically make a positive contribution to conventional GDP (cleanup costs are a plus; environmental damage deemed not assessable). Costanza *et al.* (2001) have attempted to assess the 'GDP-equivalent' of the totality of the planet's ecosystem services. Their guesstimate is that such services have a value roughly equal to global GDP as conventionally assessed. Any calculation of this kind is beset with many uncertainties, and some would argue that you simply cannot put a price upon a service which is essential to life. But I find it helpfully indicative.

One important step in the direction of a more explicit and rigorous characterization of the components of ecosystem services is to develop indicators. It can be argued that ecologists and conservation biologists could learn from economists' long-standing set of common and clear indicators, used to track and influence the development of markets. Some recent work of this kind uses composite indicators from time-series data on populations of birds or other vertebrates (see Balmford *et al.*, 2005). The UK uses one such indicator, the UK Wild Bird Index, to help evaluate the performance of its environmental policies (Gregory *et al.*, 2004). There is clear need for further theoretical development of such measures of trends in biodiversity and general ecosystem health, carefully tested against relevant data (Crane 2003). And this practice is being extended within the European Union (Gregory *et al.*, 2005).

As human numbers continue to grow, however, we need deeper understanding of how humans may alter habitats and ecosystems to provide for their needs, but do so subject to constraints which preserve both particular individual species and key elements of ecosystems. As discussed in Chapter 13, such 'co-use' will be no easy trick, involving detailed ecological understanding case by case. The alternative, however, would seem too often to be a mosaic of degraded habitat, increasingly threatening dedicated reserves, with all the tensions that entails. Terborgh's (1983) *Five New World Primates* is a pioneering work in this arena. It identifies a specific subset of tree species which would need to be kept in order for more intensive human exploitation of forests in the region of his study site to be reconciled with the continued survival there of five species of New World monkeys, all omnivores with a mixed diet of different fruits and small prey items. A more wide-ranging discussion of these issues, in an African context, is given by Western *et al.* (1994).

In essence, the broadly utilitarian argument recognizes that we do not know how much biological diversity we can lose, yet still keep ecosystem services on which humans depend. In this situation, as emphasized by one of the founders of the Conservation Movement, Aldo Leopold, 'the first rule of intelligent tinkering is to keep all the pieces'. But maybe we could be clever enough to survive in a greatly biologically impoverished world. It would, very likely, be a world akin to that of the cult movie *Blade Runner*. The question arises, who would want to live in such a world? This takes us to the third argument.

15.7 Ethical considerations

The ethical argument is simply put: we have a responsibility to hand on to future generations a planet as rich in natural wonders as the one we inherited. Narrowly utilitarian considerations urge us to preserve individual species, many of them not yet recorded much less studied, because tomorrow's biotechnology may find their genes useful. Broadly utilitarian considerations worry about preserving ecosystems because we depend upon their services. Some would say ethical considerations are more vague; I find them more compelling.

Some of the complexities of the ethical responsibilities of human stewardship were set out eloquently by Aldo Leopold. Mourning the death in the Cinncinnati Zoo in 1917 of Martha, the last passenger pigeon, he wrote: 'We grieve because no living man will see again the onrushing phalanx of victorious birds sweeping a path for Spring across the March skies, chasing the defeated winter from all the woods and prairies....Our grandfathers, who saw the glory of the fluttering hosts, were less well-housed, well-fed, well-clothed than we are. The strivings by which they bettered our lot are also those which deprived us of pigeons. Perhaps we now grieve because we are not sure, in our hearts, that we have gained by the exchange.... The truth is our grandfathers, who did the actual killing, were our agents. They were our agents in the sense they shared the conviction, which we have only now begun to doubt, that it is more important to multiply people and comforts than to cherish the beauty of the land in which they live.' This not only gives poetic expression to how many of us feel, but I think it also raises the question of whether I would feel the same way if I were a poor farmer in a drought-stricken developing country, striving to feed my family.

Such questions of environmental responsibility in relation to present circumstances can be put in sharper form, by returning to the EFs defined in

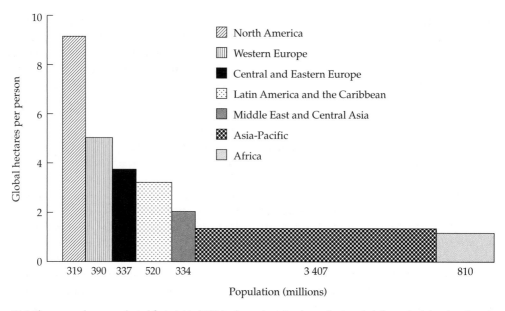

Figure 15.2 The average human ecological footprint in 2001 is shown, in units of area (hectare, ha), for each of the planet's major geographical regions. The corresponding populations are indicated on the x-axis, and the resulting rectangular areas represent the total ecological footprint by region. Adding all these together gives the total human ecological footprint for 2001, as shown in Figure 15.1.

connection with Figure 15.1. Figure 15.2 shows, for the planet's major regions, the average EF per person (the height of the respective area) and the total population (the width of the area). Obviously the total footprint of each region is the area, as shown in Figure 15.2. The total human footprint in 2001 is the sum of these areas. The resulting global average EF per person in 2001 is 2.2 ha whereas, returning to Figure 15.1, the WWF estimates the planet's BC per person that year—the EF per person that the planet could sustain, for a given value of the total human population—as 1.8 ha. Implicit in such calculations are some interesting paradoxes. Looking at individual countries, we find (WWF, 2004) that for Sweden the average person's EF is 7.0, in a country where per-capita BC is 9.8. The corresponding numbers for the average inhabitant of Egypt are EF 1.5 and BC 0.5. So the average Swede is living below the country's sustainable limits, the average Egyptian well above. Conversely, the average person's EF in Sweden is around five times that in Egypt. How do you parse ethical responsibilities here? And these two countries are not extremes in the matrix of EFs in relation to biological capacities of nations.

Similar questions can be framed on an even larger canvas. Table 15.2 shows how, in 2000, the world's population, wealth (measured by GDP), energy consumption, and inputs of carbon dioxide into the atmosphere are partitioned among the three worlds: rich (essentially the OECD countries), transitional, and poor countries. Observe that the rich world, with 13% of the global population, has more than half the global GDP and likewise consumes roughly half the energy and generates roughly half the greenhouse gas inputs.

These questions of ethical responsibilities to future generations, and how to meet them in appropriately equitable ways—given the past and currently diverse patterns of population numbers and resource consumption among nation states—lead us back to Nowak and Sigmund's early chapter (Chapter 2) on problems and paradoxes of cooperation. It is ironic to see fundamental theoretical questions about the evolution and maintenance of cooperative behaviour within human and other animal communities, which date back to Darwin, so entwined with pressing practical problems of how humankind, partitioned among squabbling nations, can affectively address its

Table 15.2 The three worlds in 2000.

	Poor	Transition	Rich
Population (billions)	4.1	1.2	0.8
Gross domestic product (trillion ppp$)	11	11	23
Industrial energy (TW)	2.9	3.2	6.3
Biomass energy (TW)	1.4	0.2	0.2
Fossil carbon (billion t of C/year)	1.6	1.7	3.1

future on a finite planet (Ehrlich and Levin, 2005). It may even be that, over the millennia since agriculture was invented, the answer shaped by evolutionary processes to the problem of building complex but stable human societies was to favour acquiescence in authoritarian hierarchies, with their concomitant rigidities. If so, our present predicament is an unusual illustration of Fisher's Fundamental Theorem[1], with its inherent tension between adaptedness and adaptability. I realize that to end on this note is interpreting theoretical ecology, with its many interesting and important questions, unusually broadly. But not, in my view, inappropriately so.

---

[1] Broadly, Fisher's Fundamental Theorem states that a population's potential rate of change of gene frequency (which measures its ability to adapt to changing circumstances) is proportional to the variance in gene frequency, which will be small if essentially all individuals are well-adapted to their current environment.

References

Aarssen, L.W. and Epp, G.A. 1990. Neighbor manipulations in natural vegetation—a review. *Journal of Vegetation Science* **1**: 13–30.

Abramson, G. and Kuperman, M. 2001. Social games in a social network. *Physical Review E* **63**: 030901R (1–4).

African Agricultural Technology Foundation. 2005. *A New Bridge to Sustainable Agricultural Development in Africa: Inaugural Report, May 2002–December 2004.* African Agricultural Technology Foundation, Nairobi, Kenya.

Agnew, D.J., Beddington, J.R., and Hill, S.L. 2002. The potential use of environmental information to manage squid stocks. *Canadian Journal of Fisheries and Aquatic Science* **59**: 1851–1857.

Agren, G.I. and Fagerstrom, T. 1984. Limiting dissimilarity in plants—randomness prevents exclusion of species with similar competitive abilities. *Oikos* **43**: 369–375.

Algar, A.C., Kerr, J.T., and Currie, D.J. 2006. A test of Metabolic Theory as the mechanism underlying broad-scale species richness gradients. *Global Ecology and Biogeography*, in press.

Allee, W.C., Park, O., Emerson, A.E., Park, T., and Schmidt, K.P. 1949. *Principles of Animal Ecology.* Saunders, Philadelphia.

Allen, A.P., Gillooly, J.F., and Brown, J.H. 2002. Global biodiversity, biochemical kinetics, and the energetic-equivalence rule. *Science* **297**: 1545–1548.

Allen, J.C. 1975. Mathematical models of species interactions in time and space. *American Naturalist* **109**: 319–342.

Anderson, K.J. and Jetz, W. 2005. The broad-scale ecology of energy expenditure of endotherms. *Ecology Letters* **8**: 310–318.

Anderson, R.M. and May, R.M. 1978. Regulation and stability of host-parasite population interactions-I. Regulatory processes. *Journal of Animal Ecology* **47**: 219–247.

Anderson, R.M. and May, R.M. 1979. Population biology of infectious diseases: Part I. *Nature* **280**: 361–367.

Anderson, R.M. and May, R.M. 1991. *Infectious Diseases of Humans: Dynamics and Control.* Oxford University Press, Oxford.

Anderson, R.M., Grenfell, B.T., and May, R.M. 1984. Oscillatory fluctuations in the incidence of infectious disease and the impact of vaccination: time series analysis. *Journal of Hygiene (Cambridge)* **93**: 587–608.

Andrewartha, H.G. and Birch, L.C. 1954. *The Distribution and Abundance of Animals.* University of Chicago Press, Chicago.

Araujo, M.B. and Pearson, R.G. 2005. Equilibrium of species' distributions with climate. *Ecography* **28**: 693–695.

Araujo, M.B., Whittaker, R.J., Ladle, R.J., and Erhard, M. 2005. Reducing uncertainty in projections of extinction risk from climate change. *Global Ecology and Biogeography* **14**: 529–538.

Armstrong, R.A. and McGehee, R. 1976a. Coexistence of two competitors on one resource. *Journal of Theoretical Biology* **56**: 499–502.

Armstrong, R.A. and McGehee, R. 1976b. Coexistence of species competing for shared resources. *Theoretical Population Biology* **9**: 317–328.

Armstrong, R.A. and McGehee, R. 1980. Competitive exclusion. *American Naturalist* **115**: 151–170.

Askew, R.R. 1971. *Parasitic Insects.* Heinemann, London.

Augspurger, C.K. and Franson, S.E. 1987. Wind dispersal of artificial fruits varying in mass, area, and morphology. *Ecology* **68**: 27–42.

Axelrod, R.M. 1984. *The Evolution of Cooperation.* Basic Books, New York.

Axelrod, R. and Hamilton, W.D. 1981. The evolution of cooperation. *Science* **211**: 1390–1396.

Babad, H.R., Nokes, D.J., Gay, N.J., Miller, E., Morgan-capner, P., and Anderson, R.M. 1995. Predicting the impact of measles vaccination in england and wales—model validation and analysis of policy options. *Epidemiology and Infection* **114**: 319–344.

Bailey, V.A., Nicholson, A.J., and Williams, E.J. 1962. Interactions between hosts and parasites where some host individuals are more difficult to find than others. *Journal of Theoretical Biology* **3**: 1–18.

Baillie, J.E.M., Hilton-Taylor, C. and Stuart, S.N. 2004. *A Global Species Assessment.* IUCN, Gland, Switzerland.

Bais, H.P., Vepachedu, R., Gilroy, S., Callaway, R.M., and Vivanco, J.M. 2003. Allelopathy and exotic plant invasion: from molecules and genes to species interactions. *Science* **301**: 1377–1380.

Balmford, A., Bruner, A., Cooper, P., Costanza, R., Farber, S., Green, R.E. *et al.* 2002. Economic reasons for conserving wild nature. *Science* **297**: 950–953.

Balmford, A., Gaston, K.J., Blyth, S., James, A., and Kapos, V. 2003. Global variation in terrestrial conservation costs, conservation benefits, and unmet conservation needs. *Proceedings of the National Academy of Sciences USA* **100**: 1046–1050.

Balmford, A., Bennun, L., ten Brink, B., Cooper, D., Côté, I.M., Crane, P. *et al.* 2005. The convention on biological diversity's 2010 target. *Science* **307**: 212–213.

Barabasi, A.-L. 2002. *Linked: The New Science of Networks.* Perseus, New York.

Baranov, F.I. 1918. On the question of the biological basis of fisheries. *Nauchnyi issledovatelskii ikhtiologicheskii Institut Isvestia* **1**: 81–128.

Barenblatt, G.I. 1996. *Scaling, Self-Similarity, and Intermediate Asymptotics,* Cambridge Texts in Applied Mathematics 14. Cambridge University Press, Cambridge.

Bartlett, M.S. 1956. Deterministic and stochastic models for recurrent epidemics. In *Proceedings of the Third Berkeley Symposium on Mathematical Statistics and Probability,* vol. 4, pp. 81–109. University of California Press, Berkeley, CA.

Bartlett, M.S. 1957. Measles periodicity and community size. *Journal of the Royal Statistical Society Series A* **120**: 48–70.

Bartlett, M.S. 1960. The critical community size for measles in the U.S. *Journal of the Royal Statistical Society Series A* **123**: 37–44.

Barton, N. 1993. The probability of fixation of a favoured allele in a subdivided population. *Genetic Research* **62**: 149–158.

Bascompte, J. and Melián, C.J. 2005. Simple trophic modules for complex food webs. *Ecology* **86**: 2868–2873.

Bascompte, J., Melián, C.J., and Sala, E. 2005. Interaction strength combinations and the overfishing of a marine food web. *Proceedings of the National Academy of Sciences USA* **102**: 5443–5447.

Bazzaz, F.A. 1975. Plant species diversity in old-field successional ecosystems in Southern Illinois. *Ecology* **56**: 485–488.

Beard, J.S. 1944. Climax vegetation in tropical America. *Ecology* **25**: 127–158.

Beard, J.S. 1955. The classification of tropical American vegetation-types. *Ecology* **36**: 89–100.

Beard, J.S. 1983. Ecological control of the vegetation of Southwestern Australia: moisture versus nutrients. In *Mediterranean-Type Ecosystems* (Kruger, F.J., Mitchell, D.T., and Jarvis, J.U.M., eds), pp. 66–73. Springer-Verlag, Berlin.

Beddington, J.R. and May, R.M. 1977. Harvesting natural populations in a randomly fluctuating environment. *Science* **197**: 463–465.

Beddington, J.R. and May, R.M. 1980. Maximum sustainable yields in systems subject to harvesting at more than one trophic level. *Mathematical Biosciences* **51**: 261–281.

Beddington, J.R. and Basson, M. 1994. The limits to exploitation on land and sea. *Philosophical Transactions of the Royal Society Series B* **343**: 87–92.

Beddington, J.R. and Kirkwood, G.P. 2005. The estimation of potential yield and stock status using life-history parameters. *Philosophical Transactions of the Royal Society Series B* **360**: 163–170.

Begon, M., Sait, S.M., and Thompson, D.J. 1995. Persistence of a parasitoid-host system—refuges and generation cycles. *Proceedings of the Royal Society of London Series B Biological Sciences* **260**: 131–137.

Begon, M., Harper, J.L., and Townsend, C.R. 1996a. *Ecology. Individuals, Populations and Communities,* 3rd edn. Blackwell Scientific Publications, Oxford.

Begon, M., Sait, S.M., and Thompson, D.J. 1996b. Predator-prey cycles with period shifts between two- and three-species systems. *Nature* **381**: 311–315.

Begon, M., Bennett, M., Bowers, R.G., French, N.P., Hazel, S.M., and Turner, J. 2002. A clarification of transmission terms in host-microparasite models: numbers, densities and areas. *Epidemiology and Infection* **129** 147–153.

Begon, M., Townsend, C.R., and Harper, J.L. 2006. *Ecology: >From Individuals to Communities,* 4th edn. Blackwell Scientific Publications, Oxford.

Bell, D.J., Roberton, S., and Hunter, P.R. 2005. Animal origins of SARS Corona virus: possible links with the international trade in small carnivores. In *SARS: A Case Study in Emerging Infections* (McLean, A.R. *et al.*, eds) pp. 51–60. Oxford University Press, Oxford.

Bell, G. 2001. Neutral macroecology. *Science* **293**: 2413–2418.

Benton, T.G., Bryant, D.M., Cole, L., and Humphrey, Q.P. 2002. Farmland biodiversity: is habitat heterogeneity the key? *Trends in Ecology and Evolution* **8**: 182–188.

Berlow, E.L., Neutel, A.-M., Cohen, J.E., de Ruiter, P.C., Ebenman, B., Emmerson, M. *et al.* 2004. Interaction strengths in food webs: issues and opportunities. *Journal of Animal Ecology* **73**: 585–598.

Bernoulli, D. 1760. Essai d'une nouvelle analyse de la mortalité causée par la petite vérole et des advantages de l'inoculation pour la prévenir. *Academie Royale des Sciences: Histoire et Memoires de Mathematique et de Physique,* 1–45.

Bernstein C., Kacelnik, A., and Krebs, J.R. 1988. Individual decisions and the distribution of predators in a patchy environment. *Journal of Animal Ecology* **57**: 1007–1026

Bernstein, C., Kacelnik, A., and Krebs, J.R. 1991. Individual decisions and the distribution of predators in a patchy environment. II. The influence of travel costs and the structure of the environment. *Journal of Animal Ecology* **60**: 205–225.

Bersier, L.-F., Dixon, P., and Sugihara, G. 1999. Scale-invariant or scale-dependent behaviour of the link-density property in food webs: a matter of sampling effort? *American Naturalist* **153**: 666–682.

Bertignac, M., Lehody, P., and Hampton, J. 1998. A spatial population dynamics simulation model of tropical tunas using a habitat index based on environmental parameters. *Fisheries Oceanography* **7**: 326–334.

Beverton, R.J.H. and Holt, S.J. 1957. On the dynamics of exploited fish populations. *MAFF Fisheries Investigation of London Series 2* **19**: 1–533.

Bjørnstad, O.N. and Grenfell, B.T. 2001. Noisy clockwork: time series analysis of population fluctuations in animals. *Science* **293**: 638–643.

Bjørnstad, O.N., Finkenstädt, B.F., and Grenfell, B.T. 2002. Dynamics of measles epidemics. I. estimating scaling of transmission rates using a time series SIR model. *Ecological Monographs* **72**: 169–184.

Black, J.N. 1964. An analysis of the potential production of swards of subterranean clover (*Trifolium subterraneum* L.) at Adelaide, South Australia. *Journal of Applied Ecology* **1**: 3–18.

Blackburn, T.M. and Hawkins, B.A. 2004. Bermann's rule and the mammal fauna of northern North America. *Ecography* **27**: 715–724.

Blouin-Demers, G. and Weatherhead, P.J. 2001. Thermal ecology of black rat snakes (*Elaphe obsolete*) in a thermally challenging environment. *Ecology* **82**: 3025–3043.

Bolker, B.M. and Grenfell, B.T. 1993. Chaos and biological complexity in measles dynamics. *Proceedings of the Royal Society of London Series B Biological Sciences* **251**: 75–81.

Bolker, B.M. and Pacala, S.W. 1999. Spatial moment equations for plant competition: understanding spatial strategies and the advantages of short dispersal. *American Naturalist* **153**: 575–602.

Bollobas, B. 2001. *Random Graphs*, 2nd edn. Academic Press, San Diego, CA.

Bomhard, B., Richardson, D.M., Donaldson, J.S., Hughes, G.O., Midgley, G.F., Raimondo, D.C. *et al.* 2005. Potential impacts of future land use and climate change on the Red List status of the Proteaceae in the Cape Floristic Region, South Africa. *Global Change Biology* **11**: 1452–1468.

Bond, E.M. and Chase, J.M. 2002. Biodiversity and ecosystem functioning at local and regional spatial scales. *Ecology Letters* **5**: 467–470.

Bonnie, R., Schwartzman, S., Oppenheimer, M., and Bloomfield, J. 2000. Counting the cost of deforestation. *Science* **288**: 1763–1764.

Bonsall, M.B. and Hassell, M.P. 2000. The effects of metapopulation structures on indirect effects in host-parasitoid assemblages. *Proceedings of the Royal Society of London Series B Biological Sciences* **267**: 2207–2212.

Bonsall, M.B. and Hastings, A. 2004. Demographic and environmental stochasticity in predator-prey metapopulation dynamics. *Journal of Animal Ecology* **73**: 1043–1055.

Bonsall, M.B. and Benmayor, R. 2005. Multiple infections alter density dependence in host-pathogen interactions. *Journal of Animal Ecology* **74**: 937–945.

Bonsall, M.B. and Hassell, M.P. 2005. Understanding ecological concepts: the role of laboratory systems. *Advances in Ecological Research* **37**: 1–36.

Bonsall, M.B., French, D.A., and Hassell, M.P. 2002. Metapopulation structures affect persistence of predator-prey interactions. *Journal of Animal Ecology* **71**: 1075–1084.

Bonsall, M.B., van der Meijden, E., and Crawley, M.J. 2003. Contrasting dynamics in the same plant-herbivore interaction. *Proceedings of the National Academy of Sciences USA* **100**: 14932–14936.

Bonsall, M.B., Hassell, M.P., Reader, P.M., and Jones, T.H. 2004a. Coexistence of natural enemies in a multitrophic host-parasitoid system. *Ecological Entomology* **29**: 639–647.

Bonsall, M.B., Jansen, V.A.A., and Hassell, M.P. 2004b. Life history trade-offs assemble ecological guilds. *Science* **306**: 111–114.

Bonsall, M.B., Bull, J.C., Pickup, N.J., and Hassell, M.P. 2005. Indirect effects and spatial scaling affects the persistence of multispecies metapopulations. *Proceedings of the Royal Society of London Series B Biological Sciences* **272**: 1465–1471.

Bowles, S. and Gintis, H. 2004. The evolution of strong reciprocity: cooperation in heterogeneous populations. *Theoretical Population Biology* **65**: 17–28.

Boyce, M.S. 1992. Population viability analysis. *Annual Review of Ecology and Systematics* **23**: 481–506.

Boyd, R. and Richerson, P.J. 2002. Group beneficial norms can spread rapidly in a structured population. *Journal of Theoretical Biology* **215**: 287–296.

Brandt, H. and Sigmund, K. 2004. The logic of reprobation: assessment and action rules for indirect reciprocity. *Journal of Theoretical Biology* **231**: 475–486.

Brandt, H. and Sigmund, K. 2005. Indirect reciprocity, image scoring, and moral hazard. *Proceedings of the National Academy of Sciences USA* **102**: 2666–2670.

Brooks, T.M., Pimm, S.L., and Collar, N.J. 1997. Deforestation predicts the number of threatened birds in insular southeast Asia. *Conservation Biology* **11**: 382–394.

Brown, J.H. 1984. On the relationship between abundance and distribution of species. *American Naturalist* **124**: 255–279.

Brown, J.H., Stevens, G.C., and Kaufman, D.M. 1996. The geographic range: size, shape, boundaries, and internal structure. *Annual Review of Ecology and Systematics* **66**: 597–623.

Brown, J.H., Gillooly, J.F., Allen, A.P., Savage, V.M., and West, G.B. 2004. Toward a metabolic theory of ecology. *Ecology* **85**: 1771–1789.

Brown, M.W. and Cameron, E.A. 1979. Effects of disparlure and egg mass size on parasitism by the gypsy moth egg parasite, *Ooencyrtus kuwanai*. *Environmental Entomology* **8**: 77–80.

Buckley, L.B. and Roughgarden, J. 2004. Biodiversity conservation: effects of changes in climate and land use. *Nature* **430**: 2 p following 33.

Bull, B., Francis, R.I.C.C., Dunn, A., McKenzie, A., Gilbert, D.J., and Smith, M.H. 2005. *CASAL User Manual*. NIWA Technical Report **126**.

Burel, F., Baudry, J., Butet, A., Clergeau, P., Delettre, Y., Le Cœur, D. *et al.* 1998. Comparative biodiversity along a gradient of agricultural landscapes. *Acta Oecologica* **19**: 47–60.

Burgman, M.A., Ferson, S., and Akçakaya, H.R. 1993. *Risk Assessment in Conservation Biology*. Chapman and Hall, London.

Burness, G.P., Diamond, J., and Flannery, T. 2001. Dinosaurs, dragons, and dwarfs: the evolution of maximal body size. *Proceedings of the National Academy of Sciences USA* **98**: 14518–14523.

Burns, C.E., Johnston, K.M., and Schmitz, O.J. 2003. Global climate change and mammalian species diversity in U.S. national parks. *Proceedings of the National Academy of Sciences USA* **100**: 11474–11477.

Butterworth, D.S., Cochrane, K.L., and De Olivera, J.A.A. 1997. Management procedures: a better way to manage fisheries? The South African experience. In *Global Trends: Fisheries Management. American Fisheries Society Symposium 20* (Pikitch, E.K., Huppert, D.D., and Sissenwine, M.P., eds), pp. 83–90. American Fisheries Society, Bethesda, MD.

Camm, J.D., Polasky, S., Solow, A., and Csuti, B. 1996. A note on optimal algorithms for reserve site selection. *Biological Conservation* **78**: 353–355.

Carl, E.A. 1971. Population control in Arctic ground squirrels. *Ecology* **52**: 395–413.

Carmel, Y. and Flather, C.H. 2006. Constrained range expansion and climate change assessments. *Frontiers in Ecology and the Environment* **4**: 178–179.

Carpenter, S.R. and Cottingham, K.L. 1997. Resilience and restoration of lakes. *Conservation Ecology* **1**: 2. www.consecol.org/vol1/iss1/art2.

Carpenter, S.R., Cottingham, K.L., and Stow, C.A. 1994. Fitting predator-prey models to time series with observation errors. *Ecology* **75**: 1254–1264.

Casdagli, M. 1992. Chaos and determinstic versus stochastic nonlinear modeling. *Journal of the Royal Statistics Society B* **54**: 303–328.

Case, T.J. 2000. *An Illustrated Guide to Theoretical Ecology*. Oxford University Press, New York.

Casper, B.B. and Jackson, R.B. 1997. Plant competition underground. *Annual Review of Ecology and Systematics* **28**: 545–570.

Cassman, K.G. 1999. Ecological intensification of cereal production systems: yield potential, soil quality, and precision agriculture. *Proceedings of the National Academy of Sciences USA* **96**: 5952–5959.

Caswell, H. 1989. *Matrix Populaton Models*. Sinauer Associates, Sunderland, MA.

Caswell, H. 2001. *Matrix Population Models: Construction, Analysis and Interpretation*, 2nd edition. Sinauer Associates, Sunderland, MA.

Cattin, M.-F., Bersier, L.-F., Banasek-Richter, C., Baltensperger, M., and Gabriel, J.-P. 2004. Phylogenetic constraints and adaptation explain food-web structure. *Nature* **427**: 835–839.

Caudill, C.C. 2005. Trout predators and demographic sources and sinks in a mayfly metapopulation. *Ecology* **86**: 935–946.

Chambers, R. 1997. *Whose Reality Counts? Putting the First Last*. Intermediate Technology Publications, London.

Chambers, R. 2005. *Ideas for Development*. Earthscan, London.

Chape, S., Harrison, J., Spalding, M., and I. Lysenko. 2005. Measuring the extent and effectiveness of protected areas as an indicator for meeting global biodiversity targets. *Philosophical Transactions of the Royal Society* **360**: 443–455.

Chapin, F.S., Zavaleta, E.S., Eviner, V.T., Naylor, R.L., Vitousek, P.M., Reynolds, H.L. *et al.* 2000. Consequences of changing biodiversity. *Nature* **405**: 234–242.

Charnov, E.L. 1993. *Life History Invariants*. Oxford University Press, New York.

Chase, J.M., Amarasekare, P., Cottenie, K., Gonzalez, A., Holt, R.D., Holyoak, M. *et al.* 2005. Competing theories for competitive metacommunities. In *Metacommunities: Spatial Dynamics and Ecological Communities* (Holyoak, M., Leibold, M.A., and Holt, R.D., eds), pp. 335–354. University of Chicago Press, Chicago.

Chesson, P.L. 1986. Environmental variation and the coexistence of species. In *Community Ecology* (Diamond, J. and Case, T., eds), pp. 240–256. Harper and Row, New York.

Chesson, P. 1994. Multispecies competition in variable environments. *Theoretical Population Biology* **45**: 227–276.

Chesson, P. 2000. Mechanisms of maintenance of species diversity. *Annual Review of Ecology and Systematics* **31**: 343–366.

Chesson, P.L. and Warner, R.R. 1981. Environmental variability promotes coexistence in lottery competitive systems. *American Naturalist* **117**: 923–943.

Chesson, P.L. and Murdoch, W.W. 1986. Aggregation of risk: relationships among host-parasitoid models. *American Naturalist* **127**: 696–715.

Chesson, P. and Huntly, N. 1997. The roles of harsh and fluctuating conditions in the dynamics of ecological communities. *American Naturalist* **150**: 519–553.

Christensen, V. and Pauly, D. 1992. ECOPATH II—a software for balancing steady-state models and calculating network characteristics. *Ecological Modeling* **61**: 169–185.

Christensen, V., Walters, C.J., and Pauly, D. 2000. *ECO-PATH with ECOSIM: A User's Guide, October 2000 Edition*. Fisheries Centre, University of British Columbia, Vancouver.

Clark, C.W. 1985. *Mathematical Bioeconomics*. Wiley Interscience, New York.

Clark, C.W., Munro, G.R., and Sumaila, U.R. 2005. Subsidies, buybacks, and sustainable fisheries. *Journal of Environment, Economics and Management* **50**: 47–58

Clark, J.A. and May, R.M. 2002. Taxonomic bias in conservation research. *Science* **297**: 191–192.

Clements, F.E. 1936. Nature and structure of the climax. *Journal of Ecology* **24**: 252–284. Reprinted in *Foundations of Ecology: Classic Papers with Commentaries* (Real, L.A. and Brown, J.H., eds), 1991. University of Chicago Press, Chicago.

Cliff, A.D. and Haggett, P. 1988. *Atlas of Disease Distributions: Analytic Approaches to Epidemiologic Data*. Blackwell, Oxford.

Cliff, A.D., Haggett, P., and Smallman-Raynor, M. 1993. *Measles: an Historical Geography of a Major Human Viral Disease from Global Expansion to Local Retreat, 1840–1990*. Blackwell, Oxford.

Cocks, K.D. and Baird, I.A. 1989. Using mathematical programming to address the multiple reserve selection problem: an example from the Eyre Peninsula, South Australia. *Biological Conservation* **49**: 113–130.

Cohen, J.E. 1995. *How Many People Can the Earth Support?* Norton, New York.

Cohen, J.E., Briand, F., and Newman, C.M. 1990. *Community Food Webs: Data and Theory*. Springer-Verlag, Berlin.

Collie, J.S. and Gislason, H. 2001. Biological reference points for fish stocks in a multispecies context. *Canadian Journal of Fisheries and Aquatic Science* **58**: 2233–2246.

Collingham, Y.C. and Huntley, B. 2000. Impacts of habitat fragmentation and patch size upon migration rates. *Ecological Applications* **10**: 131–144.

Comins, H.N. and Hassell, M.P. 1996. Persistence of multispecies host-parasitoid interactions in spatially distributed models with local dispersal. *Journal of Theoretical Biology* **183**: 19–28.

Comins, H.N., Hassell, M.P., and May, R.M. 1992. The spatial dynamics of host-parasitoid systems. *Journal of Animal Ecology* **61**: 735–748.

Connell, J.H. 1971. On the role of natural enemies in preventing competitive exclusion in some marine animals and in rain forest trees. In *Dynamics of Populations. Proceedings of the Advanced Study Institute on Dynamics of Numbers in Populations* (den Boer, B.J. and Gradwell, G.R., eds), pp. 298–310. Centre for Agricultural Publishing and Documentation, Wageningen.

Connell, J.H. 1980. Diversity and the coevolution of competitors, or the Ghost of Competition Past. *Oikos* **35**: 131–138.

Connell, J. 1983. On the prevalence and relative importance of interspecific competition: evidence from field experiments. *American Naturalist* **122**: 661–696.

Conway, G.R. 1976. Man versus pests. In *Theoretical Ecology: Principles and Applications* (May, R.M., ed.), pp. 257–281. Blackwell Scientific Publications, Oxford.

Conway, G.R. 1985. Agroecosystem analysis. *Agricultural Administration* **20**: 31–55.

Conway, G.R. 1987. The properties of agroecosystems. *Agricultural Systems* **24**: 95–117.

Conway, G. 1997. *The Doubly Green Revolution. Food for all in the 21st Century*. Penguin Books, London.

Conway, G. 2005. GM crops—an international perspective on the economic and environmental benefits. In *GMOs—Ecological Dimensions*, conference proceedings, University of Reading, Berkshire, 9 September 2002 (van Emden, H.F. and Gray, A.J., eds). Association of Applied Biologists, Wellesbourne.

Conway, G.R. and J.N. Pretty, 1991. *Unwelcome Harvest: Agriculture and Pollution*. Earthscan, London.

Cook, R. 1980. The biology of seeds in the soil. In *Demography and Evolution in Plant Populations* (Solbrig, O.T., ed.), pp. 107–131. Blackwell Scientific Publications, Oxford.

Cook, R.M. and Armstrong, W. 1984. Density effects in the recruitment and growth of North Sea haddock and whiting. *ICES Council Meeting Papers* **G68**: 21.

Cooke, J.G. and Beddington, J.R. 1985. The relationship between catch rates and abundance in fisheries. *Journal of Mathematical Biology* **1**: 391–406.

Costantino, R.F., Desharnais, R.A., Cushing, J.M., and Dennis, B. 1997. Chaotic dynamics in an insect population. *Science* **275**: 389–391.

Costanza, R., d'Arge, R., de Groot, R., Farber, S., Grasso, M., Hannon, B. *et al.* 1997. The value of the world's ecosystem services and natural capital. *Nature* **387**: 253–260.

Cottam, D.A., Whittaker, J.B., and Malloch, A.J.C. 1986. The effects of Chrysomelid beetle grazing and plant competition on the growth of *Rumex obtusifolius*. *Oecologia* **70**: 452–456.

Cottingham, K.L., Brown, B.L., and Lennon, J.T. 2001. Biodiversity may regulate the temporal variability of ecological systems. *Ecology Letters* **4**: 72–85.

Coulson, T., Catchpole, E.A., Albon, S.D., Morgan, B.J.T., Pemberton, J.M., Clutton-Brock, T.H. *et al.* 2001. Age, sex, density, winter weather, and population crashes in Soay sheep. *Science* **292**: 1528–1531.

Coulson, T., Rohani, P., and Pascual, M. 2004. Skeletons, noise and population growth: the end of an old debate? *Trends in Ecology and Evolution* **19**: 359–364.

Cowan, I.M.C., Chapman, D.G., Hoffman, R.S., McCullough, D.R., Swanson, G.A., and Weeden, R.B. 1974. *Report of Committee on the Yellowstone Grizzlies*. National Academy of Science, Washington DC.

Craighead, J.J., Hornocker, M.G., and Craighead, J., F.C. 1969. Reproductive biology of young female grizzly bears. *Journal of Reproduction & Fertility, Supplement* **6**: 447–475.

Craighead, J.J., Sumner, J.S., and Mitchell, J.A. 1995. *The Grizzly Bears of Yellowstone*. Island Press, Washington DC.

Crane, P. (ed.) 2003. *Measuring Biodiversity for Conservation*, Royal Society Policy Document 11–03. Royal Society, London.

Crawley, M.J. 1983. *Herbivory: the Dynamics of Animal-Plant Interactions*. Blackwell Scientific Publications, Oxford.

Crawley, M.J. 1990. Rabbit grazing, plant competition and seedling recruitment in acid grassland. *Journal of Applied Ecology* **27**: 803–820.

Crawley, M.J. 1997. Plant-herbivore dynamics. In *Plant Ecology* (Crawley, M.J., ed.), pp. 401–474. Blackwell Scientific Publications, Oxford.

Crawley, M.J. 2000. Seed predators and population dynamics. In *Seeds: the Ecology of Regeneration in Plant Communities* (Fenner, M., ed.), pp. 167–182. CAB International, Wallingford.

Crawley, M.J. and May, R.M. 1987. Population dynamics and plant community structure—competition between annuals and perennials. *Journal of Theoretical Biology* **125**: 475–489.

Crawley, M.J. and Long, C.R. 1995. Alternate bearing, predator satiation and seedling recruitment in *Quercus robur* L. *Journal of Ecology* **83**: 683–696.

Crawley, M.J., Albon, S.D., Bazely, D.R., Milner, J.M., Pilkington, J.G., and Tuke, A.L. 2004. Vegetation and sheep population dynamics. In *Soay Sheep. Dynamics and Selection in an Island Population* (Clutton-Brock, T.H. and Pemberton, J.M., eds.), pp. 89–112. Cambridge University Press, Cambridge.

Crawley, M.J., Johnston, A.E., Silvertown, J., Dodd, M., de Mazancourt, C., Heard, M.S. *et al.* 2005. Determinants of species richness in the Park Grass Experiment. *American Naturalist* **165**: 348–362.

Crofton, H.D. 1971. A model of host-parasite relationships. *Parasitology* **63**: 343–364.

Crowley, T.J. 1990. Are there any satisfactory geologic analogs for a future greenhouse warming? *Journal of Climate* **3**: 1282–1292.

Crozier, L., Dwyer, G. 2006. Combining population-dynamic and ecophysiological models to predict climate-induced range shifts. *American Naturalist* **187**: 853–866.

Csada, R.D., James, P.C., and Espie, R.H.M. 1996. The 'file drawer problem' of non-significant results: does it apply to biological research? *Oikos* **76**: 591–593.

Csuti, B., Polasky, S., Williams, P.H., Pressey, R.L., Camm, J.D., Kershaw, M. *et al.* 1997. A comparison of reserve selection algorithms using data on terrestrial vertebrates in Oregon. *Biological Conservation* **80**: 83–97.

Currie, D.J. 1991. Energy and large-scale patterns of animal and plant species richness. *American Naturalist* **137**: 27–49.

Currie, D.J., Mittelbach, G.G., Cornell, H.V., Field, R., Guegan, J.F., Hawkins, B.A. *et al.* 2004. Predictions and tests of climate-based hypotheses of broad-scale variation in taxonomic richness. *Ecology Letters* **7**: 1121–1134.

Czeck, B. and Krausman, P.R. 1997. Distribution and causation of species endangerment in the United States. *Science* **277**: 1116–1117.

Da Fonseca, G.A.B., Sechrest, W., and Oglethorpe, J. 2005. Managing the matrix. In *Climate Change and Biodiversity* (Lovejoy, T.E. and Hannah, L., eds), pp. 346–358. Yale University Press, New Haven, CT.

Daily, G.C. (ed.) 1997 *Nature's Services. Societal Dependence on Natural Ecosystems*. Island Press, Washington DC.

Daily, G.C., Alexander, S., Ehrlich, P.R., Goulder, L., Lubchenco, J., Matson, P.A. *et al.* 1997. Ecosystem services: benefits supplied to human societies by natural ecosystems. *Issues in Ecology* **2**: 1–16.

Daily, G.C., Soderqvist, T., Aniyer, S., Arrow, K., Dasgupta, P., Ehrlich, P.R. *et al.* 2000. The value of nature and the nature of value. *Science* **289**: 395–396.

Darwin, C. 1859. *The Origin of Species by Means of Natural Selection*. Reprinted by The Modern Library, Random House, New York.

Daufresne, T. and Hedin, L.O. 2005. Plant coexistence depends on ecosystem nutrient cycles: extension of the resource-ratio theory. *Proceedings of the National Academy of Sciences USA* **102**: 9212–9217.

Davis, M.B. and Shaw, R.G. 2001. Range shifts and adaptive responses to quaternary climate change. *Science* **292**: 673–679.

Deevey, E.S. 1960. The human population. *Scientific American* **203**: 195–204.

Deguise, I. and Kerr, J.T. 2006. Protected areas and prospects for endangered species conservation in Canada. *Conservation Biology* **20**: 48–55.

de Kroon, H., Huber, H., Stuefer, J., and van Groenendael, J. 2005. A modular concept of phenotypic plasticity in plants. *New Phytologist* **166**: 73–82.

Delcourt, H.R. and Delcourt, P.A. 1984. Ice Age haven for hardwoods. *Natural History* September: 22–28.

Delgado, C.C., Hopkins, J. and Kelly, V.A. 1998. *Agricultural Growth Linkages in Sub-Saharan Africa*, IFPRI Research Report 107. International Food Policy Research Institute, Washington DC.

Dennis, B., Desharnais, R.A., Cushing, J.M., and Costantino, R.F. 1995. Non-linear demographic dynamics—mathematical models, statistical methods and biological experiments, *Ecological Monographs* **65**: 261–281.

Dennis, R.L.H. and Shreeve, T.G. 1991. Climatic change and the British butterfly fauna: opportunities and constraints. *Biological Conservation* **55**: 1–16.

Department for International Development. 2005. *Growth and Poverty Reduction: the Role of Agriculture*. Department for International Development, London.

de Ruiter, P.C., Neutel, A.-M., and Moore, J.C. 1995. Energetics, patterns of interaction strengths, and stability in real ecosystems. *Science* **269**: 1257–1260.

Desrocher, A.E., Lunn, N.J., and Stirling, I. 2004. Polar bears in a warming climate. *Integrative and Comparative Biology* **44**: 163–176.

de Valpine, P. and Hastings, A. 2002. Fitting population models incorporating process noise and observation error. *Ecological Monographs* **72**: 57–76.

Diamond, J.M. and May, R.M. 1977. Species turnover rates on islands dependence on census interval. *Science* **197**: 266–270.

Diamond, J.M. and May, R.M. 1981. Island biogeography and the design of natural reserves. In *Theoretical Ecology* (May, R.M., ed.), pp. 228–252. Blackwell Scientific Publications, Oxford.

Dietz, K. 1976. The incidence of infectious diseases under the influence of seasonal fluctuations. *Lecture Notes in Biomathematics* **11**: 1–15.

Dinerstein, E. and Wikramanayake, E.D. 1993. Beyond "hotspots": how to prioritize investments to conserve biodiversity in the Indo-Pacific region. *Conservation Biology* **7**: 53–65.

Doak, D., Bigger, D., Harding, E., Marvier, M., O'Malley, R., and Thomson, D. 1998. The statistical inevitability of stability-diversity relationships in community ecology. *American Naturalist* **151**: 264–276.

Dobson, A.P. 2005. Monitoring global rates of biodiversity change: challenges that arise in meeting the 2010 goals. *Philosophical Transactions of the Royal Society Series B* **360**: 229–244.

Dobson, A. and Crawley, M. 1995. Pathogens and the structure of plant communities. *Trends in Ecology & Evolution* **9**: 393–398.

Dobson, A.P., Bradshaw, A.D., and Baker, A.J.M. 1997a. Hopes for the future: Restoration ecology and conservation biology. *Science* **277**: 515–521.

Dobson, A.P., Rodriguez, J.P., Roberts, W.M., and Wilcove, D.S. 1997b. Geographic distribution of endangered species in the United States. *Science* **275**: 550–553.

Dobson, A.P., Kinnaird, M.F., and O'Brien, T.G. 1997c. *An Analysis of Grizzly Bear Demography in the Yellowstone and Northern Continental Divide Populations*. EEB Department, Princeton University, Princeton, NJ.

Dobson, A.P., Kinnaird, M.F., and O'Brien, T.G. 1991. *An Analysis of Grizzly Bear Demography in the Yellowstone and Northern Continental Divide Populations*. A report for the Wilderness Society EEB Department, Princeton University, Princeton, NJ.

Dobson, A.P., Lodge, D.M., Alder, J., Cumming, G., Keymer, J.E., Mooney, H.A. *et al.* 2006. Habitat loss, trophic collapse and the decline of ecosystem services. *Ecology* **87**: 1915–1924.

Donald, C.M. 1951. Competition among pasture plants. I. Intra-specific competition among annual pasture plants. *Australian Journal of Agricultural Research* **2**: 355–376.

Dornelas, M., Connelly, S.R., and Hughes, T.P. 2006. Coral reef diversity refutes the neutral theory of biodiversity. *Nature* **440**: 80–82.

Drechsler, M. 2005. Probabilistic approaches to scheduling reserve selection. *Biological Conservation* **122**: 253–262.

Dunne, J.A. 2006. The network structure of food webs. In *Ecological Networks: Linking Structure to Dynamics in Food Webs* (Pascual, M. and Dunne, J.A., eds), pp. 27–86. Oxford University Press, Oxford.

Dunne, J.A., Williams, R.J., and Martinez, N.D. 2002. Food-web structure and network theory: the role of connectance and size. *Proceedings of the National Academy of Sciences USA* **99**: 12917–12922.

Dunne, J.A., Williams, R.J., and Martinez, N.D. 2004. Network structure and robustness of marine food webs. *Marine Ecology Progress Series* **273**: 291–302.

Durrett, R. and Levin, S.A. 1994a. The importance of being discrete (and spatial). *Theoretical Population Biology* **46**: 363–394.

Durrett, R. and Levin, S.A. 1994b. Stochastic spatial models: a user's guide to ecological applications. *Philosophical Transactions of the Royal Society Series B* **343**: 329–350.

Dyar, H.G. 1890. The number of moults of Lepidopterous larvae. *Psyche* **5**: 420–422.

Earn, D.J.D., Rohani, P., Bolker, B.M., and Grenfell, B.T. 2000. A simple model for complex dynamical transitions in epidemics. *Science* **287**: 667–670.

Easterling, M.R., Ellner, S.P., and Dixon, P.M. 2000. Size-specific sensitivity: applying a new structured population model. *Ecology* **81**: 694–708.

Eastwood, R. and Fraser, A.M. 1999. Associations between lycaenid butterflies and ants in Australia. *Australian Journal of Ecology* **24**: 503–537.

Ebel, H. and Bornholdt, S. 2002. Coevolutionary games on networks. *Physical Review E* **66**: 056118 (1–8).

Eberhardt, L.L., Knight, R.R., and Blanchardt, B.M. 1986. Monitoring grizzly bear population trends. *Journal of Wildlife Management* **50**: 613–618.

Eckburg, P.B. 2005. Diversity of the human intestinal microbial flora. *Science* **308**: 1635–1638.

Edwards, A.W.F. 1972. *Likelihood*. John Hopkins University Press, Baltimore, MD.

Ehler, L.E. 1987. Patch-exploitation efficiency in a torymid parasite of a gall midge. *Environmental Entomology* **16**: 198–201.

Ehrlich, P.R. and Ehrlich, A.H. 1981. *Extinction, The Causes and Consequences of the Disappearance of Species*. Random House, New York.

Ehrlich, P.R. and Levin, S.A. 2005. The evolution of norms. *PLoS Biology* **3**: 0943–0948.

Ellenberg, H. 1953. Physiologisches und okologisches Verhalten derselben Pflanzenarten. *Bericht der Deutschene Botanischen Gesellschaft* **65**: 351–361.

Ellner, S.P. and Turchin, P. 1995. Chaos in a 'noisy' world: new methods and evidence from time series analysis. *American Naturalist* **145**: 343–375.

Ellner, S.P., McCauley, E., Kendall, B.E., Briggs, C.J., Hosseini, P.R., Wood, S.N. *et al.* 2001. Habitat structure and population persistence in an experimental community. *Nature* **412**: 538–542.

Elton, C.S. 1924. Fluctuations in the numbers of animals: their cause and effects. *British Journal of Experimental Biology* **2**: 119–163.

Elton, C.S. 1927. *Animal Ecology*. Sidgewick and Jackson, London.

Elton, C.S. 1958. *The Ecology of Invasions by Animals and Plants*. London, Methuen.

Elton, C. and Nicholson, M. 1942. Fluctuations in the numbers of muskrat (*Ondatra zibethica*) in Canada. *Journal of Animal Ecology* **11**: 96–126.

Eshel, I. 1972. Neighbor effect and the evolution of altruistic traits. *Theoretical Population Biology* **3**: 258–277.

Essington, T.E., Beaudreau, A.H., and Weidenmann, J. 2006. Fishing through marine food webs. *Proceedings of the National Academy of Sciences USA* **103**: 3171–3175.

Fagan, W.F. 1997. Omnivory as a stabilizing feature of natural communities. *American Naturalist* **150**: 554–567.

Faith, D.P. 1992. Conservation evaluation and phylogenetic diversity. *Biological Conservation* **61**: 1–10.

Faith, D.P., Carter, G., Cassis, G., Ferrier, S., and Wilkie, L. 2003. Complementarity, biodiversity viability analysis, and policy-based algorithms for conservation. *Environmental Science & Policy* **6**: 311–328.

Fehr, E. and S. Gachter. 2002. Altruistic punishment in humans. *Nature* **415**: 137–140.

Fehr, E. and Fischbacher, U. 2003. The nature of human altruism. *Nature* **425**: 785–791.

Feigenbaum, M.J. 1978. Quantitative universality for a class of non-linear transformations. *Journal of Statistical Physics* **19**: 25–52.

Feinsinger, P., Whelan, R.J., and Kiltie, R.A. 1981. Some notes on community composition: assembly by rules or by dartboards? *Bulletin of the Ecological Society of America* **62**: 19–23.

Fenchel, T. and Finlay, B.J. 2004. The ubiquity of small species: patterns of local and global diversity. *BioScience* **54**: 777–784.

Ferguson, N.M., Donnelly, C.A., and Anderson, R.M. 2001a. The foot-and-mouth epidemic in Great Britain: pattern of spread and impact of interventions. *Science* **292**: 1155–1160.

Ferguson, N.M., Donnelly, C.A., and Anderson, R.M. 2001b. Transmission intensity and impact of control policies on the foot and mouth epidemic in Great Britain. *Nature* **413**: 542–548.

Ferguson, N.M., Cummings, D.A.T., Cauchemez, S., Fraser, C., Riley, S., Meeyai, A. *et al.* 2005. Strategies for containing an emerging influenza pandemic in Southeast Asia. *Nature* **437** 209–214.

Fine, P.E.M. and Clarkson, J.A. 1982. Measles in England and Wales-I: an analysis of factors underlying seasonal patterns. *International Journal of Epidemiology* **11**: 5–15.

Finkenstadt, B. and Grenfell, B. 1998. Empirical determinants of measles metapopulation dynamics in England and Wales. *Proceedings of the Royal Society of London Series B Biological Sciences* **265**: 211–220.

Finlay, B.J., Thomas, J.A., McGavin, G.C., Fenchel, T., and Clarke, R.T. 2006. Self-similar patterns of nature: insect diversity at local to global scales. *Proceedings of the*

Royal Society of London Series B Biological Sciences **273**: 1935–1941.

Fischer, D.T. and Church, R.L. 2005. The SITES reserve selection system: a critical review. *Environmental Modeling and Assessment* **10**: 215–228.

Fisher, R.A. and Ford, E.B. 1950. The Sewall Wright effect. *Heredity* **4**: 117–119.

Fishman, M.A. 2003. Indirect reciprocity among imperfect individuals. *Journal of Theoretical Biology* **225**: 285–292.

Forbes, S.A. 1887. The lake as a microcosm. *Bulletin of the Peoria Scientific Association*: 77–27. Reprinted in *Foundations of Ecology: Classic Papers with Commentaries* (Real, L.A. and Brown, J.H., eds), 1991. University of Chicago Press, Chicago.

Fowler, N.L. 1986. Density-dependent population regulation in a Texas grassland. *Ecology* **67**: 545–554.

Fox, G.A. and Gurevitch, J. 2000. Population numbers count: tools for near-term demographic analysis. *American Naturalist* **156**: 242–256.

Frank, S.A. 1998. *Foundations of Social Evolution.* Princeton University Press, Princeton, NJ.

Frank, S.A. 2002. *Immunology and Evolution of Infectious Disease.* Princeton: Princeton University Press, NJ.

Fraser, C., Riley, S., Anderson, R.M., and Ferguson, N.M. 2004. Factors that make an infectious disease outbreak controllable. *Proceedings of the National Academy of Sciences USA* **101**: 6146–6151.

Freckleton, R.P. and Watkinson, A.R. 2002. Are weed population dynamics chaotic? *Journal of Applied Ecology* **39**: 699–707.

Free, C.A., Beddington, J.A., and Lawton, J.H. 1977. On the inadequacy of simple models of mutual interference for parasitism and predation. *Journal of Animal Ecology* **46**: 543–554.

Freemark, K. and Boutin, C. 1995. Impacts of agricultural herbicide use on terrestrial wildlife in temperate landscapes: a review with special reference to North America. *Agriculture Ecosystems and Environment* **52**: 67–91.

Frost, T.M., Carpenter, S.R., Ives, A.R., and Kratz, T.K. 1995. Species compensation and complementarity. In *Linking Species and Ecosystems* (Jones, C.G. and Lawton, J.H., eds), pp. 224–239. Chapman and Hall, New York.

Frost, T.M., Montz, P.K., Kratz, T.K., Badillo, T., Brezonik, P.L., Gonzalez, M.J. *et al.* 1999. Multiple stresses from a single agent: diverse responses to the experimental acidification of Little Rock Lake, Wisconsin. *Limnology and Oceanography* **44**: 784–794.

Fudenberg, D. and Maskin, E. 1990. Evolution and cooperation in noisy repeated games. *American Economics Review* **80**: 274–279.

Gaines, S. and Roughgarden, J. 1985. Larval settlement rate: a leading determinant of structure in an ecological community of the marine intertidal zone. *Proceedings of the National Academy of Sciences USA* **82**: 3707–3711.

Gallagher, K.D., Kenmore, P.E., and Sogawa, K. 1994. Judicial use of insecticides deter planthopper outbreaks and extend the life of resistant varieties in Southeast Asian rice. In *Ecology and Management of Planthoppers* (Denno, R.F. and Perfect, T.J. eds), pp. 599–614. Chapman and Hall, London.

Garcia, S.M. and Newton, C.H. 1997. Current situation, trends and prospects in world capture fisheries. In *Global Trends: Fisheries Management. American Fisheries Society Symposium 20* (Pikitch, E.K., Huppert, D.D., and Sissenwine, M.P., eds), pp. 3–27. American Fisheries Society, Bethesda, MD.

Garcia, S.M. and Grainger, J.R. 2005. Gloom and doom? The future of marine capture fisheries. *Philosophical Transactions of the Royal Society Series B* **360**: 21–46.

Gaston, K.J. 1991. How large is a species' geogrpahic range? *Oikos* **61**: 434–438.

Gaston, K.J. 1994. *Rarity.* Chapman and Hall, New York.

Gaston, K.J. and May, R.M. 1992. Taxonomy of taxonomists. *Nature* **356**: 281–282.

Gaston, K.J., Blackburn, T.M., and Goldewijk, K.K. 2003. Habitat conversion and global avian biodiversity loss. *Proceedings of the Royal Society of London B Biological Sciences* **270**: 1293–1300.

Gause, G.F. 1934. *The Struggle for Existence.* Waverly Press, Baltimore, MD.

Gilpin, M.E. and Diamond, J.M. 1980. Subdivision of nature reserves and the maintenance of species diversity. *Nature* **285**: 567–568.

Gilpin, M. and Hanski, I. (eds) 1991. *Metapopulation Dynamics: Empirical and Theoretical Investigations.* Academic Press, San Diego.

Gleason, H.A. 1926. The individualistic concept of the plant association. *Bulletin of the Torrey Botanical Society* **53**: 7–26. Reprinted in *Foundations of Ecology: Classic Papers with Commentaries* (Real, L.A. and Brown, J.H., eds), 1991. University of Chicago Press, Chicago.

Gleeson, S. and Tilman, D. 1990. Allocation and the transient dynamics of succession on poor soils. *Ecology* **71**: 1144–1155.

Godfray, H.C.J. 1994. *Parasitoids.* Princeton University Press, Princeton, NJ.

Godfray, H.C.J. and Hassell, M.P. 1989. Discrete and continuous insect populations in tropical environments. *Journal of Animal Ecology* **58**: 153–174.

Godfray, H.C.J. and Knapp, S (eds) 2004. Taxonomy for the twenty-first century. *Philosophical Transactions of the Royal Society* **358**: 557–739.

Goldberg, D.E. and Barton, A. 1992. Patterns and consequences of interspecific competition in natural communities—a review of field experiments with plants. *American Naturalist* **139**: 771–801.

Gordon, D.M., Nisbet, R.M., de Roos, A., Gurney, W.S.C., and Stewart, R.K. 1991. Discrete generations in host-parasitoid models with contrasting life-cycles. *Journal of Animal Ecology* **60**: 295–308.

Grabherr, G., Gottfried, M., and Pauli, H. 1994. Climate effects on mountain plants. *Nature* **369**: 448.

Graham, A.L., Allen, J.E., and Read, A.F. 2005. Evolutionary causes and consequences of immunopathology. *Annual Review of Ecology Evolution and Systematics* **36**: 373–397.

Graham, R.W. 1992. Late Pleistocene faunal changes as a guide to understanding effects of greenhouse warming on the mammalian fauna of North America. In *Global Warming and Biological Diversity* (Peters, R.L. and Lovejoy, T.E., eds), pp. 76–87. Yale University Press, New Haven, CT.

Graham, R.W. and Grimm, E.C. 1990. Effects of global climate change on the patterns of terrestrial biological communities. *Trends in Ecology and Evolution* **5**: 289–292.

Graham, R.W., Lundelius Jr, E.L., Graham, M.A., Schroeder, E.K., Toomey, III, R.S., Anderson, E. *et al.* 1996. Spatial response of mammals to late quaternary environmental fluctuations. *Science* **272**: 1601–1606.

Gravel, D., Canham, C.D., Beaudet, M., and Messier, C. 2006. Reconciling niche and neutrality: the continuum hypothesis. *Ecology Letters* **9**: 399–409.

Green, R.E., Cornell, S.J., Scharlemann, J.P.W., and Balmford, A. 2005. Farming and the fate of wild nature. *Science* **307**: 550–554.

Gregory, R.D., Noble, D.A., and Custance, J. 2004. The state of play of farmland birds: population trends and conservation status of farmland birds in the United Kingdom. *Ibis* **146** (suppl. 2): 1–13.

Gregory, R.D., van Strien, A., Vorisek, P., Gmelig Meyling, A.W., Noble, D.G., Foppen, R.P.B., and Gibbons, D.W. 2005. Developing indicators for European birds. *Philosophical Transactions of the Royal Society Series B* **360**: 269–288.

Grenfell, B.T. and Bolker, B.M. 1998. Cities and villages: infection hierarchies in a measles metapopulation. *Ecological Letters* **1**: 63–70.

Grenfell, B.T. and Dobson, A.P. (eds) 1995. *Ecology of Infectious Diseases in Natural Populations*. Cambridge University Press, Cambridge.

Grenfell, B.T., Kleczkowski, A. Ellner, S.P., and Bolker, B.M. 1994. Measles as a case-study in nonlinear forecasting and chaos. *Philosophical Transactions of the Royal Society Series A* **348**: 515–530.

Grenfell, B.T., Bjørnstad, O.N., and Kappey, J. 2001. Travelling waves and spatial hierarchies in measles epidemics. *Nature* **414**: 716–723.

Grenfell, B.T., Bjørnstad, O.N., and Finkenstädt, B.F. 2002. Dynamics of measles epidemics. II. Scaling noise, determinism and predictability with the time series SIR model. *Ecological Monographs* **72**: 185–202.

Grenfell, B.T., Pybus, O.G., Gog, J.R., Wood, J.L.N., Daly, J., Mumford, J.A., and Holmes, E.C. 2004. Unifiying the ecological and evolutionary dynamics of pathogens. *Science* **303**: 327–332.

Grimm, V. and Wissel, C. 1997. Babel, or the ecological stability discussions: an inventory and analysis of terminology and a guide for avoiding confusion. *Oecologia* **109**: 323–334.

Gross, K. and Ives, A.R. 1999. Inferring host-parasitoid stability from patterns of parasitism among patches. *American Naturalist* **154**: 489–496.

Gross, S.J. and Price, T.D. 2000. Determinants of the northern and southern range limits of a warbler. *Journal of Biogeography* **27**: 869–878.

Grossman, Z. 1980. Oscillatory phenomena in a model of infectious diseases. *Theoretical Population Biology* **18**: 204–243.

Grover, J.P. 1997. *Resource Competition*. Chapman & Hall, London.

Gueron, S. and Levin, S.A. 1993. Self-organization of front patterns in large wildebeest herds. *Journal of Theoretical Biology* **165**: 541–552.

Gueron, S., Levin, S.A., and Rubenstein, D.I. 1996. The dynamics of herds: from individuals to aggregations. *Journal of Theoretical Biology* **182**: 85–98.

Guthrie, R.D. 2003. Rapid size decline in Alaskan Pleistocene horse before extinction. *Nature* **426**: 169–171.

Guthrie, R.D. 2006. New carbon dates link climatic change with human colonization and Pleistocene extinctions. *Nature* **441**: 207–209.

Hails, R.S. and Crawley, M.J. 1992. Spatial density dependence in populations of a cynipid gall-former *Andricus quercuscalicis*. *Journal of Animal Ecology* **61**: 567–583.

Hairston, N.G., Allan, J.D., Colwell, R.K., Futuyma, D.J., Howell, J., Lubin, M.D. *et al.* 1968. The relationship between species diversity and stability: an experimental approach with protozoa and bacteria. *Ecology* **49**: 1091–1101.

Halloran, M.E., Longini, I.M., Nizam, A., and Yang, Y. 2002. Containing bioterrorist smallpox. *Science* **298**: 1428–1432.

Hamilton, W.D. 1964a. The genetical evolution of social behaviour I. *Journal of Theoretical Biology* **7**: 1–16.

Hamilton, W.D. 1964b. The genetical evolution of social behaviour II. *Journal of Theoretical Biology* **7**: 17–52.

Hamilton, W.D. 1998. *Narrow Roads of Gene Land: the Collected Papers of W.D. Hamilton, Volume 1: Evolution of Social Behaviour.* Oxford University Press, New York.

Hammond, P.M. 1995. The current magnitude of biodiversity. In *Global Biodiversity Assessment* (Heywood, V. H., ed.), pp. 113–128. Cambridge University Press, Cambridge.

Hampe, A. and Petit, R.J. 2005. Conserving biodiversity under climate change: the rear edge matters. *Ecology Letters* **8**: 461–467.

Hannah, L. and Hansen, L. 2005. Designing landscapes and seascapes for change. In *Climate Change and Biodiversity* (Lovejoy, T.E. and Hannah, L., eds), pp. 329–342. Yale University Press, New Haven, CT.

Hannah, L., Midgley, G.F., Lovejoy, T., Bond, W.J., Bush, M., Lovette, J.C. *et al.* 2002a. Conservation of biodiversity in a changing climate. *Conservation Biology* **16**: 264–268.

Hannah, L., Midgley, G.F., and Millar, D. 2002b. Climate change-integrated conservation strategies. Global Ecology and Biogeography **11**: 485–495.

Hannah, L., Midgley, G., Hughes, G., and Bomhard, B. 2005. The view from the Cape: extinction risk, protected areas, and climate change. *BioScience* **55**: 231.

Hansen, S.R. and Hubbell, S.P. 1980. Single-nutrient microbial competition: qualitative agreement between experimental and theoretically forecast outcomes. *Science* **207**: 1491–1493.

Hanski, I. 1982. Dynamics of regional distribution: the core and satellite species hypothesis. *Oikos* **38**: 210–221.

Hanski, I. 1994. Patch-occupancy dynamics in fragmented landscapes. *Trends in Ecology and Evolution* **9**: 131–135.

Hanski, I. 1996. Metapopulation ecology. In *Population Dynamics in Ecological Space and Time* (Rhodes, O.E., Chesser, R.K., and Smith, M.A., eds), pp. 11–43. University of Chicago Press, Chicago.

Hanski, I. 1999. *Metapopulation Ecology.* Oxford University Press, Oxford.

Hanski, I. and Gilpin, M. (eds) 1997. *Metapopulation Biology. Ecology, Evolution and Genetics.* Academic Press, London.

Hanski, I. and Gaggiotti, O.E. 2004. *Ecology, Genetics and Evolution of Metapopulations.* Academic Press, London.

Hanski, I., Poyry, J., Pakkala, T., and Kuussaari, M. 1995. Multiple equilibria in metapopulation dynamics. *Nature* **377**: 618–621.

Hardin, G. 1960. The competitive exclusion principle. *Science* **131**: 1292–1297.

Harlan, H.V. and Martini, M.L. 1938. The effect of natural selection on a mixture of barley varieties. *Journal of Agricultural Research* **57**: 189–199.

Harper, J.L. 1977. *Population Biology of Plants.* Academic Press, London.

Harrison, S. and Taylor, A.D. 1997. Empirical evidence for metapopulation dynamics. In *Metapopulation Biology* (Hanski, I. and Gilpin, M.E., eds), pp. 27–39. Academic Press, San Diego.

Harte, J., Torn, M., and Jensen, D. 1992. The nature and consequences of indirect linkages between climate change and biological diversity. In *Climate Change and Biodiversity* (Lovejoy, T.E. and Hannah, L., eds), pp. 325–343. Yale University Press, New Haven, CT.

Harte, J., Kinzig, A., and Green, J. 1999. Self-similarity and species-area relations. *Science* **284**: 334–336.

Harte, J., Ostling, A., Green, J.L., and Kinzig, A. 2004. Biodiversity conservation: climate change and extinction risk. *Nature* **430**: 3 p following 33.

Harvey, P.H., Colwell, R.K., Silvertown, J.W., and May, R.M. 1983. Null models in ecology. *Annual Review of Ecology and Systematics* **14**: 189–211.

Haskell, J.P., Ritchie, J.D., and Olff, H. 2002. Fractal geometry predicts varying body size scaling relationships for mammal and bird home ranges. *Nature* **418**: 527–530.

Hassanali, H., Herren, H., Khan, Z.R., Pickett, J.A., and Woodcock, C.M. 2006. Integrated pest management: the push-pull approach for controlling insect pests and weeds of cereals, and its potential for other agricultural systems including animal husbandry. *Philosophical Transactions of the Royal Society of London B*, in press.

Hassell, M.P. 1978. *The Dynamics of Arthropod Predator-Prey Systems.* Princeton University Press, Princeton, NJ.

Hassell, M.P. 1982. Patterns of parasitism by insect parasitoids in patchy environments. *Ecological Entomology* **7**: 365–377.

Hassell, M.P. 2000. *The Spatial and Temporal Dynamics of Host-Parasitoid Interactions.* Oxford University Press, Oxford.

Hassell, M.P. and May, R.M. 1973. Stability in insect host-parasite models. *Journal of Animal Ecology* **42**: 693–726.

Hassell, M.P. and May, R.M. 1974. Aggregation of predators and insect parasites and its effect on stability, *Journal of Animal Ecology* **43**: 567–594.

Hassell, M.P. and Comins, H.N. 1976. Discrete time models for two-species competition. *Theoretical Population Biology* **12**: 202–221.

Hassell, M.P. and May, R.M. 1988. Spatial heterogeneity and the dynamics of parasitoid-host systems. *Annales Zoologici Fennici* **25**: 55–61.

Hassell, M.P. and Anderson, R.M. 1989. Predator-prey and host-pathogen interactions. In *Ecological Concepts, the Contribution of Ecology to an Understanding of the Natural World* (Cherrett, J.M., ed.), pp. 147–196, The 29th Symposium of The British Ecological Society. Blackwell Scientific Publications, Oxford.

Hassell, M.P. and Pacala, S. 1990. Heterogeneity and the dynamics of host-parasitoid interactions. *Philosophical Transactions of the Royal Society Series B Biological Sciences* **330**: 203–220.

Hassell, M.P., Lawton, J.H., and May, R.M. 1976. Patterns of dynamical behavior in single-species populations. *Journal of Animal Ecology* **45**: 471–486.

Hassell, M.P., Pacala, S., May, R.M., and Chesson, P.L. 1991a. The persistence of host-parasitoid associations in patchy environments. I. A general criterion. *American Naturalist* **138**: 568–583.

Hassell, M.P., Comins, H.N., and May, R.M. 1991b. Spatial structure and chaos in insect population dynamics. *Nature* **353**: 255–258.

Hassell, M.P., Comins, H.N., and May, R.M. 1994. Species coexistence and self-organizing spatial dynamics. *Nature* **370**: 290–292.

Hastie, T.J. and Tibshirani, R.J. 1990. *Generalized Additive Models*. Chapman and Hall, London.

Hastings, A. 1980. Disturbance, coexistence, history and the competition for space. *Theoretical Population Biology* **18**: 363–373.

Hastings, A. 1983. Age-dependent predation is not a simple process. I. Continuous time models. *Theoretical Population Biology* **23**: 347–362.

Hastings, A. 1984. Age-dependent predation is not a simple process. II. Wolves, ungulates and a discrete-time model for predation on juveniles with a stabilizing tail. *Theoretical Population Biology* **26**: 271–282.

Hastings, A. 2003. Metapopulation persistence with age-dependent disturbance or succession. *Science* **301**: 1525–1526.

Hastings, A., Harrison, S., and McCann, K. 1998. Unexpected spatial patterns in an insect outbreak match a predator diffusion model. *Proceedings of the Royal Society of London Series B Biological Sciences* **264**: 1837–1840.

Hastings, H.M. and Sugihara, G. 1993. *Fractals: A Users' Guide for the Natural Sciences*. Oxford University Press, Oxford.

Hauert, C. and Doebeli, M. 2004. Spatial structure often inhibits the evolution of cooperation in the snowdrift game. *Nature* **428**: 643–646.

Hauert, C., De Monte, S., Hofbauer, J., and Sigmund, K. 2002. Volunteering as red queen mechanism for cooperation in public goods games. *Science* **296**: 1129–1132.

Hawkins, B.A., Field, R., Cornell, H.V., Currie, D.J., Guegan, J.F., Kaufman, D.M. *et al*. 2003. Energy, water, and broad-scale geographic patterns of species richness. *Ecology* **84**: 3105–3117.

Heads, P.A. and Lawton, J.H. 1983. Studies on the natural enemy complex of the holly leaf miner—the effects of scale on the detection of aggregative responses and the implications for biological control. *Oikos* **40**: 267–276.

Henno, J. and Baanante, C. 1999. *Estimating Rates of Nutrient Depletion in Soils of Agricultural Lands of Africa*. International Fertilizer Development Center, Muscle Shoals, AL

Herlihy, C.R. Eckert, C.G. 2005. Evolution of self-fertilization at geographical range margins? A comparison of demographic, floral, and mating system variables in central vs. peripheral populations of *Aquilegia Canadensis* (Ranunculaceae). *American Journal of Botany* **92**: 744–751.

Herren, H.R. 1996. Cassava and cowpea in Africa. In *Biotechnology and Integrated Pest Management* (Persley, G.J., ed.), pp. 136–149. CAB International, Wallingford.

Herz, A.V.M. 1994. Collective phenomena in spatially extended evolutionary games. *Journal of Theoretical Biology* **169**: 65–87.

Hethcote, H.W., Yorke, J.A., and Nold, A. 1982. Gonorrhea modeling—a comparison of control methods. *Mathematical Biosciences* **58**: 93–109.

Hewitt, N. 1998. Seed size and shade-tolerance: a comparative analysis of North American temperate trees. *Oecologia* **114**: 432–440.

Higgs, A.J. and Usher, M.B. 1980. Should nature reserves be large or small? *Nature,Lond.* **285**: 568–569.

Hilborn, R. and Mangel, M. 1998. *The Ecological Detective*. Princeton University Press, Princeton, NJ.

Hilborn, R. and Walters C.J. 1992. *Quantitative Fisheries Stock Assessment: Choice, Dynamics and Uncertainty*. Chapman and Hall, London.

Hilborn, R., Maunder, M., Parma, A., Ernst, B., Payne, J., and Starr, P. 2003. *COLERAINE. A Generalized Age-structured Stock Assessment Model. User's Manual Version 2.0*. School of Aquatic and Fishery Sciences, University of Washington.

Hilborn, R., Stokes. K., Maguire, J.-J., Smith, T., Botsford, L.W., Mangel, M. *et al*. 2004. When can marine reserves improve fisheries management? *Ocean & Coastal Management* **47**: 197–205.

Hill, J.K., Thomas, C.D., and Huntley, B. 1999. Climate and habitat availability determine 20th century changes in a butterfly's range margin. *Proceedings of the Royal Society of London B Biological Sciences* **266**: 1197–1206.

Hill, J., Collingham, Y., Thomas, C., Blakeley, D., Fox, R., Moss, D., and Huntley, B. 2001. Impacts of landscape structure on butterfly range expansion. *Ecology Letters* **4**: 313–321.

Hill, J.K., Thomas, C.D., Fox, R., Telfer, M., Willis, S., Asher, J., and Huntley, B. 2002. Responses of butterflies to twentieth century climate warming: Implications for future ranges. *Proceedings of the Royal Society of London B Biological Sciences* **269**: 2163–2171.

Hobbs R.J. and Hopkins, A.J.M. 1991. The role of conservation corridors in a changing climate. In *Nature Conservation: the Role of Corridors* (Saunders, D.A. and Hobbs, R.J., eds), pp. 281–290. Surrey Beatty and Sons, Chipping Norton, NSW.

Hochberg, M.E., Hassell, M.P., and May, R.M. 1990. The dynamics of host-parasitoid-pathogen interactions. *American Naturalist* **135**: 74–94.

Hofbauer, J. and Sigmund, K. 1998. *Evolutionary Games and Population Dynamics*. Cambridge University Press, Cambridge.

Holling, C.S. 1959a. The components of predation as revealed by a study of small mammal predation of the European pine sawfly. *Canadian Entomologist* **91**: 293–320.

Holling, C.S. 1959b. Some characteristics of simple types of predation and parasitism. *Canadian Entomologist* **91**: 385–398.

Holling, C.S. 1973. Resilience and stability of ecological systems. *Annual Review of Ecology and Systematics* **4**: 1–23.

Hollowed, A.B., Bax, N., Beamish, R., Collie, J.S., Fogarty, M, Livingston, P.A. *et al.* 2000. Are multispecies models an improvement on single-species models for measuring fishing impacts on marine ecosystems? *ICES Journal of Marine Science* **57**: 707–719.

Holyoak, M. and Lawler, S.P. 1996. Persistence of an extinction-prone predator-prey interaction through metapopulation dynamics. *Ecology* **77**: 1867–1879.

Hoogerbrugge, I.D. and Fresco, L. 1993. *Homegarden Systems: Agricultural Characteristics and Challenges*, Gatekeeper Series no. 39. Sustainable Agriculture Programme, International Institute for Environment and Development, London.

Hooper, D.U. and Vitousek, P.M. 1997. The effects of plant composition and diversity on ecosystem processes. *Science* **277**: 1302–1305.

Hooper, D.U., Chapin, III, F.S., Ewel, J.J., Hector, A., Inchausti, P., Lavorel, S. *et al.* 2005. Effects of biodiversity on ecosystem functioning: a consensus of current knowledge. *Ecological Monographs* **75**: 3–35.

Hooper, W.D. and Ash, H.B. 1935. *Marcus Porcius Cato on Agriculture. Marcus Terentius Varro on Agriculture.* Loeb Classical Library, Harvard University Press, Cambridge and William Heinemann, London.

Hopkins, A.D. 1920. The bioclimatic law. *Monthly Weather Review* **48**: 355.

Hoppensteadt, F.C. and Keller, J.B. 1976. Synchronization of periodical cicada emergences. *Science* **194**: 335–337.

Horn, H.S. and MacArthur, R.H. 1972. Competition among fugitive species in a harlequin environment. *Ecology* **53**: 749–752.

Horn, H.S. and May, R.M. 1977. Limits to similarity among coexisting competitors. *Nature* **270**: 660–661.

Hotelling, H. 1929. Stability in Competition. *Economics Journal* **39**: 41–57.

Howard, L.O. and Fiske, W.F. 1911. The importation into the United States of the parasites of gypsy moth and the brown-tailed moth. *Bulletin of the Bureau of Entomology of the US Agricricultural Department* **91**: 1–312.

Hsu, S.B., Hubbell, S.P., and Waltman, P. 1977. A mathematical theory for single-nutrient competition in continuous cultures of microorganisms. *SIAM Journal of Applied Mathematics* **32**: 366–383.

Huang, J., Rozelle, S., Pray, C., and Q. Wang, Q. 2002. Plant biotechnology in China. *Science* **295**: 674–677.

Hubbell, S.P. 2001. *The Unified Neutral Theory of Biodiversity and Biogeography*. Monographs in Population Biology, no. 32. Princeton Univeristy Press, Princeton, NJ.

Hubbell, S.P. 2006. Neutral theory and the evolution of ecological equivalence. *Ecology* **87**: 1387–1398.

Hubbell, S.P. and Foster, R.B. 1986. Biology, chance, and history and the structure of tropical rain forest tree communities. In *Community Ecology* (Diamond, J. and Case, T.J., eds), pp. 314–329. Harper and Row, New York.

Hudson, P.J., Dobson, A.P., and Newborn, D. 1998. Prevention of population cycles by parasite removal. *Science* **282**: 2256–2258.

Hudson, P.J., Rizzoli, A., Grenfell, B.T., Heesterbeek, J.A.P., and Dobson, A.P. (eds) 2001. *Ecology of Wildlife Diseases*. Oxford University Press, Oxford.

Huffaker, C.B. 1958. Experimental studies on predation: dispersion factors and predator-prey oscillations. *Hilgardia* **27**: 343–383.

Huffaker, C.B. and Kennett, C.E. 1956. Experimental studies on predation: predation and cyclamen mite populations on strawberries in California. *Hilgardia* **24**: 191–222.

Huffaker, CB., Shea, K.P., and Herman, S.G. 1963. Experimental studies on predation: complex dispersion and levels of food in an acarine predator-prey interaction. *Hilgardia* **34**: 305–329.

Hughes, C.L., Hill, J.K., and Dytham, C. 2003. Evolutionary trade-offs between reproduction and dispersal in populations at expanding range boundaries. *Proceedings of the Royal Society of London B Biological Sciences* **270**: 147–150.

Huisman, J. and Weissing, F.J. 1994. Light-limited growth and competition for light in well-mixed aquatic environments—an elementary model. *Ecology* **75**: 507–520.

Huisman, J. and Weissing, F.J. 1999. Biodiversity of plankton by species oscillations and chaos. *Nature* **402**: 407–410.

Huisman, J., Thi, N.N.P., Karl, D.M., and Sommeijer, B. 2006. Reduced mixing generates oscillations and chaos in the oceanic deep chlorophyll maximum. *Nature* **439**: 322–325.

Humphries, M.M., Boutin, S., Thomas, D.W., Ryan, J.D., Selman, C., McAdam, A.G. *et al.* 2005. Expenditure freeze: the metabolic response of small mammals to cold environments. *Ecology Letters* **8**: 1326–1333.

Hunter M.L., Jacobson, G.L., and Webb, T. 1988. Paleoecology and the coarse-filter approach to maintaining biological diversity. *Conservation Biology* **2**: 375–385.

Hutchings, J.A. 1999. Influence of growth and survival costs of reproduction on Atlantic cod, *Gadus morhua*, population growth rate. *Canadian Journal of Fisheries and Aquatic Science* **56**: 1612–1623.

Hutchinson, G.E. 1953. The concept of pattern in ecology. *Proceedings of the National Academy of Sciences USA* **105**: 1–12.

Hutchinson, G.E. 1957. Concluding remarks. *Cold Spring Harbor Symposium in Quantitative Biology* **22**: 415–457.

Hutchinson, G.E. 1959. Hommage to Santa Rosalia, or why are there so many kinds of animals? *American Naturalist* **93**: 145–159.

Hutchinson, G.E. 1961. The paradox of the plankton. *American Naturalist* **95**: 137–145.

Hutchinson, G.E. 1965. *The Ecological Theatre and the Evolutionary Play*. Yale University Press, New Haven, CT.

Hutchinson, G.E. and MacArthur, R.H. 1959. A theoretical ecological model of size distributions among species of animals. *American Naturalist* **93**: 117–125.

Illoldi-Rangel, P., Sanchez-Cordero, V., and Peterson, A.T. 2004. Predicting distributions of Mexican mammals using ecological niche modeling. *Journal of Mammalogy* **85**: 658–662.

International Livestock Research Institute. 1999a. *Livestock and Nutrient Cycling*. International Livestock Research Institute, Nairobi.

International Livestock Research Institute. 1999b. *Improving Smallholder Farming Through Animal Agriculture*. International Livestock Research Institute, Nairobi.

IPCC. 2001. Climate change 2001: the scientific basis. In *Contribution of Working Group 1 in the Third Assessment Report of the Intergovernmental Panel on Climate Change*. Cambridge University Press, Cambridge.

Ismael, Y.R., Bennet, R., and Morse, S. 2001. Farm level impact of Bt cotton in South Africa. *Biotechnology and Development Monitor* **48**: 15–19.

IUCN. 2004. *Red Data Book 2004*. www.redlist.org/info/tables.html.

Iverson, L.R., Schwartz, M.W., and Prasad, A.M. 2004. How fast and far might tree species migrate in the eastern United States due to climate change? *Global Ecology and Biogeography* **13**: 209–219.

Ives, A.R. 1992. Density-dependent and density-independent parasitoid aggregation in model host-parasitoid systems. *American Naturalist* **140**: 912–937.

Ives, A.R. 1995. Measuring resilience in stochastic systems. *Ecological Monographs* **65**: 217–233.

Ives, A.R. 2005. Community diversity and stability: changing perspectives and changing definitions. In *Ecological Paradigms Lost: Routes to Theory Change* (Cuddington, K. and Beisner, B., eds), pp. 159–182. Academic Press, Amsterdam.

Ives, A.R. and Cardinale, B.J. 2004. Food-web interactions govern the resistance of communities after non-random extinctions. *Nature* **429**: 174–177.

Ives, A.R., Klug, J.L., and Gross, K. 2000. Stability and species richness in complex communities. *Ecology Letters* **3**: 399–411.

Ives, A.R., Dennis, B., Cottingham, K.L., and Carpenter, S.R. 2003. Estimating community stability and ecological interactions from time-series data. *Ecological Monographs* **73**: 301–330.

Jackson, D.B., Fuller, R.J., and Campbell, S.T. 2004. Long-term population changes among breeding shorebirds in the Outer Hebrides, Scotland, in relation to introduced hedgehogs (*Erinaceus europaeus*). *Biological Conservation* **117**: 151–166.

James, C. 2005. *Global Status of Biotech/GM Crops: 2005, ISAAA Briefs no 34*. International Service for the Acquisition of Agri-Biotech Applications, Ithaca, NY

Janzen, D.H. 1970. Herbivores and the number of tree species in tropical forests. *American Naturalist* **104**: 501–508.

Jeffers, P. 2001. *Maize Pathology Research: Increasing Maize Productivity and Sustainability in Biologically Stressed Environments*. International Maize and Wheat Improvement Center (CIMMYT), Mexico

Jensen, A.L. 1996. Beverton and Holt life history invariants result from optimal trade-off of reproduction and survival *Canadian Journal of Fisheries and Aquatic Science* **53**: 820–822.

Jepson, P. 2001. Global biodiversity plan needs to convince local policy-makers. *Nature* **409**: 12–12.

Johnson, D.M. 2004. Source-sink dynamics in a temporally, heterogeneous environment. *Ecology* **85**: 2037–2045.

Jones, M.P. 1999. Basic breeding strategies for high yielding rice varieties at WARDA. *Japanese Journal of Crop Science* **67**: 133–136.

Jones, T.H. and Hassell, M.P. 1988. Patterns of parasitism by *Trybliographa rapae*, a cynipid parasitoid of the cabbage root fly, under laboratory and field conditions. *Ecological Entomology* **13**: 309–317.

Kaspari, M., Ward, P.S., and Yuan, M. 2004. Energy gradients and the geographic distribution of local ant diversity. *Oecologia* **140**: 407–413.

Kawanabe, H., J.E. Cohen and K. Iwasaki, eds. 1993. *Mutualism and Community Organization: behavioural, theoretical, and food-web approaches*. Oxford University Press, Oxford.

Keeling, M.J. 2005. Models of Foot-and-Mouth Disease. *Proceedings of the Royal Society of London Series B Biological Sciences* **272**: 1195–1202.

Keeling, M.J. and Grenfell, B.T. 1997. Disease extinction and community size: modeling the persistence of measles. *Science* **275**: 65–67.

Keeling, M.J. and Eames, K.T.D. 2005. Networks and epidemic models. *Journal of the Royal Society, Interface* **2** 295–307.

Keeling, M.J., Rohani, P., and Grenfell, B.T. 2001a. Seasonally forced disease dynamics explored as switching between attractors. *Physica D* **148**: 317–335.

Keeling, M.J., Woolhouse, M.E.J., Shaw, D.J., Matthews, L., Chase-Topping, M., Haydon, T.D. *et al.* 2001b. Dynamics of the UK Foot and Mouth epidemic: stochastic dispersal in a heterogeneous landscape. *Science* **294**: 813–817.

Keeling, M.J., Woolhouse, M.E.J., May, R.M., Davies, G., and Grenfell, B.T. 2003. Modelling vaccination strategies against foot-and-mouth disease. *Nature* **421**: 136–142.

Keeling, M.J., Brooks, S.P., and Gilligan, C.A. 2004. Using conservation of pattern to estimate spatial parameters from a single snapshot. *Proceedings of the National Academy of Sciences of the United States of America* **101**: 9155–9160.

Keeling, M.J., Grenfell, B.T., Woolhouse, M.E., Brooks, S.P., Deardon, R., Shaw, D.J. *et al.* 2006. Optimal reactive vaccination strategies for a foot-and-mouth outbreak in the UK. *Nature* **440**: 83–86.

Keith, L.B. 1983. Role of food in hare population cycles. *Oikos* **40**: 385–395.

Kell, L.T., Pilling, G.M., Kirkwood, G.P., Pastoors, M., Mesnil, B., Korsbrekke, K. *et al.* 2005. An evaluation of the implicit management procedure used for some ICES roundfish stocks. *ICES Journal of Marine Science* **62**: 750–759.

Keller, E.F. 2005. Revisiting 'scale-free' networks. *BioEssays* **27**: 1060–1068.

Keller, L. (ed.) 1999. *Levels of Selection in Evolution*. Princeton University Press, Princeton, NJ.

Kempton, R.A. 1979. The structure of species abundance and measurement of diversity. *Biometrics* **35**: 307–321.

Kendall, B.E., Briggs, C.J., Murdoch, W.W., Turchin, P., Ellner, S.P., McCauley, E. *et al.* 1999. Why do populations cycle? A synthesis of statistical and mechanistic modeling approaches. *Ecology* **80**: 1789–1805.

Kendall, C.K. and Roberts, E.K. 2001. Whitebark Pine Decline: Infection, Mortality, and Population Trends. In *Whitebark Pine COmmunities. Ecology and Restoration*. (Tomback, D.F., Arno, S.E., and Keane, R.E., eds), pp. 221–242. Island Press, Washington DC.

Kenmore, P. 1991. *How Rice Farmers Clean Up the Environment Conserve Biodiversity, Raise More Food, Make Higher Profits: Indonesia's IPM—a Model for Asia*. Food and Agriculture Organisation, Manila, Philippines.

Kerr, B. and Godfrey-Smith, P. 2002. Individualist and multi-level perspectives on selection in structured populations. *Biology & Philosophy* **17**: 477–517.

Kerr, J.T. and Currie, D.J. 1999. The relative importance of evolutionary and environmental controls on broad-scale patterns of species richness in North America. *Ecoscience* **6**: 329–337.

Kerr, J.T. and Packer, L. 1999. The environmental basis of North American species richness patterns among *Epicauta* (Coleoptera: Meloidae). *Biodiversity and Conservation* **8**: 617–628.

Kerr, J.T. and Cihlar, J. 2003. Land use and cover with intensity of agriculture for Canada from satellite and census data. *Global Ecology and Biogeography* **12**: 161–172.

Kerr, J.T. and Ostrovsky, M. 2003. From space to species: ecological applications for remote sensing. *Trends in Ecology and Evolution* **18**: 299–305.

Kerr, J.T. and Cihlar, J. 2004. Patterns and causes of species endangerment in Canada. *Ecological Applications* **14**: 743–753.

Kerr, J.T. and Deguise, I. 2004. Habitat loss and the limits to endangered species recovery. *Ecology Letters* **7**: 1163–1169.

Kerr, J.T., Vincent, R., and Currie, D.J. 1998. Lepidopteran richness patterns in North America. *Ecoscience* **5**: 448–453.

Kerr, J.T., Southwood, T.R.E., and Cihlar, J. 2001. Remotely sensed habitat diversity predicts butterfly species richness and communtiy similarity in Canada. *Proceedings of the National Academy of Sciences USA* **98**: 11365–11370.

Kerr, S.R. and. Dickie, L.M. 2001. *The Biomass Spectrum: A Predator-Prey Theory of Aquatic Production*. Columbia University Press, Columbia.

Keymer, J.E., Marquet, P.A., Velasco-Hernandez, J.X., and Levin, S.A. 2000. Extinction thresholds and meta-population persistence in dynamic landscapes. *American Naturalist* **156**: 478–494.

Killingback, T. and Doebeli, M. 1996. Spatial evolutionary game theory: Hawks and Doves revisited. *Proceedings of the Royal Society of London Series B Biological Sciences* **263**: 1135–1144.

Kingsland, S.E. 1991. Foundational papers. In *Foundations of Ecology: Classic Papers with Commentaries* (Real, L.A. and Brown, J.H., eds), pp. 1–23. University of Chicago Press, Chicago.

Kinzig, A.P., Pacala, S.W. and Tilman, D. 2001. *The Functional Consequences of Biodiversity. Empirical Progress and Theoretical Extensions*, Monographs in Population Biology. Princeton University Press, Princeton, NJ.

Kirkpatrick, J.B. 1983. An iterative method for establishing priorities for the selection of nature reserves: an example from Tasmania. *Biological Conservation* **25**: 127–134.

Kirkpatrick, M. and Barton, N.H. 1997. Evolution of a species' range. *American Naturalist* **150**: 1–23.

Kirkwood, G.P. 1997. The revised management procedure of the International Whaling Commission. In *Global Trends: Fisheries Management. American Fisheries Society Symposium 20* (Pikitch, E.K., Huppert, D.D., and Sissenwine, M.P., eds), pp. 91–99. American Fisheries Society, Bethesda, MD.

Kleidon, A. and Mooney, H.A. 2000. A global distribution of biodiversity inferred from cliamtic constraints: results from a process-based modelling study. *Global Change Biology* **6**: 507–523.

Klinkhamer, P.G.L., de Jong, T.J., Metz, J.A.J., and Val, J. 1987. Life-history tactics of annual organisms—the joint effects of dispersal and delayed germination. *Theoretical Population Biology* **32**: 127–156.

Klironomos, J.N. 2002. Feedback with soil biota contributes to plant rarity and invasiveness in communities. *Nature* **417**: 67–70.

Knight, R.R. and Eberhardt, L.L. 1985. Population dynamics of Yellowstone grizzly bears. *Ecology* **66**: 323–334.

Koelle, K., Rodo, X., Pascual, M., Yunus, M., and Mostafa, G. 2005. Refractory periods and climate forcing in cholera dynamics. *Nature* **436**: 696–700.

Konvicka, M., Maradova, M., Benes, J., Fric, Z., and Kepka, P. 2003. Uphill shifts in distribution of butterflies in the Czech Republic: effects of changing clmate detected on a regional scale. *Global Ecology and Biogeography* **12**: 403–410.

Kot, M. 1992. Discrete-time travelling waves: ecological examples. *Journal of Mathematical Biology* **30**: 413–436.

Kotiaho, J.S., Kaitala, V., Komonen, A., and Paivinen, J. 2005. Predicting the risk of extinction from shared ecological characteristics. *Proceedings of the National Academy of Sciences USA* **102**: 1963–1967.

Krebs, C.J., Boutin, S., Boonstra, R., Sinclair, A.R.E., Smith, J.N.M., Dale, M.R.T., and Turkington, M.R. 1995. Impact of food and predation on snowshoe hare cycle. *Science* **269**: 1112–1115.

Krebs, J.R. and Davies, N.B. 1993. *An Introduction to Behavioural Ecology*, 3rd edn. Blackwell Scientific Publications, Oxford.

Krebs, J.R. and Davies, N.B. (eds) 1997. *Behavioural Ecology: an Evolutionary Approach*, 4th edn. Blackwell Science, Malden, MA.

Kremen, C., Niles, J.O., Dalton, M.G., Daily, G.C., Ehrlich, P.R., Fay, J.P. *et al.* 2000. Economic incentives for rain forest conservation across scales. *Science* **288**: 1828–1832.

Kremen, C., Williams, N.M., and Thorp, R.W. 2002. Crop pollination from native bees at risk from agricultural intensification. *Proceedings of the National Academy of Sciences USA* **99**: 16812–16816.

Kremer, M. and Glennerster, R. 2004. *Strong Medicine: Creating Incentives for Pharmaceutical Research on Neglected Diseases*. Princeton University Press, Princeton, NJ.

Kuiken, T., Leighton, F.A., Fouchier, R.A.M., LeDuc, J.W., Peiris, J.S.M., Schudel, A. *et al.* 2005. Public health—pathogen surveillance in animals. *Science* **309**: 1680–1681.

Kukal, O., Ayres, M.P., and Scriber, J.M. 1991. Cold tolerance of pupae in relation to the distribution of swallowtail butterflies. *Canadian Journal of Zoology* **69**: 3028–3037.

Ladle, R.J., Jepson, P., Araujo, M.B., and Whittaker, R.J. 2004. Dangers of crying wolf over risk of extinctions. *Nature* **428**: 799.

Laevastu, T. 1992. Interactions of size-selective fishing with variations in growth rates and effects on fish stocks. *ICES Council Meeting Papers* **G**: **5**: 13.

Laine, A.L. 2004. Resistance variation within and among host populations in a plant-pathogen metapopulation: implications for regional pathogen dynamics. *Journal of Ecology* **92**: 990–1000.

Lande, R. 1987. Extinction thresholds in demographic models of territorial populations. *American Naturalist* **130**: 624–635.

Lande, R. 1988a. Demographic models of the northern spotted owl. *Oecologia* **75**: 601–607.

Lande, R. 1988b. Genetics and demography in biological conservation. *Science* **241**: 1455–1460.

Lande, R., Engen, S., Sæther, B.-E., Filli, F., Matthysen, E., and Weimerskirch, H. 2002. Estimating density dependence from population time series using demographic theory and life-history data. *American Naturalist* **159**: 321–337.

Lande, R., Engen, S., and Sæther, B.-E. 2003. *Stochastic Population Dynamics in Ecology And Conservation*. Oxford University Press, Oxford.

Langat, M., Mukwana, E., and Woomer, P.L. 2000. *MBILI Update: Testing an Innovative Cropping Arrangement*. Sustainable Agriculture Centre for Research and Development in Africa, Bungoma, Kenya.

Law, R. and Rowell, C.A. 1993. Cohort-structured populations, selection responses, and exploitation of the North Sea cod. In *The Exploitation of Evolving Resources*. Lecture Notes in Biomathematics, vol. 99 (Stokes, T.K., McGlade, J.M., and Law, R., eds), pp. 155–173. Springer-Verlag, Berlin.

Lawler, J.J., White, D., and Master, L.L. 2003. Integrating representing and vulnerability: two approaches for prioritizing areas for conservation. *Ecological Applications* **13**: 1762.

Lawton, J.H. 1993. Range, population abundance and conservation. *Trends in Ecology and Evolution* **8**: 409–413.

Lawton, J.H. and May, R.M. 1983. The birds of Selborne. *Nature* **306**: 732–733.

Lawton, J.H. and May, R.M. 1995. *Extinction Rates*. Oxford University Press, Oxford.

Le Galliard, J.-F., Ferrière, R., and Dieckmann, U. 2003. The adaptive dynamics of altruism in spatially heterogeneous populations. *Evolution* **57**: 1–17.

Lederhouse, R.C., Aynes, M.P., and Scriber, J.M. 1995. Physiological and behavioural adaptations to variable thermal environments in North American Swallowtail butterflies. In *Swallowtail Butterflies: Ecology and Evolution* (Scriber, J.M., Tsubaki, Y., and Lederhouse, R.C., eds), pp. 71–82. Scientific Publishers, Gainseville, GA.

Lehman, C.L. and D. Tilman. 2000. Biodiversity, stability, and productivity in competitive communities. *American Naturalist* **156**: 534–552.

Lehmann-Ziebarth, N. and Ives, A.R. 2006. The structure and stability of model ecosystems assembled in a variable environment. *Oikos* **114**: 451–464.

Lehmann-Ziebarth, N., Heideman, P.P., Shapiro, R.A., Stoddart, S.L., Hsiao, C.C.L., Stephenson, G.R. *et al.* 2005. Evolution of periodicity in periodical cicadas *Ecology* **86**: 3200–3211.

Lei, G.C. and Hanski, I. 1998. Spatial dynamics of two competing specialist parasitoids in a host metapopulation. *Journal of Animal Ecology* **67**: 422–433.

Leigh, E.G. 1983. When does the good of the group override the advantage of the individual? *Proceedings of the National Academy of Sciences USA* **80**: 2985–2989.

Leimar, O. and Hammerstein, P. 2001. Evolution of cooperation through indirect reciprocation. *Proceedings of the Royal Society of London Series B Biological Sciences* **268**: 745–753.

Leopold, A. 1948. The land ethic. In A Sand County Almanac, and Sketches Here and There. Oxford University Press, New York.

Leslie, P.H. 1945. On the use of matrices in certain population mathematics. *Biometrika* **33**: 183–212.

Lessells, C.M. 1985. Parasitoid foraging: should parasitism be density dependent? *Journal of Animal Ecology* **54**: 27–41.

Levin, B.R. and Kilmer, W.L. 1974. Intermedic selection and the evolution of altruism: a computer simulation study. Evolution *28*: 527–545.

Levin, B.R., F.M. Stewart, and L. Chao. 1977. Resource-limited growth, competition, and predation: a model and experimental studies with bacteria and bacteriophage. *American Naturalist* **111**: 3–24.

Levin, S.A. 1970. Community equilibria and stability, and an extension of the competitive exclusion principle. American Naturalist. **104**: 413–423.

Levin, S.A., and R.T. Paine. 1974. Disturbance, patch formation, and community structure. *Proceedings of the National Academy of Sciences U.S.A.* **71**: 2744–2747.

Levin, S.A., Cohen, D., and Hastings, A. 1984. Dispersal strategies in patchy environments. *Theoretical Population Biology* **26**: 165–191.

Levins, R. 1969. Some demographic and genetic consequences of environmental heterogeneity for biological control. *Bulletin of Entomological Research* **15**: 237–240.

Levins, R. 1970. Extinction. In *Some Mathematical Problems in Biology* (Gesternhaber, M., ed.), pp. 77–107. American Mathematical Society, Providence, RI.

Levins, R. 1979. Coexistence in a variable environment. *American Naturalist* **114**: 765–783.

Levins, R. and Culver, D. 1971. Regional coexistence of species and competition between rare species. *Proceedings of the National Academy of Sciences USA* **68**: 1246–1248

Lewis, M.A. 1997. Variability, patchiness, and jump dispersal in the spread of an invading population. In D.

Tilman and P. Kareiva, eds. *Spatial ecology: the role of space in population dynamics and interspecific interactions*, pp. 46–74. Princeton University Press, New Jersey.

Lewontin, R.C. and Cohen, D. 1969. On population growth in a randomly varying environment. *Proceedings of the National Academy of Sciences USA* **62**: 1056–1060.

Li, T.Y. and Yorke, J.A. 1975. Period 3 implies chaos. *American Mathematical Monthly* **82**: 985–992.

Lieberman, E., Hauert, C., and Nowak, M.A. 2005. Evolutionary dynamics on graphs. *Nature* **433**: 312–316.

Lloyd, A.L. and May, R.M. 1996. Spatial heterogeneity in epidemic models. *Journal of Theoretical Biology* **179**: 1–11.

Lloyd-Smith, J.O., Schreiber, S.J., Kopp, P.E., and Getz, W.M. 2005. Superspreading and the effect of individual variation on disease emergence. *Nature* **438**: 355–359.

Longley, M. and Sotherton, N.W. 1997. Factors determining the effects of pesticides upon butterflies inhabiting arable farmland. *Agriculture, Ecosystems and Environment* **61**: 1–12.

Loreau, M., Naeem, S., Inchausti, P., Bengtsson, J., Grime, J.P., Hector, A. *et al.* 2001. Biodiversity and ecosystem functioning: current knowledge and future challenges. *Science* **294**: 804–808.

Lorenzen, K. 1995. Population dynamics and management of culture-based fisheries. *Fisheries Management and Ecology* **2**: 61–73.

Lorenzen, K. 2005. Population dynamics and potential of fisheries stock enhancement: practical theory for assessment and policy analysis. *Philosophical Transactions of the Royal Society Series B* **360**: 171–189.

Lorenzen, K. and Enberg, K. 2002. Density-dependent growth as a key mechanism in the regulation of fish populations: evidence from among-population comparisons. *Proceedings of the Royal Society of London Series B Biological Sciences* **269**: 49–54.

Lotka, A. 1925. *Elements of Physical Biology*. Williams and Wilkins Co., Baltimore, MD.

Lubchenco, J. 1978. Plant species diversity in a marine intertidal community: importance of herbivore food preference and algal competitive abilities. *American Naturalist* **112**: 23–39.

Luce, R.D. and Riaffa, H. 1957. *Games and Decisions*. John Wiley, New York.

Ludwig, D. and Walters C.J. 1985. Are age-structured models appropriate for catch-effort data? *Canadian Journal of Fisheries and Aquatic Science* **42**: 1066–1072.

Lyons, S.K. 2003. A quantitative assessment of the range shifts of Pleistocene mammals. *Journal of Mammalogy* **84**: 385–402.

MacArthur, R.H. 1955. Fluctuations of animal populations and a measure of community stability. *Ecology* **36**: 533–536.

MacArthur, R.H. 1969. Species packing, and what interspecies competition minimizes. *Proceedings of the National Academy of Sciences USA* **64**: 1369–1371.

MacArthur, R.H. 1972. *Geographical Ecology: Patterns in the Distribution of Species*. Princeton University Press, Princeton, NJ.

MacArthur, R.H. and Wilson, E.O. 1963. An equilibrium theory of insular zoogeography. *Evolution* **17**: 373–387.

MacArthur, R. and Levins, R. 1964. Competition, habitat selection, and character displacement in a patchy environment. *Proceedings of the National Academy of Sciences USA* **51**: 1207–1210.

Macarthur, R.H. and Wilson, E.O. 1967. *The Theory of Island Biogeography*. Princeton University Press, Princeton, NJ.

MacDougall, A.S. 2005. Responses of diversity and invasibility to burning in a northern oak savanna. *Ecology* **86**: 3354–3363.

Mace, P.M. 1994. Relationships between common biological reference points used as threshold and targets of fisheries management strategies. *Canadian Journal of Fisheries and Aquatic Science* **51**: 110–122.

Mace, P.M. and Doonan, I.J. 1988. *A Generalized Bioeconomic Simulation Model*. NZ Fisheries Assessment Research Document no. 88/4.

Mack, R.N. and Harper, J.L. 1977. Interference in dune annuals: spatial pattern and neighbourhood effects. *Journal of Ecology* **65**: 345–363.

Magurran, A.E. 2004. *Measuring Biological Diversity*. Blackwell, Oxford.

Malcolm, J.R., Markham, A., Neilson, R.P., and Garaci, M. 2002. Estimated migration rates under scenarios of global climate change. *Journal of Biogeography* **29**: 835–849.

Malthus, T. 1798. *An Essay on the Principle of Population*. J. Johnson, London.

Mangel, M. and Levin, P.S. 2005. Regime, phase and paradigm shifts: making community ecology the basic science for fisheries. *Philosophical Transactions of the Royal Society Series B* **360**: 95–106.

Mann, C.C. 1999. Crop scientists seek a new revolution, *Science* **283**: 311–314.

Manne, L.L., Brooks, T.M., and Pimm, S.L. 1999. Relative risk of extinction of passerine birds on continents and islands. *Nature* **399**: 258–261.

Margules, C.R., Nicholls, A.O., and Pressey, R.L. 1988. Selecting networks of reserves to maximize biological diversity. *Biological Conservation* **43**: 63–76.

Maron J.L. and Harrison S. 1997. Spatial pattern formation in an insect host-parasitoid system. *Science* **278**: 1619–1621.

Maruyama, T. 1970. Effective number of alleles in a subdivided population. *Theoretical Population Biology* **1**: 273–306.

Matessi, C. and Jayakar, S.D. 1976. Conditions for the evolution of altruism under Darwinian selection. *Theoretical Population Biology* **9**: 360–387.

Matthiessen, P. 1987. *Wildlife in America*. Penguin, New York.

Mattson, D.J., Blanchard, B.M., and Knight, R.R. 1992. Yellowstone grizzly bear mortality, human habituation, and whitebark pine seed crops. *Journal of Wildlife Management* **56**: 432–442.

Mattson, D.J., Herrero, S., Wright, R.G., and Pease, C.M. 1996. Science and management of Rocky mountain grizzly bears. *Conservation Biology* **10**: 1013–1025.

May, R.M. 1972. Will a large complex system be stable? *Nature* **238**: 413–414.

May, R.M. 1973a. *Stability and Complexity in Model Ecosystems*. Princeton University Press, Princeton, NJ.

May, R.M. 1973b. On the relationships among various types of population models. *American Naturalist* **107**: 46–57.

May, R.M. 1974a. Ecosystem properties in randomly fluctuating environments. In *Progress in Theoretical Biology* (Rosen, R. and Snell, F.M., eds), pp. 1–20. Academic Press, New York.

May, R.M. 1974b. *Stability and Complexity in Model Ecosystems*, 2nd edn. Princeton University Press, Princeton, NJ.

May, R.M. 1974c. Biological populations with non-overlapping generations: stable points, stable cycles and chaos. *Science* **186**: 645–647.

May, R.M. 1975. Patterns of species abundance and diversity. In *Ecology and Evolution of Communities* (Diamond, J.M. and Cody, M., eds), pp. 81–120. Harvard University Press, Cambridge, MA.

May, R.M. 1976a. Simple mathematical models with very complicated dynamics. *Nature* **261**: 459–467.

May, R.M. 1976b. Models for single populations. In *Theoretical Ecology: Principles and Applications* (May, R.M., ed.), pp. 4–25. W.B. Saunders Company, Philadelphia.

May, R.M. 1977a. Togetherness among schistosomes: its effects on the dynamics of the infection. *Mathematical Biosciences* **35**: 301–343.

May, R.M. 1977b. Dynamical aspects of host-parasite associations: Crofton's model revisited. *Parasitology* **75**: 259–276.

May, R.M. 1977c. Thresholds and breakpoints in ecosystems with a multiplicity of stable states. *Nature* **269**: 471–477.

May, R.M. 1978a. Host-parasitoid systems in patchy environments: a phenomenological model. *Journal of Animal Ecology* **47**: 833–843.

May, R.M. 1978b. The dynamics and diversity of insect faunas. In *Diversity of Insect Faunas* (Mound, L.A. and Waloff, N., eds), pp. 188–204. Blackwell, Oxford.

May, R.M. 1980. Population biology of microparasitic infections. In *Mathematical Ecology*, vol. 17 (Hallam, T.G. and Levin, S.A., eds), pp. 405–441. Springer-Verlag, Berlin.

May, R.M. 1986. The search for patterns in the balance of nature: advances and retreats. *Ecology* **67**: 1115–1126.

May, R.M. 1987. More evolution of cooperation. *Nature* **327**: 15–17.

May, R.M. 1988. How many species are there on earth? *Science* **241**: 1441–1449.

May, R.M. 1990a. How many species? *Philosophical Transactions of the Royal Society B Biological Sciences* **330**: 293–304.

May, R.M. 1990b. Taxonomy as destiny. *Nature* **347**: 129–130.

May, R.M. 1992. How many species inhabit the earth? *Scientific American* October:42–48.

May, R.M. 1994. The effects of spatial scale on ecological questions and answers. In *Large-scale Ecology and Conservation Biology* (Edwards, P.J., May, R.M., and Webb, N.R., eds), pp. 1–17. Blackwell Scientific Publications, Oxford.

May, R.M. 1999. The dimensions of life on Earth. In *Nature and Human Society*, pp. 30–45. National Academy of Sciences Press, Washington DC.

May, R.M. 2001. *Stability and Complexity in Model Ecosystems: Princeton Landmarks in Biology*. Princeton University Press, Princeton, NJ.

May, R.M. 2004. Uses and abuses of mathematics in biology. *Science* **303**: 790–793.

May, R.M. 2006. Network structure and the biology of populations. *Trends in Ecology and Evolution* **21**: 394–399.

May, R.M. and MacArthur, R.H. 1972. Niche overlap as a function of environmental variability. *Proceedings of the National Academy of Sciences USA* **69**: 1109–1113.

May, R.M. and Leonard, W.J. 1975. Nonlinear aspects of competition between three species. *Journal of Applied Mathematics* **29**: 243–253.

May, R.M. and Anderson, R.M. 1978. Regulation and stability of host-parasite population interactions-II. Destabilizing processes. *Journal of Animal Ecology* **47**: 248–267.

May, R.M. and Anderson, R.M. 1979. Population biology of infectious diseases: part II. *Nature* **280**: 455–461.

May, R.M. and Anderson, R.M. 1984. Spatial heterogeneity and the design of immunization programs. *Mathematical Biosciences* **72**: 83–111.

May, R.M. and Nowak, M.A. 1994. Superinfection, metapopulation dynamics, and the evolution of diversity. *Journal of Theoretical Biology* **170**: 95–114.

May, R.M. and Stumpf, M.P.H. 2000. Species-area relations in tropical forests. *Science* **290**: 2084–2086.

May, R.M., Beddington, J.R., Clark, C.W., Holt, S.J., and Laws, R.M. 1979. Management of multispecies fisheries. *Science* **205**: 267–279.

May, R.M., Lawton, J.H., and Stork, N.E. 1995. Assessing extinction rates. In *Extinction Rates* (Lawton, J.H. and May, R.M., eds), pp. 1–24. Oxford University Press, Oxford.

Maynard Smith, J. 1964. Group selection and kin selection. *Nature* **63**: 20–29.

Maynard Smith, J. 1976. Group selection. *Quarterly Review of Biology* **201**: 145–147.

McAllister, M.K. and Ianelli, J.N. 1997. Bayesian stock assessment using catch-age data and the sampling-importance sampling algorithm. *Canadian Journal of Fisheries and Aquatic Science* **54**: 284–300.

McAllister, M.K. and Kirkwood, G.P. 1998. Bayesian stock assessment: a review and example application using the logistic model. *ICES Journal of Marine Science* **55**: 1031–1060.

McCann, S.K. 2000. The diversity-stability debate. *Nature* **405**: 228–233.

McGill, B.J. 2003. A test of the unified neutral theory of biodiversity. *Nature* **424**: 1006–1007.

McKibben, B. 1999. Nature without people? *New York Review of Books* 12 August: 44–48.

McLachlan, J.S., Clark, J.S., and Manos, P.S. 2005. Molecular indicators of the migration capacity under rapid climate change. *Ecology* **86**: 2088–2098.

McNaughton, S.J. 1977. Diversity and stability of ecological communities: a comment on the role of empiricism in ecology. *American Naturalist* **111**: 515–525.

McNaughton, S.J. 1993. Biodiversity and function of grazing ecosystems. In *Biodiversity and Ecosystem Function* (Schulze, E.-D. and Mooney, H.A., eds), pp. 361–383. Springer-Verlag, Berlin.

Meir, E., Andelman, S.J., and Possingham, H.P. 2004. Does conservation planning matter in a dynamic and uncertain world? *Ecology Letters* **7**: 615–622.

Menge, B.A. and Sutherland, J.P. 1987. Community regulation: variation in disturbance, competition, and predation in relation to environmental stress recruitment. *American Naturalist* **130**: 730–757.

Methot, R.D. 1990. Synthesis model: an adaptive framework for analysis of diverse stock assessment data. *International North Pacific Fisheries Commission Bulletin* **50**: 259–277.

Meynecke, J. 2004. Effects of global climate change on geographic distributions of vertebrates in North Queensland. *Ecological Modelling* **174**: 347–357.

Michod, R.E. 1999. *Darwinian Dynamics: Evolutionary Transitions in Fitness and Individuality*. Princeton University Press, Princeton, NJ.

Milgram, S. 1967. The small world problem. *Psychology Today* **2**: 60–67.

Milinski, M., Semmann, D., and Krambeck, H.-J. 2002. Reputation helps solve the 'tragedy of the commons'. *Nature* **415**: 424–426.

Milinski, M., Semmann, D., Bakker, T.C.M., and Krambeck, H.-J. 2001. Cooperation through indirect reciprocity: image scoring or standing strategy? *Proceedings of the Royal Society of London Series B Biological Sciences* **268**: 2495–2501.

Millennium Ecosystem Assessment. 2005. *Ecosystems and Human Well-being: Synthesis*. Island Press, Washington DC.

Miller, T.E., Burns, J.H., Munguia, P., Walters, E.L., Kneitel, J.M., Richards, P.M. *et al.* 2005. A critical review of twenty years' use of the resource-ratio theory. *American Naturalist* **165**: 439–448.

Millien, V., Lyons, S.K., Olson, L., Smith, F.A., Wilson, A. B., and Yom-Tov, Y. 2006. Ecotypic variation in the context of global climate change: revisiting the rules. *Ecology Letters* **9**: 853–869.

Mills, K.E. and Bever, J.D. 1998. Maintenance of diversity within plant communities: soil pathogens as agents of negative feedback. *Ecology* **79**: 1595–1601.

Milo, R., Shen-Orr, S., Itzkovitz, S., Kashtan, N., Chklovskii, D., and Alou, U. 2002. Network motifs: simple building blocks of complex networks. *Science* **298**: 824–827.

Mitchell, C.E. and Power, A.G. 2003. Release of invasive plants from fungal and viral pathogens. *Nature* **421**: 625–627.

Mitteldorf, J. and Wilson, D.S. 2000. Population viscosity and the evolution of altruism. *Journal of Theoretical Biology* **204**: 481–496.

Mittermeier, R.A., Mittermeier, C.G., Brooks, T.M., Pilgrim, J.D., Konstant, W.R., da Fonseca, G.A.B., and Kormos, C. 2003. Wilderness and biodiversity conservation. *Proceedings of the National Acadaemy of Sciences USA* **100**: 10309–10313.

Molander, P. 1985. The optimal level of generosity in a selfish, uncertain environment. *Journal of Conflict Resolution* **29**: 611–618.

Mollison, D. (ed.) 1995. *Epidemic Models: their Structure and Relation to Data*. Cambridge University Press, Cambridge

Montoya, J.M., Pimm, S.L., and Solè, R.V. 2006. Ecological networks and their fragility. *Nature* **442**: 259–264.

Mooney, H.A. 1977. *Convergent Evolution in Chile and California*. Dowden, Hutchinson & Ross, Stroudsburg, PA.

Morin, P.J. and McGrady-Steed, J. 2004. Biodiversity and ecosystem functioning in aquatic microbial systems: a new analysis of temporal variation and species richness-predictability relations. *Oikos* **104**: 458–466.

Morris, M. 1993. Telling tails explain the discrepancy in sexual partner reports. *Nature* **365**: 437–440.

Morris, R.S., Wilesmith, J.W., Stern, M.W., Sanson, R.L., and Stevenson, M.A. 2001. Predictive spatial modelling of alternative control strategies for the foot-and-mouth disease epidemic in Great Britain, 2001. *Veterinary Record* **149**: 137–145.

Mougeot, F., Evans, S.A., and Redpath, S.M. 2005. Interactions between population processes in a cyclic species: parasites reduce autumn territorial behaviour of male red grouse. *Oecologia* **144**: 289–298.

Muller-Landau, H.C., Condit, R.S., Chave, J., Thomas, S.C., Bohlman, S.A., Bunyavejchewin, S., *et al.* 2006. Testing metabolic ecology theory for allometric scaling of tree size, growth and mortality in tropical forests. *Ecology Letters* **9**: 575–588.

Murdoch, W.W., Briggs, C.J., and Nisbet, R.M. 1997. Dynamical effects of host size- and parasitoid state-dependent attacks by parasitoids. *Journal of Animal Ecology* **66**: 542–556.

Murdoch, W.W., Briggs, C.J., and Nisbet, R.M. 2003. *Resource-Consumer Dynamics*. Princeton University Press, Princeton, NJ.

Murdoch, W.W., Reeve, J.D., Huffaker, C.B., and Kennett, C.E. 1984. Biological control of olive scale and its relevance to ecological theory. *American Naturalist* **123**: 371–392.

Murdoch, W.W., Nisbet, R.M., Blythe, S.P., Gurney, W.S.C., and Reeve, J.D. 1987. An invulnerable age class and stability in delay-differential host-parasitoid models. *American Naturalist* **134**: 288–310.

Murphy, H.T., VanDerWal, J., and Lovett-Doust, J. 2006. Distribution of abundance across the range in eastern North American trees. *Global Ecology and Biogeography* **15**: 63–71.

Murray, J.D. 1989. *Mathematical Biology*. Springer-Verlag, London.

Murrell, D.J., Dieckmann, U., and Law, R. 2004. On moment closures for population dynamics in continuous space. *Journal of Theoretical Biology* **229**: 421–432.

Mwangi, W.W. 1997. Low use of fertilizers and low productivity in sub-Saharan Africa. *Nutrient Cycling in Agroecosystems* **47**: 135–147.

Myers, N. 1992. Synergisms: joint effects of climate change and other forms of habitat destruction. In *Climate Change and Biodiversity* (Lovejoy, T.E. and Hannah, L., eds), pp. 344–354. Yale University Press, New Haven, CT.

Myers, N., Mittermeier, R.A., Mittermeier, C.G., da Fonseca, G.A.B., and Kent, J. 2000. Biodiversity hotspots for conservation priorities. *Nature* **403**: 853–858.

Myers, R.A. and Worm, B. 2003. Rapid world-wide depletion of predatory fish communities. *Nature* **423**: 280–283.

Myers, R.A. and Worm, B. 2005. Extinction, survival or recovery of large predatory fishes. *Philosophical Transactions of the Royal Society Series B* **360**: 13–20.

Myers, R.A., Bridson, J., and Barrowman, N.J. 1995. *Summary of worldwide spawner and recruitment data*. Canadian Technical Report on Fisheries and Aquatic Science no. 2024.

Myers, R.A., Bowen, K.G., and Barrowman, N.J. 1999. Maximum reproductive rate of fish at low population sizes. *Canadian Journal of Fisheries and Aquatic Science* **56**: 2404–2419.

Myers, R.A., Barrowman, N.J., Hilborn, R., and Kehler, D.G. 2001. Inferring the Bayes prior with limited direct data with applications for risk analysis and reference points. *North American Journal of Fisheries Management* **22**: 351–364.

Naeem, S. (ed.) 2006. Special feature—neutral community ecology. *Ecology* **87**: 1368–1431.

Nakamaru, M. and Iwasa, Y. 2005. The evolution of altruism by costly punishment in lattice-structured populations: Score-dependent viability versus score-dependent fertility. *Evolutionary Ecology Research* 7:853–870.

Nakamaru, M., Matsuda, H., and Iwasa, Y. 1997. The evolution of cooperation in a lattice-structured population. *Journal of Theoretical Biology* **184**: 65–81.

Nakamaru, M., Nogami, H., and Iwasa, Y. 1998. Score-dependent fertility model for the evolution of cooperation in a lattice. *Journal of Theoretical Biology* **194**: 101–124.

Nee, S. 1994. How populations persist. *Nature* **367**: 123–124.

Nee, S. 2000. Mutualism, parasitism and cooperation in the evolution of coviruses. *Philosophical Transactions of the Royal Society B Biological Sciences* **355**: 1607–1613.

Nee, S. 2004. More than meets the eye. Earth's real biodiversity is invisible, whether we like it or not. *Nature* **429**: 804–805.

Nee, S. 2005. The neutral theory of biodiversity: do the numbers add up? *Functional Ecology* **19**: 173–176.

Nee, S. and Colgrave, N. 2006. Paradox of the clumps. *Nature* **441**: 417–418.

Nee, S. and May, R.M. 1992. Dynamics of metapopulations: habitat destruction and competitive coexistence. *Journal of Animal Ecology* **61**: 37–40.

Nee, S. and May, R.M. 1993. Population-level consequences of conspecific brood parasitism in birds and insects. *Journal of Theoretical Biology* **161**: 95–109.

Nee, S. and May, R.M. 1997. Extinction and the loss of evolutionary history. *Science* **278**: 692–694.

Nee, S., Harvey, P.H., and May, R.M. 1991. Lifting the veil on abundance patterns. *Proceedings of the Royal Society of London Series B Biological Sciences* **243**: 161–163.

Nee, S., May, R.M., and Hassell, M.P. 1997. Two-species metapopulation models. In *Metapopulation Biology* (Hanski, I. and Gilpin, M.E., eds), pp. 123–147. Academic Press, London.

Nee, S., Colegrave, N., West, S.A., and Grafen, A. 2005. The illusion of invariant quantities in life histories. *Science* **309**: 1236–1239.

Neubert, M.G., Kot, M., and Lewis, M.A. 1995. Dispersal and pattern formation in a discrete-time predator-prey model. *Theoretical Population Biology* **48**: 7–43.

Neuhauser, C. 2001. Mathematical challenges in spatial ecology. *North American Mathematical Society* **48**: 1304–1314.

Neutel, A.M., Heesterbeek, J.A.P., and de Ruiter, P.C. 2002. Stability in real food webs: weak links in long loops. *Science* **296**: 1120–1123.

Newman, M.E.J. 2003. The structure and function of complex networks. *SIAM Review* **45**: 167–256.

Nicholson, A.J. 1933. The balance of animal populations. *Journal of Animal Ecology* **2**: 132–178.

Nicholson, A.J. 1957. The self adjustment of populations to change. *Cold Spring Harbour Symposia on Quantitative Biology* **XXII**: 153–173.

Nicholson, A.J. and Bailey, V.A. 1935. The balance of animal populations. Part I. *Proceedings of the Zoological Society of London* **3**: 551–598.

Nisbet, R.M. and Gurney, W.S.C. 1976. Population dynamics in a periodically varying environment. *Journal of Theoretical Biology* **56**: 459–475.

Nitecki, M. (ed.) 1984. *Extinctions*. Chicago Press, Chicago.

Noss, R.F. 2001. Beyond Kyoto: forest management in a time of rapid climate change. *Conservation Biology* **15**: 578–590.

Nowak, M.A. and Sigmund, K. 1992. Tit for tat in heterogeneous populations. *Nature* **355**: 250–253.

Nowak, M. and Sigmund, K. 1993. A strategy of win-stay, lose-shift that outperforms tit-for-tat in the prisoner's dilemma game. *Nature* **364**: 56–58.

Nowak, M.A. and May, R.M. 1992. Evolutionary games and spatial chaos. *Nature* **359**: 826–829.

Nowak, M.A. and May, R.M. 1994. Superinfection and the evolution of parasite virulence. *Proceedings of the Royal Society of London Series B* **255**: 81–89.

Nowak, M.A. and Sigmund, K. 1998a. Evolution of indirect reciprocity by image scoring. *Nature* **393**: 573–577.

Nowak, M.A. and Sigmund, K. 1998b. The dynamics of indirect reciprocity. *Journal of Theoretical Biology* **194**: 561–574.

Nowak, M.A. and May, R.M. 2000. *Virus Dynamics*. Oxford University Press, Oxford.

Nowak, M.A. and Sigmund, K. 2005. Evolution of indirect reciprocity. *Nature* **437**: 1291–1298.

Noy-Meir, I. 1975. Stability of grazing systems: an application of predator-prey graphs. *Journal of Ecology* **63**: 459–481.

Nuffield Council on Bioethics. 2003. *The Use of Genetically Modified Crops in Developing Countries: a follow-up discussion paper*. Nuffield Council on Bioethics, London.

Ødegaard, F. 2000. How many species of arthropods? Erwin's estimate revised. *Biological Journal of the Linnean Society* **71**: 583–597.

Odum, E.P. 1953. *Fundamentals of Ecology*. W.B. Saunders, Philadelphia, PA.

O'Grady, J.J., Reed, D.H., Brook, B.W., and Frankham, R. 2004. What are the best correlates of predicted extinction risk? *Biological Conservation* **118**: 513–520.

Ohtsuki, H. and Iwasa, Y. 2004. How should we define goodness? Reputation dynamics in indirect reciprocity. *Journal of Theoretical Biology* **231**: 107–120.

Ohtsuki, H. and Iwasa, Y. 2005. The leading eight: social norms that can maintain cooperation by indirect reciprocity. *Journal of Theoretical Biology* **239**: 435–444.

Ohtsuki, H., Hauert, C., Lieberman, E., and Nowak, M.A. 2006. A simple rule for the evolution of cooperation on graphs. *Nature* 441: 502–505.

Olsen, L.F. and Schaffer, W.M. 1990. Chaos versus noisy periodicity—alternative hypotheses for childhood epidemics. *Science* **249**: 499–504.

Olsen, L.F., Truty, G.L., and Schaffer, W.M. 1988. Oscillations and chaos in epidemics: a nonlinear dynamic study of six childhood diseases in Copenhagen, Denmark. *Theoretical Population Biology* **33**: 344–370.

Olson, D.M. and Dinerstein, E. 1998. The Global 200. A representation approach to conserving the Earth's most biologically valuable ecoregions. *Conservation Biology* **12**: 502–515.

Olson, J.S. 1958. Rates of succession and soil changes on southern Lake Michigan sand dunes. *Botanical Gazette* **119**: 125–169.

Overpeck, J., Cole, J., and Bartlein, P. 2005. A 'paleo-perspective' on climate variability and change. In *Climate Change and Biodiversity* (Lovejoy, T.E. and Hannah, L., eds), pp. 91–108. Yale University Press, New Haven, CT.

Pacala, S.W. 1986. Neighborhood models of plant population dynamics: 2. Multispecies models of annuals. *Theoretical Population Biology* **29**: 262–292.

Pacala, S.W. 1987. Neighborhood models of plant population dynamics: 3. Models with spatial heterogeneity in the physical environment. *Theoretical Population Biology* **31**: 359–392.

Pacala, S.W. 1997. Dynamics of plant communities. In *Plant Ecology* (Crawley, M.J., ed.), pp. 532–555. Blackwell Scientific Publications, Oxford.

Pacala, S. and Hassell. M.P. 1991. The persistence of host-parasitoid associations in patchy environments. II. Evaluation of field data. *American Naturalist* **138**: 584–605.

Pacala, S.W. and Crawley, M.J. 1992. Herbivores and plant diversity. *American Naturalist* **140**: 243–260.

Pacala, S.W. and Levin, S.A. 1997. Biologically generated spatial pattern and the coexistence of competing species. In *Spatial Ecology: the Role of Space in Population Dynamics and Interspecific Interactions* (Tilman, D. and Kareiva, P., eds), pp. 204–232. Princeton University Press, Princeton, NJ.

Pacala, S.W., Hassell, M.P., and May, R.M. 1990. Host-parasitoid associations in patchy environments. *Nature* **344**: 150–153.

Pacala, S.W., Canham, C.D., Saponara, J., Silander, J.A., Kobe, R.K., and Ribbens, E. 1996. Forest models defined by field measurements: estimation, error analysis and dynamics. *Ecological Monographs* **66**: 1–43.

Packard, N.H., Crutchfield, J.P., Framer, J.D., and Shaw, R.S. 1980. Geometry from a time series. *Physical Review Letters* **45**: 712–716.

Paine, R.T. 1966. Food web complexity and species diversity. *American Naturalist* **100**: 65–75.

Paine, R.T. 1969. A note on trophic complexity and community stability. *American Naturalist* **103**: 91–93.

Paine, R.T. 1988. Food webs: road map of interactions or grist for theoretical development? *Ecology* **69**: 1648–1654.

Panchanathan, K. and Boyd, R. 2003. A tale of two defectors: the importance of standing for evolution of indirect reciprocity. *Journal of Theoretical Biology* **224**: 115–126.

Pandolfi, J.M. 2006. Corals fail a test of neutrality. *Nature* **440**: 35–36.

Parmesan, C. 1996. Climate and species' range. *Nature* **382**: 765–766.

Parmesan, C. and Galbraith, H. 2004. *Observed impacts of global change in the U.S.* Pew Center on Global Climate Change, Arlington, VA.

Parmesan, C., Ryrholm, N., Stefanescu, C., Hill, J., Thomas, C., Descimon, H. *et al.* 1999. Poleward shifts in geographical ranges of butterfly species associated with regional warming. *Nature* **399**: 579–584.

Parmesan, C., Gaines, S., Gonzalez, L., Kaufman, D.M., Kingsolver, J., Peterson, A.T., and Sagarin, R. 2005. Empirical perspectives on species borders: from traditional biogeography to global change. *Oikos* **108**: 58–75.

Pascual, M. and Dunne, J.A. 2006. *Ecological Networks: Linking Structure to Dynamcis in Food Webs.* Oxford University Press, Oxford.

Pastor, J., Aber, J.D., McClaugherty, C.A., and Melillo, J.M. 1982. Geology, soils, and vegetation of Blackhawk Island, Wisconsin. *American Midland Naturalist* **108**: 266–277.

Pastor, J., Aber, J.D., McClaugherty, C.A., and Melillo, J.M. 1984. Aboveground production and N and P cycling along a nitrogen mineralization gradient on Blackhawk Island, Wisconsin. *Ecology* **65**: 256–268.

Patrick, R. 1973. The use of algae, especially diatoms, in the assessment of water quality. *American Society of Testing and Materials* **528**: 76–95.

Patrick, R. 1975. Stream communities. In *Ecology and Evolution of Communities* (Cody, M.L. and Diamond, J.M., eds), pp. 445–459. Harvard University Press, Cambridge, MA.

Patz, J.A., Campbell-Lendrum, D., Holloway, T., and Foley, J.A. 2005. Impact of regional climate change on human health. *Nature* **438**: 310–317.

Paulsson, J. 2002. Multileveled selection on plasmid replication. *Genetics* **161**: 1373–1384.

Pauly, D., Christensen, V., Dalsgaard, A., Froese, R., and Torres, J. 1998. Fishing down marine food webs. *Science* **279**: 860–863.

Pauly, D., Christensen, V., and Walters, C.J. 2000. ECOPATH, ECOSYM and ECOSPACE as tools for evaluating ecosystem impact on fisheries. *ICES Journal of Marine Science* **57**: 697–706.

Pavoine, S., Ollier, S., and Dufour, A.-B. 2005. Is the originality of a species measurable? *Ecology Letters* **8**: 579–586.

Pearson, R.G. 2006. Climate change and the migration capacity of species. *Trends in Ecology and Evolution* **21**: 111–113.

Pearson, R.G. and Dawson, T.P. 2003. Predicting the impacts of climate change on the distribution of species: are bioclimate envelope models useful? *Global Ecology and Biogeography* **12**: 361–371.

Pease, C.M. and Mattson, D.J. 1999. Demography of the Yellowstone grizzly bears. *Ecology* **80**: 957–975.

Pella, J.J. and Tomlinson P.K. 1969. A generalized stock production model. *Inter-American Tropical Tuna Commission Bulletin* **16**: 419–496.

Petchey, O.L., Casey, T., Jiang, L., McPhearson, P.T., and Price, J. 2002. Species richness, environmental fluctuations, and temporal change in total community biomass. *Oikos* **99**: 231–240.

Peterman, R.M., Clark, W.C., and Hollings, C.S. 1979. The dyamics of resilience: shifting stability domains in fish and insect systems. In *Population Dynamics* (Anderson, R.A., Turner, B.D., and Taylor, L.R., eds), pp. 321–341. Blackwell Scientific Publications, Oxford.

Peters, C.M., Gentry, A.H., and Mendelsohn, R.O. 1989. Valuation of an Amazonian rainforest. *Nature* **339**: 655–656.

Peters, R.L. 1992. Conservation of biological diversity in the face of climate change. In *Climate Change and Biodiversity* (Lovejoy, T.E. and Hannah, L., eds), pp. 15–30. Yale University Press, New Haven, CT.

Peters, R.L. and Darling, J.D.S. 1985. The greenhouse effect and nature reserves. *BioScience* **35**: 707–717.

Peterson, A.T., Martinez-Meyer, E., Gonzalez-Salazar, C., and Hall, P.W. 2004. Modeled climate change effects on distributions of Canadian butterfly species. *Canadian Journal of Zoology* **82**: 851–858.

Peterson, R.O. 1999. Wolf-moose interaction on Isle Royale: the end of natural regulation? *Ecological Applications* **9**: 10–16.

Pimentel, D. 1961. Species diversity and insect population outbreaks. *Annals of the Entomological Society of America* **54**: 76–86.

Pimm, S.L. 1984. The complexity and stability of ecosystems. *Nature* **307**: 321–326.

Pimm, S.L. 1991. *The Balance of Nature?* Chicago University Press, Chicago.

Pimm, S.L. and Askins, R.A. 1995. Forest losses predict bird extinctions in eastern North America. *Proceedings of the National Academy of Sciences USA* **92**: 9343–9347.

Pimm, S.L. and Raven, P. 2000. Extinction by numbers. *Nature* **403**: 843–845.

Pimm, S.L., Russell, G.J., Gittleman, J.L., and Brooks, T.M. 1995. The future of biodiversity. *Science* **269**: 347–350.

Pingali, P.L. and Rosegrant, M.W. 1998. *Intensive Food Systems in Asia: Can the Degradation Problems be Reversed?* Paper presented at the Pre-Conference Workshop Agricultural Intensification, Economic Development and the Environment of the Annual Meeting of the American Agricultural Economics Association, Salt Lake City, Utah, 31 July–1 Aug 1998.

Pingali, P.L. and Heisey, P.W. 1999. *Cereal Productivity in Developing Countries: Past Trends and Future Prospects,* CIMMYT Economics Paper 99–03. International Maize and Wheat Improvement Center (CIMMYT), Mexico

Plagányi, É.E. and Butterworth, D.S. 2004. A critical look at the potential of Ecopath with Ecosim to assist in practical fisheries management. *African Journal of Marine Science* **26**: 261–287.

Plotkin, J.B. Potts, M.D., Yu, D.W., Bunyavejchewin, S., Condit, R., Foster, R.B. *et al.* 2000. Predicting species diversity in tropical forests. *Proceedings of the National Academy of Sciences USA* **97**: 10850–10854.

Pope, J.G. 1972. An investigation of the accuracy of virtual population analysis using cohort analysis. *Research Bulletin of the International Commission of Northwest Atlantic Fisheries* **9**: 65–74.

Pope, J.G. 1991. The ICES Multispecies Assessment Working Group: evolution insights, and future problems. Multispecies models relevant to management of living resources. *ICES Marine Science Symposia* **193**: 22–33.

Possingham, H.P. and Wilson, K.A. 2005. Biodiversity: turning up the heat on hotspots. *Nature* **436**: 919–920.

Pounds, J.A., Bustamante, M.R., Coloma, L.A., Consuegra, J.A., Fogden, M.P.L., Foster, P.N. *et al.* 2006. Widespread amphibian extinctions from epidemic disease driven by global warming. Nature **439**: 161–167.

Pressey, R.L. and Taffs, K.H. 2001. Scheduling conservation action in production landscapes: priority areas in western New South Wales defined by irreplaceability and vulnerability to vegetation loss. *Biological Conservation* **100**: 355–376.

Pressey, R.L., Possingham, H.P., and Margules, C.R. 1996. Optimality in reserve selection algorithms: when does it matter and how much? *Biological Conservation* **76**: 259–267.

Preston, F.W. 1948. The commonness, and rarity, of species. *Ecology* **29**: 254–283.

Preston, F.W. 1962. The canonical distribution of commonness and rarity. *Ecology* **43**: 185–215 and 410–432.

Proches, S., Wilson, J.R.U., and Cowling, R.M. 2006. How much evolutionary history in a 10x10m plot? *Proceedings of the Royal Society of London Series B Biological Sciences* **273**: 1143–1148.

Proulx, S.R., Promislow, D.E.L., and Phillips, P.C. 2005. Network thinking in ecology and evolution. *Trends in Ecology and Evolution* **20**: 345–353.

Pulliam, H.R. 1988. Sources, sinks, and population regulation. *American Naturalist* **132**: 652–661.

Pulliam, H.R. 1996. Sources and sinks. In *Population Dynamics in Ecological Space and Time* (Rhodes, O.E., Chesser, R.K., and Smith, M.A., eds), pp. 41–70. University of Chicago Press, Chicago.

Punt, A.E., Smith, A.D.M., and Cui, G. 2002. Evaluation of management tools for Australia's south east fishery.

1. Modelling the south east fishery taking account of technical interactions. *Marine and Freshwater Research* **53**: 615–629.

Pyke, C.R., Andelman, S.J., and Midgley, G. 2005. Identifying priority areas for bioclimatic representation under climate change: a case study for Proteaceae in the Cape Floristic Region, South Africa. *Biological Conservation* **125**: 1–9.

Quinn, II, T.J. and Deriso, R.B. 1999. *Quantitative Fish Dynamics*. Oxford University Press, Oxford.

Quinn, II, T.J. and Collie, J.S. 2005. Sustainability in single-species population models. *Philosophical Transactions of the Royal Society Series B* **360**: 147–162.

Rand, D.A. and Wilson, H. 1991. Chaotic stochasticity: a ubiquitous source of unpredictability in epidemics. *Proceedings of the Royal Society of London Series B Biological Sciences* **246**: 179–184.

Rand, D.A. and Wilson, H.B. 1995. Using spatio temporal chaos and intermediate scale determinism to quantify spatially extended ecosystems. *Proceedings of the Royal Society of London Series B Biological Sciences* **259**: 111–117.

Randhawa, N.S. no date. *Some Concerns for Future of Punjab Agriculture*. Mimeo, New Delhi, India.

Rapoport, A. and Chammah, A.M. 1965. *Prisoner's Dilemma*. University of Michigan Press, Ann Arbor, MI.

Raup, D.M. 1978. Cohort analysis of genetic survivorship. *Paleobiology* **4**: 1–15.

Raven, P.H. 2004. Taxonomy: where are we now? *Proceedings of the Royal Society of London Series B Biological Sciences* **359**: 729–730.

Read, A.F. and Taylor, L.H. 2001. The ecology of genetically diverse infections. *Science* **292**: 1099–1102.

Reader, J. 2004. *Cities*. Heinemann, London.

Redfern, M., Jones, T.H., and Hassell, M.P. 1992. Heterogeneity and density dependence in a field study of a tephritid-parasitoid interaction. *Ecological Entomology* **17**: 255–262.

Redford, K.H., Coppolillo, P., Sanderson, E.W., Da Fonseca, G.A.B., Dinerstein, E., Groves, C. *et al.* 2003. Mapping the conservation landscape. *Conservation Biology* **17**: 116–131.

Rees, M. 1997. Seed dormancy. In *Plant Ecology* (Crawley, M.J., ed.), pp. 214–238. Blackwell Scientific Publications, Oxford.

Rees, M. and Crawley, M.J. 1989. Growth, reproduction and population dynamics. *Functional Ecology* **3**: 645–653.

Rees, M. and Crawley, M.J. 1991. Do plant-populations cycle? *Functional Ecology* **5**: 580–582.

Rees, M., Grubb, P.J., and Kelly, D. 1996. Quantifying the impact of competition and spatial heterogeneity on the structure and dynamics of a four-species guild of winter annuals. *American Naturalist* **147**: 1–32.

Reeve, J.D. 1990. Stability, variability, and persistence in host-parasitoid systems. *Ecology* **71**: 422–426.

Reeve, J.D., Kerans, B.L., and Chesson, P.L. 1989. Combining different forms of parasitoid aggregation: effects on stability and patterns of parasitism. *Oikos* **56**: 233–239.

Reeve, J.D., Cronin, J.T., and Strong, D.R. 1994. Parasitism and generation cycles in a salt-marsh planthopper. *Journal of Animal Ecology* **63**: 912–920.

Reid, W.V. 1992. How many species will there be? In *Tropical Deforestation and Species Extinction* (Whitmore, T.C. and Sayer, J.A., eds), pp. 55–73. Chapman & Hall/IUCN, London.

Repasky, R.R. 1991. Temperature and the northern distribuitons of wintering birds. *Ecology* **72**: 2274–2285.

Rex, M.A., McClain, C.R., Johnson, N.A., Etter, R.J., Allen, J.A., Bouchet, P., and Waren, A. 2005. A source-sink hypothesis for abyssal biodiversity. *American Naturalist* **165**: 163–178.

Richardson, D.M., Allsopp, N., D'Antonio, C.M., Milton, S.J., and Rejmánek, M. 2000. Plant invasions—the role of mutualisms. *Biological Review* **75**: 65–93.

Ricker, W.E. 1954. Stock and recruitment. *Journal of Fish Research Board Canada* **11**: 559–623.

Ricketts, T.H. 2001. The matrix matrix: effective isolation in fragmented landscapes. *American Naturalist* **158**: 87–99.

Ricketts, T.H., Daily, G.C., Ehrlich, P.R., and Michener, C. D. 2004. Economic value of tropical forest to coffee production. *Proceedings of the National Academy of Sciences USA* **101**: 12579–12582.

Ricketts, T.H., Dinerstein, E., Boucher, T., Brooks, T.M., Butchart, S.H.M., Hoffmann, M. *et al.* 2005. Pinpointing and preventing imminent extinctions. *Proceedings of the National Academy of Sciences USA* **102**: 18497–18501.

Rickleffs, R.E. 2006. The unified neutral theory of biodiversity: do the numbers add up? *Ecology* **87**: 1424–1431.

Rijnsdorp, A.D. 1993. Fisheries as a large-scale experiment on life-history evolution: Disentangling phenotypic and genetic effects in changes in maturation and reproduction in North Sea plaice, *Pleuronectes platessa* L. *Oecologia*, **96**: 391–401.

Roberts, C.M. and Hawkins, J.P. 2003. *Fully Protected Marine Reserves: a Guide 2003*. World Wildlife Fund. www.panda.org/resources/publications/water/mpreserves/mar/dwnld.htm.

Roberts, C.M., Hawkins, J.P., and Gell, F.R. 2005. The role of marine reserves in achieving sustainable fisheries. *Philosophical Transactions of the Royal Society Series B* **360**: 123–132.

Roberts, M.G., Smith, G., and Grenfell, B.T. 1995. Mathematical models for macroparasites of wildlife. In

Ecology of Infectious Diseases in Natural Populations (Grenfell, B.T. and Dobson, A.P., eds), pp. 177–208. Cambridge University Press, Cambridge.

Rodrigues, A.S.L. and Gaston, K.J. 2002a. Optimisation in reserve selection procedures—why not? *Biological Conservation* **107**: 123–129.

Rodrigues, A.S.L. and Gaston, K.L. 2002b. Maximizing phylogenetic diversity in selection of networks of conservation areas. *Biological Conservation* **105**: 103–111.

Rodrigues, A.S., Cerdeira, J.O., and Gaston, K.J. 2000. Flexibility, efficiency, and accountability: adapting reserve selection algorithms to more complex conservation problems. *Ecography* **23**: 565–574.

Roff, D.A. 2002. *Life History Evolution*. Sinauer Associates, Sunderland, MA.

Rohani. P., May, R.M., and Hassell, M.P. 1996. Metapopulations and equilibrium stability: the effects of spatial structure. *Journal of Theoretical Biology* **181**: 97–109.

Rooney, N., McCann, K., Gellner, G., and More, J.C. 2006. Structural asymmetry and the stability of diverse food webs. *Nature* **442**: 265–269.

Root, T.L. 1988a. Environmental factors associated with avian distributional boundaries. *Journal of Biogeography* **15**: 489–505.

Root, T.L. 1988b. Energy constraints on avian distributions and abundances. *Ecology* **69**: 330–339.

Root, T.L. and Hughes, L. 2005. Present and future phenological changes in wild plants and animals. In *Climate Change and Biodiversity* (Lovejoy, T.E. and Hannah, L., eds), pp. 61–74. Yale University Press, New Haven, CT.

Root, T.L., Price, J.T., Hall, K.R., Scheider, S.H., Rosenzweig, C., and Pounds, J.A. 2003. Fingerprints of global warming on wild animals and plants. *Nature* **421**: 57–60.

Rosenzweig, M.I. and MacArthur, R.H. 1963. Graphical representation and stability conditions of predator-prey interactions. *American Naturalist* **97**: 209–223.

Rothhaupt, K.O. 1988. Mechanistic resource competition theory applied to laboratory experiments with zooplankton. *Nature* **333**: 660–662.

Rothschild, B.J. 1986. *Dynamics of Marine Fish Populations*. Harvard University Press, Cambridge, MA.

Royal Society. 1998. *Genetically Modified Plants for Food Use*. Royal Society, London.

Royama, T. 1992. *Analytical Population Dynamics*. Chapman and Hall, London.

Saccheri, I. and Hanski, I. 2006. Natural selection and population dynamics. *Trends in Evolution and Ecology* **21**: 341–347.

Sagarin, R.D. and Gaines, S.D. 2002. Geographical abundance distributions of coastal invertebrates: using one-dimensional ranges to test biogeographic hypotheses. *Journal of Biogeography* **29**: 985–997.

Sainsbury, K.J. 1988. The ecological basis for multispecies fisheries, and management of a demersal fishery in tropical Australia. In *Fish Population Dynamics*, 2nd ed. *The Implications for Management* (Gulland, J.A., ed.), pp. 349–382. John Wiley & Sons, New York

Sainsbury, K.J., Punt, A.E., and Smith, A.D.M. 2000. Design of operational management strategies for achieving fishery ecosystem objectives. *ICES Journal of Marine Science* **51**: 731–741.

Sala, O.E., Chapin, F.S., Armesto, J.J., Berlow, E., Bloomfield, J., Dirzo, R. *et al.* 2000. Global biodiversity scenarios for the year 2100. *Science* **287**: 1770–1774.

Samson, D.A. and Werk, K.S. 1986. Size-dependent effects in the analysis of reproductive effort in plants. *American Naturalist* **127**: 667–680.

Sanchez, P.A. 2002. Soil fertility and hunger in Africa. *Science* **295**: 2019–2020.

Sanderson, E., Jaiteh, M., Levy, M.A., Redford, K.H., Wannebo, A., and Woolmer, G. 2002. The human footprint and the last of the wild. *Bioscience* **52**: 891–904.

Santos, F.C. and Pacheco, J.M. 2005. Scale-free networks provide a unifying framework for the emergence of cooperation. *Physical Review Letters* **95**: 98104–1–98104–4.

Santos, F.C., Rodrigues, J.F., and Pacheco, J.M. 2005. Graph topology plays a determinant role in the evolution of cooperation. *Proceedings of the Royal Society of London Series B Biological Sciences* **273**: 51–55.

Sasaki, A.J. 1997. Clumped distribution by neighbourhood competition. *Journal of Theoretical Biology* **186**: 415–430.

Savolainen, V., Cowan, R.S., Vogler, A.P., and Roderick, G.K. (eds) 2005. DNA barcoding of life. *Philosophical Transactions of the Royal Society* **360**: 1803–1980.

Schaefer, M.B. 1954. Some aspects of the dynamics of populations important to the management of commercial marine fisheries. *Inter-American Tropical Tuna Commission Bulletin* **1**: 25–56.

Schaffer, W.M. 1985. Order and chaos in ecological systems. *Ecology* **66**: 93–106.

Schaffer, W.M. 2000. Foreword. In *Chaos in Real Data: the Analysis of Nonlinear Dynamics from Short Ecological Time Series* (Perry, J.N., Smith, R.H., Woiwod, I.P., and Morse, D.R., eds), pp. vii–x. Kluver Academic Publications, Dordrecht.

Schaffer, W.M. and Kot, M. 1985a. Do strange attractors govern ecological-systems. *Bioscience* **35**: 342–350.

Schaffer, W.M. and Kot, M. 1985b. Nearly one dimensional dynamics in an epidemic. *Journal of Theoretical Biology* **112**: 403–427.

Scheffer, M. and van Ness, E.H. 2006. Self-organised similarity, and the evolutionary emergence of groups of similar species. *Proceedings of the National Academy of Sciences USA* **103**: 6230–6235.

Scheffer, M., Carpenter, S.R., Foley, J.A., Folke, C., and Walker, B. 2001. Catastrophic changes in ecosystems. *Nature* **413**: 591–596.

Schellnhuber, H.J., Cramer, W., Nakicenovic, N., Wigley, T., and Yohe, G. (eds) 2006. *Avoiding Dangerous Climate Change*. Cambridge University Press, Cambridge.

Schenkeveld, A.J. and Verkaar, H.J. 1984. The ecology of short-lived forbs in chalk grasslands—distribution of germinative seeds and its significance for seedling emergence. *Journal of Biogeography* **11**: 251–260.

Schenzle, D. 1984. An age-structured model of pre- and post-vaccination measles transmission. *IMA Journal of Mathematics Applied in Medicine and Biology* **1**: 169–191.

Schoener, T.W. 1983. Field experiments on interspecific competition. *American Naturalist* **122**: 240–285.

Schwinning, S. and Parsons, A.J. 1996. A spatially explicit population model of stoloniferous N-fixing legumes in mixed pasture with grass. *Journal of Ecology* **84**: 815–826.

Schwinning, S. and Weiner, J. 1998. Mechanisms determining the degree of size asymmetry. *Oecologia* **113**: 447–455.

Scoones, I. and Thompson, J. (eds) 1994. *Beyond Farmer First: Rural People's Knowledge, Agricultural Research and Extension Practice*. Intermediate Technology Publications, London.

Scott, D., Malcolm, J.R., and Lemieux, C. 2002. Climate change and modelled biome representation in Canada's national park system: implications for system planning and park mandates. *Global Ecology and Biogeography* **11**: 475–484.

Seabloom, E.W., Dobson, A.P., and Stoms, D.M. 2002. Extinction rates under nonrandom patterns of habitat loss. *Proceedings of the National Academy of Sciences USA* **99**: 11229–11234.

Seinen, I. and Schram, A. 2006. Social status and group norms: indirect reciprocity in a repeated helping experiment. *European Economics Review* **50**: 581–602.

Sepkoski, J.J. 1992. Phylogentic and ecologic patterns in the Phanerozoic history of marine biodiversity. In *Systematics, Ecology, and the Biodiversity Crisis* (Eldredge, N., ed.), pp. 77–100. Columbia University Press, Columbia.

Serrano, D., Oro, D., Esperanza, U., and Tella, J.L. 2005. Colony size selection determines adult survival and dispersal preferences: Allee effects in a colonial bird. *American Naturalist* **166**: E22–E31.

Shaffer, M.L. 1990. Population viability analysis. *Conservation Biology* **4**: 39–40.

Shapiro, B., Drummond, A.J., Rambaut, A., Wilson, M.C., Matheus, P.E., Sher, A.V. *et al.* 2004. Rise and fall of the Beringian steppe bison. *Science* **306**: 1561–1565.

Shaw, R.G. and Antonovics, J. 1986. Density-dependence in *Salvia lyrata*, a herbaceous perennial—the effects of experimental alteration of seed densities. *Journal of Ecology* **74**: 797–813.

Shepherd, J.G. 1982. A versatile new stock-recruitment relationship for fisheries and the construction of sustainable yield curves. *Journal du Conseil. Conseil Permanent International pour l'Exploration de la Mer* **40**: 67–75.

Shipley, B. and Dion, J. 1992. The allometry of seed production in herbaceous angiosperms. *American Naturalist* **139**: 467–483.

Shurin, J.B., Gruner, D.S., and Hillebrand, H. 2006. All wet or dried up? Real differences between aquatic and terrestrial food webs. *Proceedings of the Royal Society of London Series B Biological Sciences* **273**: 1–9.

Sibly, R.M. and Smith, R.H. 1998. Identifying key factors using lambda contribution analysis. *Journal of Animal Ecology* **67**: 17–24.

Sih, A., Crowley, P., McPeck, M., Petraaka, J., and Strohmeier, K. 1986. Predation, competition, and prey communities: a review of field experiments. *Annual Review of Ecology and Systematics* **16**: 269–311.

Siitonen, P., Lehtinen, A., and Siitonen, M. 2005. Effects of forest edges on the distribution, abundance, and regional persistence of wood-rotting fungi. *Conservation Biology* **19**: 250–260.

Simberloff, D. 1992. Do species-area curves predict extinction in fragmented forest? In *Tropical Deforestation and Species Extinction* (Whitmore, T.C. and Sayer, J. A., eds), pp. 75–89. Chapman & Hall/IUCN, London.

Sinha, S. and Sinha, S. 2005. Evidence of universality for the May-Wigner stability theorem for random networks with local dynamics. *Physical Review E* **71**: 020902.

Skellam, J.G. 1951. Random dispersal in theoretical populations. *Biometrika* **38**: 196–218.

Skyrms, B. and Pemantle, R. 2000. A dynamic model of social network formation. *Proceedings of the National Academy of Sciences USA* **97**: 9340–9346.

Slatkin, M. 1981. Fixation probabilities and fixation times in a subdivided population. *Evolution* **35**: 477–488.

Smith, D.L., Lucey, B., Waller, L.A., Childs, J.E., and Real, L.A. 2002. Predicting the spatial dynamics of rabies epidemics on heterogeneous landscapes. *Proceedings of the National Academy of Sciences of the United States of America* **99**: 3668–3772.

Smith, M.A. and Green, D.M. 2005. Dispersal and the metapopulation paradigm in amphibian ecology and conservation: are all amphibian populations metapopulations? *Ecography* **28**: 110–128.

Smith, R.H. and Mead, R. 1974. Age-structure and stability in models of prey-predator systems. *Theoretical Population Biology* **6**: 308–322.

Snall, T., Ehrlen, J., and Rydin, H. 2005. Colonization-extinction dynamics of an epiphyte metapopulation in a dynamic landscape. *Ecology* **86**: 106–115.

Sober, E. and Wilson, D.S. 1998. *Unto Others: the Evolution and Psychology of Unselfish Behavior*. Harvard University Press, Cambridge, MA.

Soemarwoto, O. and Conway, G.R. 1991. The Javanese homegarden. *Journal for Farming Systems Research and Extension* **2**: 95–118.

Solomon, M.E. 1949. The natural control of animal populations. *Journal of Animal Ecology* **18**: 1–35.

Solow, A.R. and Beet, A.R. 1998. On lumping species in food webs. *Ecology* **79**: 2013–2018.

Solow, A.R., Mound, L.A., and Gaston, K.J. 1995. Estimating the rate of synonymy. *Systematic Biology* **44**: 93–96.

Sommer, U. 1986. Nitrate and silicate competition among Antarctic phytoplankton. *Marine Biology* **98**: 345–351.

Sommer, T. 1990. Phytoplankton nutrient competition—from laboratory to lake. In *Perspectives on Plant Competition* (Grace, J.B. and Tilman, D. eds), pp. 193–213. Academic Press, San Diego.

Southwood, T.R.E., May, R.M., and Sugihara, G.S. 2006. Observations of related ecological exponents. *Proceedings of the National Academy of Sciences USA* **103**: 6931–6933.

Sperling, L. and Scheidegger, U. 1995. *Participatory Selection of Beans in Rwanda: Results, Methods and Institutional Issues*, Gatekeeper Series no. 51. International Institute for Environment and Development, London.

Stachowicz, J.J. 2001. Mutualism, facilitation, and the structure of ecological communities. *BioScience* **51**: 235–246.

Stamp, S.D. 1969. *Nature Conservation in Britain. New Naturalist*. Collins, London.

Stefansson, G. and Palsson, O.K. 1998. A framework for multi-species modelling of Arctic-boreal systems. *Review of Fish Biology and Fisheries* **8**: 101–104.

Stenseth, N.C., Falck, W., Bjørnstad, O.N., and Krebs, C.J. 1997. Population regulation in snowshoe hare and Canadian lynx: asymmetric food web configuration between hare and lynx. *Proceedings of the National Academy of Sciences USA* **94**: 5147–5152.

Stenseth, N.C., Mysterud, A., Ottersen, G., Hurrell, J.W., Chan, K.S., and Lima, M. 2002. Ecological effects of climate fluctuations. *Science* **297**: 1292–1296.

Stenseth, N.C., Chan, K.-S., Tavecchia, G., Coulson, T., Mysterud, A., Clutton-Brock, T.H., and Grenfell, B.T. 2004. Modelling non-additive and non-linear signals from climatic noise in ecological time series: Soay sheep as an example. *Proceedings of the Royal Society of London Series B Biological Sciences* **271**: 1985–1993.

Stephens, P.A. and Sutherland, W.J. 1999. Consequences of the Allee effect for behaviour, ecology and conservation. *Trends in Ecology and Evolution* **14**: 401–405.

Stoll, P. and Bergius, E. 2005. Pattern and process: competition causes regular spacing of individuals within plant populations. *Journal of Ecology* **93**: 395–403.

Stoll, P., Weiner, J., Muller-Landau, H., Muller, E., and Hara, T. 2002. Size symmetry of competition alters biomass-density relationships. *Proceedings of the Royal Society of London Series B Biological Sciences* **269**: 2191–2195.

Stone, L. 1995. Biodiversity and habitat destruction—a comparative-study of model forest and coral-reef ecosystems. *Proceedings of the Royal Society of London Series B Biological Sciences* **261**: 381–388.

Stone, R. 1992. Researchers score victory over pesticides—and pests- in Asia. *Science* **256**: 1272.

Storch, D., Marquet, P.A., and Brown, J.H. (eds) 2006. *Scaling Biodiversity*. Oxford University Press, Oxford.

Stouffer, D.B., Camacho, J., Guimera, R., Ng, C.A., and Nunes-Amaral, L.A. 2005. Quantitative patterns in the structure of model and empirical food webs. *Ecology* **86**: 1301–1311.

Strobeck, C. 1973. N species competition. *Ecology* **54**: 650–654.

Strong, D.R., Simberloff, D., Abele, L.G., and Thistle, A.B. 1984. *Ecological Communities: Conceptual Issues and the Evidence*. Princeton University Press, Princeton, NJ.

Stumpf, M.P.H., Wiuf, C., and May, R.M. 2005. Subnets of scale-free networks are not scale-free: sampling properties of networks, *Proceedings of the National Academy of Sciences USA* **102**: 4221–4224.

Suffling, R. and Scott, D. 2002. Assessment of climate change effects on Canada's National Park system. *Environmental Monitoring and Assessment* **74**: 117–139.

Sugden, R. 1986. *The Economics of Rights, Co-operation and Welfare*. Blackwell Scientific Publications, Oxford.

Sugihara, G. 1980. Minimal community structure: an explanation of species abundance patterns. *American Naturalist* **116**: 770–787.

Sugihara, G. 1982. *Niche Hierarchy: Structure, Organization and Assembly in Natural Communities*. PhD thesis, Princeton University, NJ.

Sugihara, G. 1984. Graph theory, homology and food webs. In *Population Biology: Proceedings of Symposia in Applied Mathematics*, vol. 30 (Levin, S.A., ed.), pp. 83–101. American Mathematical Society, Providence, RI.

Sugihara, G. and May, R.M. 1990. Nonlinear forecasting as a way of distinguishing chaos from measurement error in time series. *Nature* **344**: 734–741.

Sugihara, G., Grenfell, B., and May, R.M. 1990. Distinguishing error from chaos in ecological time series. *Philosophical Transactions of the Royal Society Series B Biological Sciences* **330**: 235–251.

Sugihara, G., Bersier, L.-F., and Schoenly, K. 1997. Effects of taxonomic and trophic aggregation on food web properties. *Oecologia* **112**: 272–284.

Sugihara, G., Bersier, L.-F., Southwood, T.R.E., Pimm, S. L., and May, R.M. 2003. Predicted correspondence between species abundances and dendrograms of niche similarities. *Proceedings of the National Academy of Sciences USA* **100**: 5246–5251.

Sukatschew, W.N. 1928. *Plant Communities* (in Russian). Nauk, Moscow.

Sutherland, W.J. 1996. *For Individual Behaviour to Population Ecology*. Oxford University Press, Oxford.

Svenning, J.C. and Skov, F. 2004. Limited filling of the potential range in European tree species. *Ecology Letters* **7**: 565–573.

Swenson, W., Wilson, D.S., and Elias, R. 2000. Artificial ecosystem selection. *Proceedings of the National Academy of Sciences USA* **97**: 9110–9114.

Symonides, E., Silvertown, J., and Andreasen, V. 1986. Population-cycles caused by overcompensating density-dependence. *Oecologia* **71**: 156–158.

Szabó, G. and Vukov, J. 2004. Cooperation for volunteering and partially random partnerships. *Physical Review E* **69**: 036107 (1–7).

Szathmary, E. and Demeter, L. 1987. Group selection of early replicators and the origin of life. *Journal of Theoretical Biology* **128**: 463–486.

Takahashi, N. and Mashima, R. 2003. *The Emergence of Indirect Reciprocity: is the Standing Strategy the Answer?*, Working Paper Series no. 29. Center for the Study of Cultural and Ecological Foundations of the Mind, Hokkaido University, Japan.

Takens, F. 1981. Detecting strange attractors in turbulence. In *Dynamical Systems and Turbulence* (Rand, D.A. and Young, L.-S., eds), pp. 366–381, Lecture Notes in Mathematics, vol. 898. Springer-Verlag, Berlin.

Taylor, A.D. 1988. Parasitoid competition and the dynamics of host-parasitoid models. *American Naturalist* **132**: 417–436.

Taylor, P.D. and Jonker, L.B. 1978. Evolutionary stable strategies and game dynamics. Mathematical Biosciences **40**: 145–156.

Terborgh, J. 1974. Preservation of natural diversity: the problem of extinction prone species. *BioScience* **24**: 715–722.

Terborgh, J.T. 1983. *Five New World Primates: a Study in Comparative Ecology*. Princeton University Press, Princeton, NJ.

Terborgh, J.T. 1999. *Requiem for Nature*. Island Press, Washington DC.

Thebault, E. and Loreau, M. 2005. Trophic interactions and the relationship between species diversity and ecosystem stability. *American Naturalist* **166**: E95–E114.

Thomas, C.D. 2000. Dispersal and extinction in fragmented landscapes. *Proceedings of the Royal Society of London B Biological Sciences* **267**: 139–145.

Thomas, C.D. 2005. Recent evolutionary effects of climate change. In *Climate Change and Biodiversity* (Lovejoy, T. E. and Hannah, L., eds), pp. 75–88. Yale University Press, New Haven, CT.

Thomas, C.D., Cameron, A., Green, R.E., Bakkenes, M., Beaumont, L.J., Collingham, Y.C. *et al.* 2004. Extinction risk from climate change. *Nature* **427**: 145–148.

Thompson, K. 2000. The functional ecology of soil seed banks. In *Seeds: the Ecology of Regeneration in Plant Communities* (Fenner, M., ed.), pp. 215–235. CAB International, Wallingford.

Thuiller, W., Araujo, M.B., Pearson, R.G., Whittaker, R.J., Brotons, L., and Lavorel, S. 2004. Biodiversity conservation: uncertainty in predictions of extinction risk. *Nature* **430**: 1 p following 33.

Thuiller, W., Lavorel, S., and Araujo, M.B. 2005. Niche properties and geophical extent as predictors of species sensitivity to climate change. *Global Ecology and Biogeography* **14**: 247–357.

Tilman, D. 1976. Ecological competition between algae: experimental confirmation of resource-based competition theory. *Science* **192**: 463–465.

Tilman, D. 1977. Resource competition between planktonic algae: an experimental and theoretical approach. *Ecology* **58**: 338–348.

Tilman, D. 1980a. Resources: a graphical-mechanistic approach to competition and predation. *American Naturalist* **116**: 362–393.

Tilman, D. 1980b. Resource competition, spatial heterogeneity, and species diversity: an equilibrium approach to plant community structure. *American Naturalist* **116**: 362–393.

Tilman, D. 1982. *Resource Competition and Community Structure*. Monographs in Population Biology, Princeton University Press, Princeton, NJ.

Tilman, D. 1988. *Plant Strategies and the Dynamics and Structure of Plant Communities*. Princeton University Press, Princeton, NJ.

Tilman, D. 1990. Mechanisms of plant competition for nutrients: the elements of a predictive theory of competition. In *Perspectives on Plant Competition* (Grace, J. and Tilman, D., eds), pp. 117–141. Academic Press, San Diego.

Tilman, D. 1994. Competition and biodiversity in spatially structured habitats. *Ecology* **75**: 2–16.

Tilman, D. 1996. Biodiversity: population versus ecosystem stability. *Ecology* **77**: 350–363.

Tilman, D. 1999. The ecological consequences of changes in biodiversity: a search for general principles. *Ecology* **80**: 1455–1474.

Tilman, D. 2004. Niche tradeoffs, neutrality, and community structure: a stochastic theory of resource competition, invasion, and community assembly. *Proceedings of the National Academy of Sciences USA* **101**: 10854–10861.

Tilman, D. and Wedin, D. 1991a. Plant traits and resource reduction for five grasses growing on a nitrogen gradient. *Ecology* **72**: 685–700.

Tilman, D. and Wedin, D. 1991b. Dynamics of nitrogen competition between successional grasses. *Ecology* **72**: 1038–1049.

Tilman, D. and Pacala, S. 1993. The maintenance of species richness in plant communities. In *Species Diversity in Ecological Communities* (Ricklefs, R.E. and Schluter, D., eds), pp. 13–25. University of Chicago Press, Chicago.

Tilman, D. and Karieva, P. (eds) 1997. *Spatial Ecology: the Role of Space in Population Dynamics and Interspecific Interactions*, Monographs in Population Biology. Princeton University Press, Princeton, NJ.

Tilman, D., Kilham, S.S., and Kilham, P. 1982. Phytoplankton community ecology: the role of limiting nutrients. *Annual Review of Ecology and Systematics* **13**: 349–372.

Tilman, D., May, R.M., Lehmanm, C.L., and Nowak, M.A. 1994. Habitat destruction and the extinction debt. *Nature* **371**: 65–66.

Tilman, D., Knops, J., Wedin, D., Reich, P., Ritchie, M., and Sieman, E. 1997. The influence of functional diversity and composition on ecosystem processes. *Science* **277**: 1300–1302.

Tilman, D., Reich, P.B., Knops, J., Wedin, D., Mielke, T., and Lehman, C. 2001a. Diversity and productivity in a long-term grassland experiment. *Science* **294**: 843–845.

Tilman, D., Fargione, J., Wolff, B., D'Antonio, C.M., Dobson, A.P., Howarth, R.W. *et al.* 2001b. Forecasting agriculturally driven global environmental change. *Science* **292**: 281–284.

Tilman, D., Reich, P.B., and Knops, J. 2006. Biodiversity and ecosystem stability in a decade-long grassland experiment. *Nature* **441**: 629–632.

Tong, H. 1990. *Non-linear Time Series: a Dynamical Systems Approach.* Oxford University Press, Oxford.

Traulsen, A. and Nowak, M.A. 2006. Evolution of cooperation by multi-level selection. *Proceedings*

of the National Academy of Sciences USA **103**: 10952–10955.

Traulsen, A., Sengupta, A.M., and Nowak, M.A. 2005. Stochastic evolutionary dynamics on two levels. *Journal of Theoretical Biology* **235**: 393–401.

Travis, J.M.J. 2003. Climate change and habitat destruction: a deadly anthropogenic cocktail. *Proceedings of the Royal Society of London* **270**: 467–473.

Tregonning, K. and Roberts, A. 1979. The robustness of natural systems. *Nature* **281**: 563–564.

Trenbath, B.R., Conway, G.R., and Craig, I.A. 1990. Threats to sustainability in intensified agricultural systems: analysis and implications for management. In *Agroecology: Researching the Ecological Basis for Sustainable Agriculture* (Gliessman, S.R., ed.), pp. 337–366. Springer-Verlag, New York.

Trivers, R.L. 1971. The evolution of reciprocal altruism. *Quarterly Review of Biology* **46**: 35–57.

Tscharntke, T. and Brandl, R. 2004. Plant-insect interactions in fragmented landscapes. *Annual Review of Entomology* **49**: 405–430.

Tuljapurkar, S.D. 1982. Population dynamics in variable environments. 3. Evolutionary dynamics of *r* selection. *Theoretical Population Biology* **21**: 141–165.

Turchin, P. 2003. *Complex Population Dynamics: a Theoretical/Empirical Synthesis.* Princeton University Press, Princeton, NJ.

Turin, H. and Den Boer, P.J. 1988. Changes in the distribution of carabid beetles in The Netherlands since 1880. II. Isolation of habitats and long-term time trends in the occurrence of carabid species with different powers of dispersal (Coleoptera, Carabidae). *Biological Conservation* **44**: 179–200.

Turnbull, L.A., Rees, M., and Crawley, M.J. 1999. Seed mass and the competition/colonization trade-off: a sowing experiment. *Journal of Ecology* **87**: 899–912.

Turnbull, L.A., Crawley, M.J., and Rees, M. 2000. Are plant populations seed-limited? A review of seed sowing experiments. *Oikos* **88**: 225–238.

Turner, II, B.L., Clark, W.C., Kates, R.W., Richards, J.F., Mathews, J.T., and Meyer, W.B. 1990. *The Earth as Transformed by Human Action.* Cambridge University Press, Cambridge.

Turner, J.R.G., Gatehouse, C.M., and Corey, C.A. 1987. Does solar energy control organic diversity? Butterflies, moths and the British climate. *Oikos* **48**: 195–205.

Turner, W.R. and Wilcove, D.S. 2006. Adaptive decision rules for the acquisition of nature reserves. *Conservation Biology* **20**: 527–537.

Turner, W.R., Wilcove, D.S., and Swain, H.M. 2006. *State of the Scrub: Conservation Progress, Management*

Responsibilities, and Land Acquisition Priorities for Imperiled Species of Florida's Lake Wales Ridge. Archbold Biological Station, Lake Placid, Florida.

UNICEF. 2001. *Progress Since the World Summit for Children: A Statistical Review.* United Nations Children's Fund, New York.

United Nations Food and Agriculture Organization. 2001. *The State of Food Insecurity in the World 2001.* UN Food and Agriculture Organization, Rome.

United Nations Food and Agriculture Organization. 2006. *FAOSTAT.* Food and Agriculture Organization, Rome.

Utida, S. 1957. Population fluctuation, an experimental and theoretical approach. *Cold Spring Harbour Symposia on Quantitative Biology* **XXII**: 139–151.

Uyenoyama, M. and Feldman, M.W. 1980. Theories of kin and group selection: a population genetics perspective. *Theoretical Population Biology* **17**: 380–414.

Vandermeer, J.H. 1969. The competitive structure of communities: an experimental approach with Protozoa. *Ecology* **50**: 362–371.

Vandermeer, J. and Perfecto, I. 1995. *Breakfast of Biodiversity. The Truth about Rainforest Destruction.* First Foof Books, Oakland, CA.

Vandermeer, J. and Perfecto, I. 2005. The future of farming and conservation. *Science* **308**: 1257.

Vane-Wright, R.I., Humphries, C.J., and Williams, P.H. 1991. What to protect? Systematics and Choice. *Biological Conservation* **55**: 235–254.

van Nouhuys, S. and Hanski, I. 2002. Colonization rates and distances of a host butterfly and two specific parasitoids in a fragmented landscape. *Journal of Animal Ecology* **71**: 639–650.

Varley, G.C. and Gradwell, G.R. 1960. Key factors in population studies. *Journal of Animal Ecology* **29**: 399–401.

Verhulst, P.F. 1838. Notice sur la loi que la population suit dans son accroissement. *Correspondances Mathématiques et Physiques* **10**: 113–121.

Vinther, M. 2001. Ad hoc multispecies VPA tuning applied for the Baltic and North Sea fish stocks. *ICES Journal of Marine Science* **58**: 311–320.

Vitousek, P.M., Aber, J.D., Howarth, R.W., Likens, G.E., Matson, P.A., Schindler, D.W. *et al.* 1997a. Human alteration of the global nitrogen cycle: sources and consequences. *Ecological Applications* **7**: 737–750.

Vitousek, P.M., Mooney, H.A., Lubchenco, J., and Melillo, J.M. 1997b. Human domination of earth's ecosystems. *Science* **277**: 494–499.

Vitousek, P.M., Ehrlich, P.R., Ehrlich, A.H., and Matson, P.A. 1986. Human appropriation of the products of photosynthesis. *BioScience* **36**: 368–373.

Volterra V. 1926. Variations and fluctuations of the number of individuals in animal species living together [in Italian]. *Memoires Accademia dei Lincei* **2**: 31–113.

Volterra, V. 1931. Variations and fluctuations of the number of individuals in animal species living together. In *Animal Ecology* (Chapman, R.N., ed.), pp. 409–448. McGraw-Hill, New York.

von Bertalanffy, L. 1938. A quantitative theory of organic growth (Inquiries on growth laws II). *Human Biology* **10**: 181–213.

Vucetich, J.A. and Peterson, R.O. 2004a. Grey wolves—Isle Royale. In *Biology and Conservation of Wild Canids* (MacDonald, D.W. and Sillero-Zubiri, C., eds), pp. 285–296. Oxford University Press, Oxford.

Vucetich, J.A. and Peterson, R.O. 2004b. The influence of prey consumption and demographic stochasticity on population growth rate of Isle Royale wolves *Canis lupus. Oikos* **107**: 209–320.

Vucetich, J.A. and Peterson, R.O. 2004c. The influence of top-down, bottom-up and abiotic factors on the moose (*Alces alces*) population of Isle Royale. *Proceedings of the Royal Society of London Series B Biological Sciences* **271**: 183–189.

Vucetich, J.A., Peterson, R.O., and Schaefer, C.L. 2002. The effect of prey and predator densities on wolf predation. *Ecology* **83**: 3003–3013.

Waage, J.K. 1979. Foraging for patchily-distributed hosts by the parasitoid *Nemeritis canescens. Journal of Animal Ecology* **48**: 353–371.

Wade, M.J. 1976. Group selection among laboratory populations of *Tribolium. Proceedings of the National Academy of Sciences USA* **73**: 4604–4607.

Walde, S. and Murdoch, W.W. 1989. Spatial density dependence in parasitoids. *Annual Review of Entomology* **33**: 441–466.

Walker, B.H. 1992. Biodiversity and ecological redundancy. *Conservation Biology* **6**: 18–23.

Walters, C.J. and Kitchell, J.F. 2001. Cultivation/depensation effects on juvenile survival and recruitment: implications for the theory of fishing. *Canadian Journal of Fisheries and Aquatic Science* **58**: 39–50.

Walters, C., Christensen. V., and Pauly, D. 1997. Structuring dynamic models of exploited ecosystems from trophic mass-balance assessments. *Review of Fish Biology* **7**: 139–172.

Walther, G., Post, E., Convey, P., Menzel, A., Parmesan, C., Beebee, T. *et al.* 2002. Ecological responses to recent climate change. *Nature* **413**: 389–396.

Wambugu, F. and Kiome, R. 2001. *The Benefits of Biotechnology for Small-Scale Banana Farmers in Kenya*, ISAAA Briefs no. 22. International Service for the Acquisition of Agri-Biotech Applications, Ithaca, NY.

Warren, M.S., Hill, J.K., Thomas, J.A., Asher, J., Fox, R., Huntley, B. *et al.* 2001. Rapid responses of British butterflies to opposing forces of climate and habitat change. *Nature* **414**: 65–68.

Watkinson, A.R. 1990. The population-dynamics of *Vulpia fasciculata*–a 9-year study. *Journal of Ecology* **78**: 196–209.

Watkinson, A.R., Lonsdale, W.M., and Andrew, M.H. 1989. Modeling the population-dynamics of an annual plant *Sorghum intrans* in the wet-dry tropics. *Journal of Ecology* **77**: 162–181.

Watling, J.I. and Donnelly, M.A. 2006. Fragments as islands: a synthesis of formal responses to habitat patchiness. *Conservation Biology* **20**: 1016–1025.

Watts, D.J. and Strogatz, S.H. 1998. Collective dynamics of 'small world' networks. *Nature* **393**: 440–442.

Wearing, H.J., Rohani, P., and Keeling, M.J. 2005. Appropriate models for the management of infectious diseases. *PLoS Medicine* **2**: 621–627.

WDPA. World Database on Protected Areas (2004). 2004 World Database on Protected Areas. CD-ROM. IUCN World Commission on Protected Areas. UNEP World Conservation Monitoring Centre. Washington, DC.

Webb, C.O., Ackerly, D.D., McPeek, M.A., and Donoghue, M.J. 2002. Phylogenies and community ecology. *Annual Review of Ecology and Systematics* **33**: 475–505.

Webb, T. 1992. Past changes in vegetation and climate: lessons for the future. In *Climate Change and Biodiversity* (Lovejoy, T.E. and Hannah, L., eds), pp. 59–75. Yale University Press, New Haven, CT.

Wedekind, C. and Milinski, M. 2000. Cooperation through image scoring in humans. *Science* **288**: 850–852.

Wedekind, C. and Braithwaite, V.A. 2002. The long term benefits of human generosity in indirect reciprocity. *Current Biology* **12**: 1012–1015.

Wedin, D. and Tilman, D. 1993. Competition among grasses along a nitrogen gradient: initial conditions and mechanisms of competition. *Ecological Monographs* **63**: 199–229.

Weiner, J. 1982. A neighborhood model of annual-plant interference. *Ecology* **63**: 1237–1241.

Weiner, J. 2004. Allocation, plasticity and allometry in plants. *Perspectives in Plant Ecology Evolution and Systematics* **6**: 207–215.

Weiner, J., Stoll, P., Muller-Landau, H., and Jasentuliyana, A. 2001. The effects of density, spatial pattern, and competitive symmetry on size variation in simulated plant populations. *American Naturalist* **158**: 438–450.

Western, D., Wright, R.M., and Strum, S.C. (eds) 1994. *Natural Connections: Perspectives in Community-Based Conservation*. Island Press, Washington DC.

Westoby, M., Leishman, M., and Lord, J. 1997. Comparative ecology of seed size and dispersal. In *Plant Life Histories: Ecology, Phylogeny and Evolution* (Silvertown, J., Franco, M., and Harper, J.L., eds), pp. 143–162. Cambridge University Press, Cambridge.

White, J. and Harper, J.L. 1970. Correlated changes in plant size and number in plant populations. *Journal of Ecology* **58**: 467–485.

White, K.A.J., Lewis, M.A., and Murray, J.D. 1998. On wolf territoriality and deer survival. In *Modelling Spatiotemporal Dynamics in Ecology*, pp 105–126 (Bascompte, J. and Sole, R.V.). Springer-Verlag, Berlin.

White, P.J. and Kerr, J.T. 2006. Contrasting spatial and temporal global change impacts on butterfly species richness during the 20th century. *Ecography*, in press.

Whitlock, M. 2003. Fixation probability and time in subdivided populations. *Genetics* **164**: 767–779.

Whitney, G. 1986. Relation of Michigan's presettlement pine forests to substrate and disturbance history. *Ecology* **67**: 1548–1559.

Whittaker, R.H. 1951. A criticism of the plant association and climatic climax concepts. *Northwest Science* **25**: 17–31.

Whittaker, R.H. 1956. Vegetation of the Great Smoky Mountains. *Ecological Monographs* **26**: 1–20.

Whittaker, R.H. 1975. *Communities and Ecosystems*. Macmillan, New York.

Wiersma, Y.F. and Nudds, T.D. 2001. Comparison of methods to estimate historic species richness of mammals for tests of faunal relaxation in Canadian parks. *Journal of Biogeography* **28**: 447–452.

Wigner, E. 1958. On the distribution of the roots of certain symmetric matrices. *Annals of Mathematics* **67**: 325–327.

Wilby, A. and Thomas, T.D. 2002. Natural enemy diversity and pest control: patterns of pest emergence with agricultural intensification. *Ecology Letters* **5**: 353–360.

Wilcove, D.S. and Master, L.L. 2005. How many endangered species are there in the United States? *Frontiers in Ecology and the Environment* **3**: 414–420.

Wilcove, D.S., McLellan, C.H., and Dobson, A.P. 1986. Habitat fragmentaion in the temperate zone. In *Conservation Biology. The Science of Scarcity and Diversity* (Soule, M.E., ed,.), pp. 237–256. Sinauer Associates, Sunderland, MA.

Willerslev, E., Hansen, A.J., Binladen, J., Brand, T.B., Gilbert, M.T.P., Shapiro, B. *et al.* 2003. Diverse plant and animal genetic records from holocene and pleistocene sediments. *Science* **300**: 791–795.

Williams, G.C. 1966. *Adaptation and Natural Selection*. Princeton University Press, Princeton, NJ.

Williams, P., Hannah, L., Andelman, S., Midgley, G., Araujo, M., Hughes, G. *et al.* 2005. Planning for climate change: identifying minimum-dispersal corridors for the Cape Proteaceae. *Conservation Biology* **19**: 1063–1074.

Williams, R.J. and Martinez, N.D. 2000. Simple rules yield complex food webs. *Nature* **404**: 180–183.

Williams, R.J., Berlow, E.L., Dunne, J.A., Barabasi, A.L., and Martinez, N.D. 2002. Two degrees of separation in complex food webs. *Proceedings of the National Academy of Sciences USA* **9**: 12913–12916.

Wills, C., Harms, K.E., Condit, R., and King, D. 2006. Nonrandom processes maintain diversity in tropical forests. *Science* **311**: 527–531.

Wilson, E.O. 1975. *Sociobiology*. Harvard University Press, Cambridge, MA.

Wilson, E.O. 1992. *The Diversity of Life*. Penguin, Harmondsworth.

Wilson, E.O. (ed) 1988. *Biodiversity*. National Academy Press, Washington DC.

Wilson, E.O. and Hölldobler, B. 2005. Eusociality: origin and consequences. *Proceedings of the National Academy of Sciences USA* **102**: 13367–13371.

Wilson, H.W., Godfray, H.C.J., Hassell, M.P., and Pacala, S. 1998. Deterministic and stochastic host-parasitoid dynamics in spatially extended systems. In *Modelling Spatiotemporal Dynamics in Ecology* (Bascompte, J. and Sole, R.V.), pp. 63–81. Springer-Verlag, Berlin.

Wilson, R.J., Gutierrez, D., Gutierrez, J., Martinez, D., Agudo, R., and Monserrat, V.J. 2005. Changes to the elevational limits and extent of speceis ranges associated with climate change. *Ecology Letters* **8**: 1138–1146.

Wilson, W.G., de Roos, A.M., and McCauley, E. 1993. Spatial instabilities within the diffusive Lokta-Volterra system—individual-based simulation results. *Theoretical Population Biology* **43**: 91–127.

Wolf, A., Swift, J.B., Swinney, H.L., and Vastano, J.A. 1985. Determining Lyapunov exponents from a time series. *Physica* **16D**: 285–317.

Wood, S., Cassman, K., and Gaskell, J. 2006. Assessing the condition and multi-scale impacts of cultivated systems. In *Millennium Ecosystem Assessment. Ecosystems and Human Well-Being*, vol. 1. *Current State and Trends*, pp. 209–242. Island Press, Washington DC.

Wood, S.N. 1994. Obtaining birth and mortality patterns from structured population trajectories. *Ecological Monographs* **64**: 23–44.

Wood, S.N. 2001. Partially specified ecological models. *Ecological Monographs* **71**: 1–25.

Woodward, F.I. 1987. *Climate and Plant Distribution*. Cambridge University Press, Cambridge.

Wooten, J.T. 2005. Field parameterization and experimental test of the neutral theory of biodiversity. *Nature* **440**: 80–82.

Wright, S. 1931. Evolution in Mendelian populations. *Genetics* **16**: 97–159.

WWF. 2004. *Living Planet Report 2004*. www.panda.org.

Wynn-Edwards, V.C. 1962. *Animal Dispersion in Relation to Social Behavior*. Oliver and Boyd, London.

Yachi, S. and Loreau, M. 1999. Biodiversity and ecosystem productivity in a fluctuating environment: the insurance hypothesis. *Proceedings of the National Academy of Sciences USA* **96**: 1463–1468.

Ye, X., Salim, A.B., Kloti, A., Zhang, J., Lucca, P., Beyer, P., and Potrykus, I. 2000. Engineering the pro-vitamin A (beta-carotene) biosynthetic pathway into rice endosperm. *Science* **287**: 393–405.

Yoda, K., Kira, T., and Hozumi, K. 1957. Intraspecific competition among higher plants. IX. Further analysis of the competitive interaction between adjacent individuals. *Journal of the Institute Polytechnic of Osaka City University* **8**: 161–178.

Yodzis, P. 1978. *Competition for Space and the Structure of Ecological Communities*. Springer-Verlag, New York.

Yodzis, P. 1998. Local trophodynamics and the interaction of marine mammals and fisheries in the Benguela region. *Journal of Animal Ecology* **69**: 635–658.

Index

Note: page numbers in *italics* refer to Figures and Tables.